Transformations Through Blockchain Technology

Sheikh Mohammad Idrees
Mariusz Nowostawski
Editors

Transformations Through Blockchain Technology

The New Digital Revolution

Springer

Editors
Sheikh Mohammad Idrees
Researcher, Department of Computer Science (IDI)
Norwegian University of Science and Technology (NTNU)
Gjøvik, Norway

Mariusz Nowostawski
Associate Professor, Department of Computer Science (IDI)
Norwegian University of Science and Technology (NTNU)
Gjøvik, Norway

ISBN 978-3-030-93343-2 ISBN 978-3-030-93344-9 (eBook)
https://doi.org/10.1007/978-3-030-93344-9

This Springer imprint is published by the registered company Springer Nature Switzerland AG
The registered company address is: Gewerbestrasse 11, 6330 Cham, Switzerland

Preface

Blockchain has come a long way ahead ever since its inception by Satoshi Nakamoto. Blockchain is more than just a novel decentralized technology. It is also an emerging technology that disrupts and revolutionizes almost all sectors of our society, in financial sector, industry, and institutions. Blockchain is a distributed software network that serves as a digital ledger and a means for transferring assets securely without involving any third party. Blockchain and decentralization can help to solve most of the industry's challenges, among others, interoperability. There are several different types of blockchain available today. Some blockchains were created to cater the demands of a small group of people with limited access to the network, while some are designed without any restriction where anyone throughout the world can join and use the blockchain network when needed. These are examples of permissioned blockchains, or private blockchains, and public blockchains. Irrespective of the type of blockchain, blockchain technology has the potential to alter centuries-old business structures, opening the path for increased government legitimacy and new possibilities for regular individuals to flourish.

This book *Transformations Through Blockchain Technology* familiarizes the reader with the detailed knowledge of blockchain technology and how this technology is transforming different sectors of society. This book provides a comprehensive knowledge about various cutting-edge research areas involving blockchain technology and decentralization. It serves as a connecting medium involving various domains and blockchain technology. This book facilitates sharing of information, case studies, theoretical and practical knowledge required for blockchain transformations in various sectors.

This book further focuses on the concepts of the technology and how blockchain technology can work in an integrated manner with other existing technologies to transform the society. This book serves as an interesting knowledge sharing medium to discuss the major areas that are impacted by blockchain technology. The book is divided into various chapters, and a brief summary of the chapters is presented as below.

Chapter: Revolutionising the Approach to Smart Campus Architecture Through IoT and Blockchain Technologies.

This chapter proposes a novel architecture framework for IoT and blockchain applications deployed within a smart campus environment, comparing the main technologies involved. This framework is tested in the usability scenario of Higher Education Smart Certification system for issuing authentic, verifiable, and sharable student credentials. The results of this study are particularly useful in the context of developing countries seeking to achieve and maintain relevant safety and security standards.

Chapter: Transforming Education Through Blockchain Technology.

The chapter defines the major issues and challenges in front of digital world along with the specific leakages all around the computing environment. The reason blockchain technology is the primary consideration for present and future digital platforms is that through blockchain one can learn how to protect all essential credentials digitally in a suitable manner and establish commutation services across the globe with the management of higher level of security, privacy, and confidentiality.

Chapter: Development of a Blockchain-Based Survival Game for Blockchain Education.

Blockchain applications provide many potential use cases for organizations and businesses. However, the complexity of blockchain technology has limited its adoption in addressing real-life problems. Gamification provides a visual approach for reducing complex concepts into easily understandable contents. This chapter of the book provides gamification of the main concepts in the blockchain using a closed system such that participants can generate and exchange assets and record their activities in a transparent and tamperproof ledger.

Chapter: Integrating Blockchain with Education: Proposed Model, Prospects, and Challenges.

Blockchain is one of the most important emerging technologies which is aimed at decentralization. In this work, we apply the concepts of blockchain to solve the problems faced by the Indian education sector. We propose a model for credit management and a blockchain-based certificate management system. The proposed work aims to incorporate the newly emerging educational institutes and various initiatives which lack any certified creditability. Using our system, such initiatives can be integrated and approved by higher authorities, thereby allowing students to claim what they have learned.

Chapter: Towards a Distributed Record of Measurement Adapters Powered by Blockchain Technology.

Scope, resolution, and coverage of distributed data collectors are a current challenge. Data interoperability in a heterogeneous context is a concern. Real-time data collection systems use measurement adapters (MA) like semantic bridges for fostering data interoperability. This work proposes a distributed record of MA based on blockchain technology for articulating direct and indirect data transmissions among measurement adapters. It provides an open-source library implementing the described functionality.

Chapter: Blockchain Technology: A Breakthrough in the Healthcare Sector.

Technological innovations rapidly changed the healthcare sector from stages of drug development to patient care. Despite these, many challenges need to be addressed. To provide solutions, blockchain technology is an effective way with its enhanced security features. This chapter deals with the latest applications of blockchain technology in various health domains, problems resolved by blockchain in the health sector, with its future perspectives.

Chapter: Digital Transformation of Healthcare Sector by Blockchain Technology.

The chapter deals with key features such as introduction to blockchain, its key elements, life cycle, and technologies using blockchain. This chapter focuses on applications, benefits, and challenges of blockchain in the healthcare sector. It enables readers to answer some significant questions such as how blockchain can transform the health sector, what is the advantage of using BCT in the healthcare sector? Is it essential to apply blockchain technology in the healthcare sector? What is the role of blockchain in a pandemic such as COVID-19?

Chapter: A Systematic Review on Blockchain in Transforming the Healthcare Sector.

This chapter presents a comparative analysis of the work carried out by various researchers to provide solutions for enhancing the healthcare system. Some have focused on providing complete user control for sharing their own health records whereas other have given importance to enhance interoperability of data globally and to secure the sensitive data of patients. Using smart contract can lead to seamless connectivity.

Chapter: Blockchain Technology for Contact Tracing During COVID-19.

This chapter analyzes how blockchain technology can be exploited to develop contact tracing applications that guarantee the security of users' privacy. First, the general concepts underlying blockchain technology are introduced, and then practical cases of use of this technology in various fields are analyzed. Finally, practical cases of application of blockchain technologies for contact tracing in the Covid-19 pandemic are analyzed.

Chapter: Transforming Healthcare Sector in India Through Blockchain Technology: Challenges and Opportunities from Legal Perspectives.

This chapter aims to address the techno-socio-legal aspects and implications of blockchain technology in the healthcare sector. It engages inter alia with diverse medical practices and major health insurance models exploring the viability of Blockchain 4.0. It further engages and evaluates the socio-legal challenges which are potential impediments for the acceptability of this technology. Finally, the chapter recommends the possible adaptation mechanism by which the technology may integrate with law and society in the healthcare sector of the Global South, especially in India.

Chapter: Smart Contracts in Blockchain Application: Review Chain.

In the form of ratings and reviews, many e-commerce and other digital sites allow for greater consumer participation. A recent study has highlighted the importance of this feedback, confirming that positive feedback increases product sales and therefore popularity. Managers of online portals are in charge of overseeing the whole evaluation process. Online operators have the ability to tamper with legitimate reviews, and in the worst-case situation, they may prevent users from leaving evaluations if it has a negative impact on sales. Our focus will be on limiting the digital site operators through the use of smart contracts and decentralized apps with storage, allowing all customers to give and observe unrestricted data.

Chapter: Blockchain Technology Transforms Digital Marketing by Growing Consumer Trust.

Today, digital marketing allows companies to enhance their marketing strategies and to establish global marketplaces, as well as drive further demand for their products and services. In this vibrant market, blockchain technology has emerged to enable better customer journeys. In this chapter, we discuss how blockchain technology can impact a company's digital marketing efforts, contributing to market engagement and empowering a customer-driven paradigm. Furthermore, we discuss how blockchain technology promotes digitalization, assists in preventing fraud, strengthens trust and transparency, and allows for better privacy protection while also enhancing security.

Chapter: Blockchain Technology: Unlocking the Business Model Maze for Evolving Businesses and Start-Up.

Blockchain technology has received notable recognition with bitcoin since 2008. Contrary to the popular notion that blockchain can only revolutionize the financial sector, it is capable enough to bring revolution to the business models irrespective of industry nature, size, ownership, geographic location, etc. This chapter aims to explore and establish successful application of blockchain solutions in developing business models for well-established business units and startups as well, in the field of health care, supply chain, energy, e-governance, and crypto currency.

Chapter: Transformation of Logistics Value Chain for Enhancing Cross-Border Trade Using Blockchain and IOT.

Cross-border shipment has always been a tedious process because of a lot of paperwork and manual human intervention by various stakeholders of the entire process, for example customs authorities, freight forwarders, buyers, sellers, transporters, etc. The authors have tried to optimize the process by integrating blockchain technology into the process of cross-border trade ultimately making the process more secure and robust.

Chapter: Rural Logistics Transformation Through Blockchain.

The concept of rural logistics encompasses transport, distribution, storage, material handling, and the packaging of goods in rural areas, as well as the flow of information and funds in support of rural production and consumption. This chapter researches the possibility of the development of sustainable rural logistics and supply chain management through decentralized data storage represented by blockchain technology. Blockchain can increase the efficiency and transparency of the supply chain and positively affect all logistic processes, from storage to delivery and payment. The goal of this chapter is to explore the possible use of blockchain technology in logistics processes and to identify the impact of blockchain technology on business transparency.

We hope you find this book interesting and enjoy delving deeper into the varied aspects of the book and enjoy reading and learning about the transformations through blockchain technology. This book would not be possible without the involvement of many people. We owe our gratitude and sincere "Thank You" to our contributors and reviewers without whose support this would not be achieved. Much appreciation goes to our authors, and we are obliged to the reviewers for their comments which improved the quality of the book. Last but not least, thanks to God for showing us the light to start this project and blessing us to complete it.

Gjøvik, Norway Sheikh Mohammad Idrees
Gjøvik, Norway Mariusz Nowostawski

Contents

Contributors

Amogh Aggarwal Graphic Era Deemed to be University, Dehradun, Uttarakhand, India

Manal Alkhammash Department of Informatics, Sussex University, Brighton, UK

Department of Computer Sciences, Jazan University, Jazan, KSA

Mona Alshahrani Department of Informatics, Sussex University, Brighton, UK

Department of Information Systems, Al Imam Mohammad Ibn Saud Islamic University, Riyadh, KSA

Sumit Badotra Department of Computer Science and Engineering, Lovely Professional University, Phagwara, Punjab, India

Natalia Beloff Department of Informatics, Sussex University, Brighton, UK

Roheet Bhatnagar Department of Computer Science and Engineering, Manipal University, Jaipur, India

Gulpreet Kaur Chadha Department of IT, Amity School of Engineering and Technology, Amity University, Noida, Uttar Pradesh, India

Shambhu Prasad Chakrabarty Centre for Regulatory Studies, Governance and Public Policy, The West Bengal National University of Juridical Sciences, Kolkata, West Bengal, India

Richa Chauhan Banasthali Vidyapith, Niwai, Rajasthan, India

Ranga Chimhundu School of Business, Faculty of Business, Education, Law and Arts, University of Southern Queensland, Toowoomba, QLD, Australia

Giuseppe Ciaburro Department of Architecture and Industrial Design, Università degli Studi della Campania Luigi Vanvitelli, Borgo San Lorenzo, Aversa (CE), Italy

Sarishma Dangi Graphic Era Deemed to be University, Dehradun, Uttarakhand, India

Mario Diván National University of La Pampa, C. Gil 353, Santa Rosa, Argentina

Girirajasekhar Dornadula Department of Pharmacy Practice, Annamacharya College of Pharmacy, Rajampeta, Andhra Pradesh, India

Lakshmi Narasimha Gunturu Department of Pharmacy Practice, Annamacharya College of Pharmacy, Rajampeta, Andhra Pradesh, India

Medini Gupta Amity Institute of Information Technology, Amity University, Noida, Uttar Pradesh, India

Rumman Hassan School of Business, Faculty of Business, Education, Law and Arts, University of Southern Queensland, Toowoomba, QLD, Australia

Varun Jaiswal School of Electrical and Computer Science Engineering, Shoolini University, Solan, Himachal Pradesh, India

Department of Food and Nutrition, College of BioNano Technology, Gachon University, Gyeonggi-do, South Korea

Vidhi Kaul Independent Researcher, New Delhi, India

Ritu Maheshwari Atal Bihari Vajpayee Institute of Good Governance and Policy Analysis (A Registered Body of Government of M.P. Under Public Service Management Department), Bhopal, Madhya Pradesh, India

Deepti Mehrotra Department of IT, Amity School of Engineering and Technology, Amity University, Noida, Uttar Pradesh, India

Souvik Mukherjee The West Bengal National University of Juridical Sciences, Kolkata, West Bengal, India

Renuka Nagpal Department of CSE, Amity School of Engineering and Technology, Amity University, Noida, Uttar Pradesh, India

Raghavendra Naveen Nimbagal Department of Pharmaceutics, Sri Adichunchanagiri College of Pharmacy, Adichunchanagiri University, Mandya, Karnataka, India

Alex Norta Department of Software Science, Tallinn University of Technology, Tallinn, Estonia

Tarun Pal Department of Biotechnology, Vignan's Foundation for Science, Technology and Research, Vadlamudi, Guntur, Andhra Pradesh, India

Zuckerberg Institute for Water Research, Blaustein Institutes for Desert Research, Ben-Gurion University of the Negev, Midreshet Ben Gurion, Israel

Fazla Rabby School of Business, Faculty of Business, Education, Law and Arts, University of Southern Queensland, Toowoomba, QLD, Australia

Ajay Rana Shobhit University, Meerut, Uttar Pradesh, India

Pranjal Rastogi Graphic Era Deemed to be University, Dehradun, Uttarakhand, India

María Laura Sánchez-Reynoso National University of La Pampa, C. Gil 353, Santa Rosa, Argentina

INCITAP, National Scientific and Technical Research Council (CONICET), Santa Rosa, Argentina

Aakarsh Shrivastava Department of IT, Amity School of Engineering and Technology, Amity University, Noida, Uttar Pradesh, India

Gaurav Srivastava Department of Computer Engineering, Poornima College of Engineering, Jaipur, India

Swati Srivastava Department of Computer Science and Engineering, Manipal University, Jaipur, India

Urvashi Srivastava Maulana Azad National Institute of Technology, Bhopal, India

Dudala Sai Sushma Department of Biotechnology, Vignan's Foundation for Science, Technology and Research, Vadlamudi, Guntur, Andhra Pradesh, India

Sarvesh Tanwar Amity Institute of Information Technology, Amity University, Noida, Uttar Pradesh, India

Chibuzor Udokwu Department of Software Science, Tallinn University of Technology, Tallinn, Estonia

Ravi Verma Department of CSE, Radharaman Institute of Technology and Science, Bhopal, India

Roxana Voicu-Dorobantu The Bucharest University of Economic Studies, Bucharest, Romania

Martin White Department of Informatics, Sussex University, Brighton, UK

Revolutionising the Approach to Smart Campus Architecture Through IoT and Blockchain Technologies

Manal Alkhammash, Mona Alshahrani, Natalia Beloff, and Martin White

1 Introduction

Information and communication technology (ICT) has a significant role in connecting objects together to enable them to communicate and exchange information in the development of smart environments that will be used to monitor, track and manage different data. Globally, governments are working towards digitalisation with the development of smart cities to accommodate the projected rise in the city's population. From 2015 to 2050, the population in the city is expected to grow from 55% to 66%, which will require the optimisation of resources and the use of technologies, such as IoT, to implement a sustainable and intelligent environment [1] for use with governance, public transport, traffic, etc. For larger organisations, such as a smart campus, which can be considered as an example of a smart city, deploying this type of technology can improve the operational efficacy of the campus in many areas such as the reduction of water usage, the efficient use of energy and becoming carbon neutral. It also has the potential to increase the user experience.

M. Alkhammash
Department of Informatics, Sussex University, Brighton, UK

Department of Computer Sciences, Jazan University, Jazan, KSA
e-mail: ma979@sussex.ac.uk

M. Alshahrani (✉)
Department of Informatics, Sussex University, Brighton, UK

Department of Information Systems, Al Imam Mohammad Ibn Saud Islamic University, Riyadh, KSA
e-mail: m.alshahrani@sussex.ac.uk

N. Beloff · M. White
Department of Informatics, Sussex University, Brighton, UK
e-mail: n.beloff@sussex.ac.uk; m.white@sussex.ac.uk

S. M. Idrees, M. Nowostawski (eds.), *Transformations Through Blockchain Technology*, https://doi.org/10.1007/978-3-030-93344-9_1

The focus on smart campuses has been fuelled with the advancement of advanced technologies such as IoT, Blockchain, and machine learning, which enable data to be captured at the level of the infrastructure, generating valuable insights for the stakeholders. However, because of the centralised architecture, there are still security and privacy issues in the current smart campus system. Therefore, this study aims to develop a comprehensive novel smart campus framework to improve the security issue. The framework combines several innovative technologies including infrastructure, storage, and communication technologies. It merges IoT and Blockchain technologies development in a smart campus environment to design various campus service systems, including smart systems for learning, library, building, parking, and waste and water management.

Blockchain recently has a tremendous impact on various fields including the higher education sector. Therefore, exploring exciting developed systems including the main processes in the higher education field is an important step towards fulfilling the purpose of this research. Moreover, this work involves addressing the current cases of utilising Blockchain in solving the issues of higher education systems. This work contains a result of ongoing research about examining the influential factors that affect the adoption of Blockchain technology in developing countries, in this study, the research's investigation involved Saudi Arabia. Additionally, this study presents a framework that is tested for the integration of Blockchain and other relevant technologies into the higher education certificating system for issuing authentic, verifiable, and sharable student credentials. Thus, the findings from this study evaluate the potential of acceptance of Blockchain technology as a viable solution to provide the higher education systems with the concept of smart certificates in the certification process. Lastly, the study proposed a complementary modelling approach for decentralised applications for smart certificates (DASC) in which Blockchain distributed technology emerges in the context of higher education.

The chapter is structured as follows. Section 2 offers a concise review of relevant literature to understand the concepts of a smart campus, IoT and Blockchain. Section 3 provides an overview of current smart campus frameworks. In Sect. 4 the chapter addresses several issues associated with current smart campus frameworks, followed by the proposal of a new one with an in-depth explanation of its IoT layers and Blockchain technology in the Sect. 5. Section 6 presents Blockchain existing applications in higher education. A Blockchain-based student certification system is designed in Sect. 7 based on the novel Blockchain adoption framework. The discussion will be addressed in Sect. 8, and the chapter ends with conclusions and suggested future work.

2 Background

This section discusses the existing literature and research about smart campus, Internet of Things (IoT), and Blockchain technology.

2.1 Smart Campus Concept

The traditional definition of a campus is that it is a piece of land on which buildings and structures are collected together, ranging from libraries to halls of residence to parking lots, to form an educational institution. In recent years, various state-of-the-art technologies have been adopted on campuses worldwide, including visual learning environments [1] and timetabling systems [2]. The purpose of adopting these technologies has been to increase service quality for stakeholders (e.g., administrators, academics, and students), as well as to control and supervise facilities. These developments ought to be developing in a continuous way to maximise efficiency, lower operational costs, promote effective decision-making, reduce effort, and improve the student experience [3]. Therefore, the phrase "smart campus" has been used to refer to campuses that have the digital infrastructure needed to collect and analyse data, and to use these for decision-making and responsivity to changes on campus without the need for human supervision [4]. According to [5], a smart campus is a setting that uses the structural features of an ambient learning environment to enable the integration of digital and social services in physical learning resources. If the smart campus is viewed as a holistic framework, it includes multiple themes such as intelligent sensor management systems, automatic security monitoring and control, smart building management, and cooperation, communication, and social networking. A range of innovative technologies have been proposed for smart campuses, including Blockchains, the IoT, the cloud, and developing subsystems that rely on mobile technologies to promote security (e.g., technologies such as RFID and ZigBee).

2.2 Internet of Things (IoT)

Due to the continual growth of the internet, connectivity and communication are ubiquitous. The integration of humans and devices has enabled connection to the internet and data transfer automatically, which has created the so-called Internet of Things (IoT). IoT technology facilitates the global connectivity of computer networks, enabling the remote control of various 'smart' objects to access specific services. IoT is an innovation that combines digital and physical components to enable novel business models and the creation of new products. With the efficient increase in broadband communication and power management, along with advances in microprocessors and increasingly reliable memory, it has become possible to digitalise functions and environments. Therefore, creating a smart world and the information thus generated may prove useful in various service domains, including smart cities, smart homes, and smart campuses.

Over the years, researchers have defined IoT from different perspectives [6]. Some definitions have concentrated on the objects that connect to networks; others have focused on aspects related to IoT such as network technology and network protocols. Some descriptions of IoT focus on semantic challenges such as big data information, storage, and search. The European Commission defined IoT as an approach to developing smart environments through merging physical and virtual worlds [7]. The term IoT usually refers to a system of smart devices connected to the internet with the ability to identify themselves and communicate with each other by collecting and sending data via the network [8]. Thus, it allows people, things, and processes to communicate and be connected in any place, at any time, using any service, and with any network.

From a technological perspective, embedded devices with sensors and chips such as tags, near-field communication (NFC), and RFID are to be identify, control, and manage the devices, they use IP addresses and other communication protocols such as MQTT to communicate without human interfaces [9]. When devices connect to the internet, they can start transmitting data to computing technologies such as the cloud, which is a powerful platform that integrates data analytics tools, data storage, and data delivery models to perform services for users and businesses. Once the data are received in the cloud, they are processed by relevant software and, in turn, valuable information is sent to the client. These data create a smart environment when they are used to make correct decisions, detect problems before they occur, and save time and money. Therefore, major IoT characteristics are perception, network, and intelligent processing.

With the growing business requirement for service applications, there is a need to develop more devices and emerging technologies to ensure availability anytime, anywhere, as well as to propose protocols to solve compatibility issues and integration among heterogeneous objects that are connected to the network. With these developments, IoT has faced several challenges in terms of security and privacy. To solve these issues, it is necessary to revise the traditional IoT architecture, which consists of three basic layers: the perception layer, network layer, and application layer [10]. For example, many developments have occurred in terms of security measures in the perception layer, including adding access control for the connected devices (i.e., to protect privacy) and providing different encryption mechanisms (i.e., to encrypt the signal from electronic tags). In addition, the network layer has been altered to enable more dynamic topologies, end-to-end authentication, security routing, and the key agreement process, among other features. One example is developing the IPSec protocol and IEEE 802.15.4 standard protocol to provide secure channels of communication between two devices. In the application layer, many mechanisms have now been developed to reduce the leakage of confidential information, as well as to bolster the robustness and security of information management (e.g., password management and resource management).

Nowadays, the amount of the connected devices is assumed to be in range between 20 and 50 billion till the year 2020 because the maximum number of devices can be supported by IoT. The exceptional growth in the system of IoT has created new opportunities by which methods allow information to be shared and accessed easily [11]. Conoscenti et al. in [14] highlights the cause of such initiatives is mainly the existence of the open data paradigm. Despite these aspects, these creative systems and methods face some significant vulnerabilities, for example, the shortage of confidence that has been shown in many scenarios (Fig. 1).

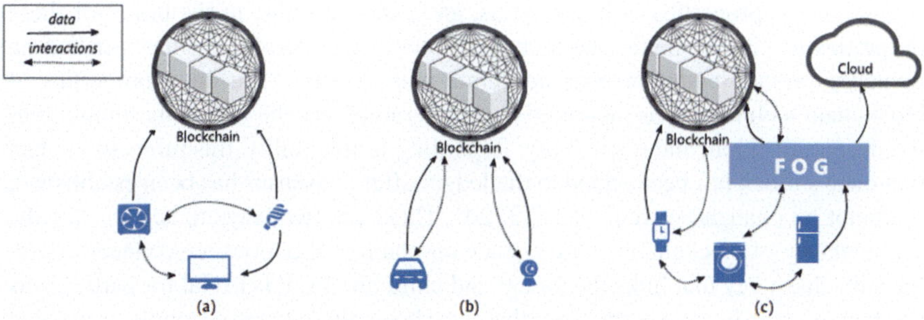

Fig. 1 Blockchain and IoT interactions [11]. (**a**) IoT-IoT, (**b**) IoT-Blockchain, (**c**) Hybrid approach

As stated by Zheng et al. in [12], Internet of Things (IoT) is considered as one of the leading fields in adopting Blockchain technology. Blockchain can increase the efficiency of the IOT by giving a sharing service that is trusted, where we can get advantage of decentralised environment, in which the information is easily traceable and reliable. Using Blockchain technology integrated with IoT will increase security, whereas in any point of time, the data's sources can be recognised, which guarantees data immutability. As result of this integration, the IoT will provide a secure environment where information can be securely shared between several participants [11]. This research will contribute to improving campus IoT security by merging it with Blockchain technology, as discussed throughout this chapter.

2.3 *Blockchain Technology*

Blockchain can be considered a revolutionary development. It is defined as a distributed record of digital events stored across all participating computers in a linked chain [13, 14]. The National Institute of Standards and Technology noted, 'Blockchains are tamper-evident and tamper-resistant digital ledgers implemented in a distributed fashion and usually without a central authority' [15]. This description of blackchin is more detailed and covers a boarder overview. This chapter will follow this definition. It describes Blockchain's vital feature as distributed technology and has an immutable and decentralised structure.

Technical world have hailed the emergence of blockchain technology as representing the arrival of the fourth industrial revolution, and it has been referred to as 'the Internet of Value Exchange' [16]. Despite the rapid growth of blockchain applications in finance, otherwise known as cryptocurrencies, there are also other valuable applications of the technology [17, 18]. One prominent alternative application area is that of supply chains [19], but it is noteworthy that the social, political, general, and educational applications of blockchain technology have also been explored [16, 20–22]. One of the classic and most well-known blockchain applications is that of Bitcoin [23], but it is important to recognise that blockchains can be deployed for diverse uses other than financial ones.

Due to the properties of blockchains, and especially due to the utility of these properties in various fields, the so-called 'blockchain revolution' has touched on numerous areas, including both academia and industry. The main properties of blockchain technology, as identified in the literature, are the following: immutability, decentralisation, and traceability. Regarding immutability, this refers to the fact that once a block has been added to the ledger after consensus has been established, it cannot be changed or removed [12, 24]. There are two ways to account for the immutability of blockchains: firstly, since on-chain transactions are connected with the same hash key that links the parent and child blocks, it is trivial for participants (nodes) to identify any transactions that have been altered; and secondly, given that blockchain is a distributed ledger technology (DLT), each node in the network contains a copy of the public ledger that is continuously synchronised across the nodes [16]. This second point implies that to change any transaction successfully, a minimum of 51% of the stored ledgers must be changed.

The third main characteristics of Blockchain is known as decentralised technology. Decentralisation property eliminates the existence of centralised organisations to proceed all the processes of transactions' validation, storage, maintenance, and transmission on Blockchain [16]. The Blockchain builds based on the structure of distributed system with lake of central controlling party to rely on [25]. As a consequence of this structure, the trust of the chain's transactions among Blockchain network's nodes can be maintained throughout mathematical methods.

An important feature that results from the DLT is traceability, which refers to the ability to trace and follow connected blocks based on their corresponding hash keys [16]. This traceability derives from the fact that transactions are ordered sequentially on a blockchain, which guarantees that every block is linked to two other blocks [12, 16, 23]. Critically, the properties of blockchains are more numerous than those discussed here, including the fact that blockchains bring transparency to every block that is appended to the chain [26].

Although Blockchain-based systems provide many benefits, Blockchain adoption has some challenges that need to be overcome to increase the acceptance of Blockchain technology in various fields. The first challenge is the scalability of network transactions. Reyna et al. note that scalability is a major concern in using Blockchain when the amount of data and the number of transactions grow very fast; for example, bitcoin transactions increase by a megabyte per block every 10 min [19]. In addition, according to Zhang et al., the Ethereum public Blockchain defines a 'gas limit' that limits the capacity of data operations to prevent attacks from manifesting through infinite looping [20]. Various studies have argued that the main reason for adopting Blockchain technology is its security characteristics. Conversely, other studies have demonstrated that security is one of the disadvantages of adopting Blockchain technology [12, 21]. Privacy can be maintained in a Blockchain by adopting both a private and a public key mechanism. Users can make transactions using both public and private keys and thereby avoid exposing their real identity. However, Huckle et al. [21] show that Blockchain cannot assure the privacy of transactions because the balances and values included in the transactions are clearly visible to the public in every public key.

In the last decade, Blockchain technology research has become a growing trend in computer science, with increasing attention from various researchers and organisations. Since 2008, it taken a place among the top five technology trends and is considered to be the next revolution in technology as it provides solutions to the issues related to classical centralised architecture [27, 28].

3 Overview of Current Smart Campus Frameworks

Many researchers have proposed different smart campus frameworks that would need to be reviewed and studied in order to develop our suitable smart campus framework. Most of these frameworks have been used IoT architecture to propose smart campus environment. The middleware layers—sitting between the hardware and sensors layer and the applications layer—in IoT architecture have almost the same functionalities; however, the underlying technologies are different. Thus, it is important to address these frameworks based on IoT architecture layers.

Some researchers have developed a smart campus framework using the traditional three-tier structure. For example, Narendrakumar and Pillai [29] proposed the *Smart Connected Campus* framework to deploy and exploit IoT technologies in a campus setting in order to remotely monitor various campus activates. The architecture generates several features reliably and easily at run-time, such as water and temperature monitoring systems, a route map and online resources, and integrates them all in a single platform. All the facilities are communicated and networked through IoT, enabling data generation at run-time and remote access to the data. The framework consists of three major components: sensor technologies, a cloud server, and Android mobile applications. In addition, authorised login credentials are used to authenticate data access. Furthermore, Hossain et al. [30] proposed a smart campus model based IoT technology, which included campus facilities such as smart parking, smart classroom, smart buildings, and cloud computing system. The model contains of three levels: a perception layer which is responsible for sensing data from IoT campus devices; a network layer which sends captured data to cloud storage and stores collected data via Wi-Fi or the internet; and lastly, an application layer, which provides services and applications to the end user.

Zhe et al. [31] proposed a smart campus-based information service. The framework integrates various service information systems with the individual's location to unified information service on campus. The architecture uses the basic three-tier structure. The presentation layer, mainly based on the mobile applications to present the information, includes messages, location and statistics to users. The application layer comprises location applications and local information services. The data layer consists of three components: data source, including location data and different applications data; basic data, which is utilised to store basic and location information; and data interface, which supports the interaction between the system data and the location service platform. The study did not, however, mention Blockchain technology.

Moreover, Yan and Hu [32] constructed a smart campus framework-based data platform, providing the association with data analysis using Apriori algorithm as association mining algorithm. The framework consists of basic three-tier structure: the user interface, which shows the results of data mining; the middle layer, where data is processed; and the data layer where the data is stored in a database. In addition, Hu and Yan [33] proposed smart campus-based big data framework to offer cooperation and interconnection between different applications systems. In addition, the framework used cloud computing as big data storage and the K-Means algorithm to process the data. However, all of these smart campuses used traditional databases, which suffer from third party issues, and Blockchain could be the solution to storing the data. The frameworks could use Blockchain technology to increase the security and protect the data.

Some studies been developed smart campus frameworks using four IoT architecture layers. For instance, Agate et al. [34] designed a smart campus-based fog computing framework to enable data collection from various smart devices. The edge of the network enhanced the services available to users and improved the user experience in the campus. This framework-based IoT architecture consists of four layers. The lowest level of the framework includes various heterogeneous sensors, which are responsible for collecting raw data and sharing it with the higher level. The next layer consists of fog entities, which received data transmission from edge devices using communication infrastructures, such as Wi-Fi, aggregating all data and sharing it with the cloud. The cloud layer computes and analyses the data, as well as being responsible for storing all information needed to display the collected data and results on the application layer. This approach focuses on integration of fog computing and IoT to improve the scalability of the system. However, such a system usually includes security and privacy issues.

Enqing et al. [35] constructed a smart campus and three-dimensional geographic-based to share data and services, as well as the management of heterogeneous and multi-source data, such as spatial real data, information database and geographic information. The framework contains of four layers. The first layer is the physical layer, which includes hardware, network devices, and the operating system. Then the data service layer has different databases, such as a spatial real database, geographic information, and other databases. Next, the platform function layer includes a basic function to support the 3D geographic information interface and the business function, which supports data interface, such as image and video queries. Lastly, the application layer is responsible for providing application and data services. This study did not cover Blockchain technology.

Lihong [36] constructed a smart campus based on IoT and deployed on Hadoop as cloud server. The framework provided resource sharing, dynamic understanding, and information released real-time. The architecture was divided into four layers: a data layer, responsible for storing and classifying data based on business requirements and data characteristics; a system layer, comprising operating system and several subsystems, a central server hardware platform and the layer supports that connect other layers; a network layer, which includes the IoT networks that support different types of sensors for coverage and transmission of various resources and information on campus; and an application layer presents obtained data to users and provides interactive services through management processes and business rules. However, the framework did not discuss the integration of Blockchain technology into its architecture.

A five-layered IoT has also been used to develop smart campus frameworks. For example, Agarwal et al. [37] proposed a five-layered IoT smart campus framework that can be implemented in various cases. The first of the five layers is a sensor and data acquisition layer, which is responsible for capturing data in real time from different sensors then sending the data to the upper layers. A compute and infrastructure layer comes next, which consists of the required infrastructures, such as Wi-fi, Zigbee, and BLE that are needed for network connectivity by gathering data from the previous layer and sending it to the cloud or backend server. Then the platform layer is responsible for developing a unified layer to support communication among heterogeneous systems, integrate the data from various systems, and define business rules. The application layer helps by defining and implementing different usage scenarios within the smart campus. Finally, a monitoring layer monitors the applications and creates alerts when needed. A dashboard and command centre can be set up to monitor applications on a regular basis. The framework of smart campus can help to implement different cases based on usage; however, it relies on IoT technologies and third parties to store the data, which leads to several issues that are discussed in more detail in the next section.

Lastly, some researchers have proposed a smart campus framework using seven IoT architecture layers. For example, Debauche et al [38]. presented (R)evoCampus as a smart campus architecture-based IoT technology. The seven layers comprise the following: infrastructure layer responsible for collecting environmental information or understanding an action; an information layer allows different objects to identify and communicate with connected devices; a communication layer, responsible for data transmission among connected devices using various technologies and protocols, such as Zigbee, 4G, and Wi-Fi; a connective layer, allowing interoperability and connectivity of exchange data among devices; a middleware layer safeguards the storing and processing of data in the cloud; a service layer delivers reliable and critical facilities for several applications, such as open data and weather station; and finally, an application layer provides various data consultations, visualisations, and presentations to end-users. However, the framework did not mention Blockchain technology.

In conclusion, all of the studies mentioned above have proposed smart campus frameworks based on IoT technologies. Each framework has been developed using different IoT architecture middleware layers, which mainly sit between the hardware and sensors layer and the applications layer. The middleware layers contain a set of functionalities, such as data storage, data management, and data processing, which are required when developing smart applications, since data is the heart of smart applications and IoT environments. Therefore, data has to be secured and must not be stored in third parties. This is discussed in the next section in order to propose our smart campus framework, which stores data in a secure way, as well as using suitable IoT middleware layers, providing appropriate technologies and tools in each layer.

4 Existing Problems for Current Smart Campus Frameworks

Although the above existing smart campus frameworks provide different architectures with a different number of layers, they all rely on an IoT centralised architecture, which is developed using a centralised server, to manage and control connected nodes. In other words, a centralised architecture is a clint-server architecture, that typically has a centralised server cloud computing and clints that are represented by IoT nodes. The centralised server has many roles, such as processing data, managing task scheduling, dealing with all requests coming from the network and storing information.

The centralised architecture provides several advantages, it allows a variety of devices to connect and communicate among each other under the management of the centralised unit and provides the needed identification and authentication for connected devices. The whole network, in this case, is easy to control and is maintained through central server. In addition, the architecture does not need to be installed in many workstations requiring software and hardware since most of the operations are done by the centralised unit, which also saves costs.

However, the centralised architecture suffers from various threats and complications. Khan [39] and Atlam and Wills [40] discussed these issues in detail as summarised in the following:

- **Security:** Many studies have been done and have confirmed that the most serious issue for the centralised system is security [41]. Since all the system operations are executed via a central unit and all data are saved in one place, it increases the risks of being a target for different types of attacks. Denial of Service (DoS) attacks and information leakage are most common. Therefore, there is a high risk in storing sensitive data in a centralised server.
- **Privacy:** Another vital issue that needs to be considered is privacy, since smart devices collect a massive quantity of confidential and personal information such as passwords and financial accounts, and is stored in centralised unit which usually belongs to third party and can easily be attacked. There are some examples of service providers attacking the privacy of users [40, 42]. For instance, some third parties such as Facebook sell user information to marketing companies, which in turn use this information to analyse user behaviour in the network [42]. In addition, data can be altered or deleted from insider attacks that will affect data integrity. Thus, a secure method is necessary to provide privacy of information.
- **Single point of failure:** When all operations, controls, and storage are achieved by a centralised server, it can create a single point of failure which means when the central unit is down, the whole system will fail and will be unavailable [40]. The common way to avoid this issue is by adding redundant switches, servers, and network connections as a backup to work alternately when the original centralised server fails. Nevertheless, this approach has many issues, such as it is

expensive to install the alternative requirements, and that there are synchronisation problems between the backup and the original server.

- **Scalability:** Another main issue for a centralised architecture is scalability [40]. The centralised unit uses a central authority to process all commands and controls which can be scaled for small networks however, it can be impractical for large organisations that are spread over different locations, which would increase the number of IoT devices needed. In this case, transporting decisions for large organisations may suffer and not operate efficiently.

Lately, Blockchain technology has been adopted in several domains such as healthcare, finance, and the supply-chain due to characteristics including resiliency, anonymity, autonomous control, support for integrity, and decentralisation [43]. Since a Blockchain network is a distributed ledger, it is not controlled by an authorised single node. Kshetri et al. [44] compared the privacy and security measures of cloud storage and Blockchain technology. In cloud storage, data is typically not secured and it can be easily accessed by central authorities, which can lead to data leakage. Blockchains, on the other hand, allow users to have complete control over their data, which offers guaranteed privacy and security. Therefore, most researchers are using Blockchain technology to improve the level of privacy and security in their environment. For example, a framework was presented by Kianmajd et al. [45] that aims to integrate Blockchain into a system for preserving user privacy while accessing community resources. In addition, Zyskind et al. [46] created a platform the enables users to manage their personal data. Their implementation used Blockchain-based protocol to secure and store sensitive data in a ledger. Instead of transferring the data to the network, encryption was used to prevent unauthorised access. Biswas and Muthukkumarasamy [47] developed a Blockchain-based framework for secure communication in a smart city. The authors stated that integrating Blockchain technology with smart city devices can create a secure environment in which all devices can transact securely. Thus, Blockchain technology can help to secure and control sensitive personal data, including student data, ensuring that it is under the user's control and not controlled by an authorised single node.

Various methods have been proposed to verify data integrity in systems such as cloud storage, including Provable Data Possession (PDP), Proof of Retrievability (POR), and Proof of Reoperation (COR) [48]. However, most of these traditional schemes rely on centralised services to perform verification operations instead of the user to alleviate the burden. As discussed previously, using a third party is an inefficient way to store data. The development of Blockchain technology has solved this issue and, as such, provides different schemes for data integrity. For example, Liu et al. [49] proposed a Blockchain-based data integrity architecture that enables data consumers and data owners to easily validate data integrity. Yue et al. [50] developed a framework based on a Merkle tree to verify data integrity using Blockchain technology for peer-to-peer cloud storage. Wang et al. [42] proposed a data audit model that replaces the centralised node with collective trust. The model allows users to gain a better understanding of their data and enables them to trace their data history. In addition, Yatskiv et al. [51] developed a novel method to ensure the integrity of video data and protect it from unauthorised changes. The concept is based on a hash function that is calculated for each frame, where its blocks are sent to cloud storage.

In addition, Blockchain technology can enhance the authorisation level since it needs to reach consensus or agreement among the majority of network peers to verify transactions. Fabiano [52] discussed various challenges in terms of data privacy in IoT systems. Controlling network access is the main issue because the existing system uses different network protocols. Blockchain technology has been used to improve the security of several IoT applications without requiring the need for centralised services to control access by using a consensus mechanism (i.e., BFT, PoS, and PoW for validation and verification criteria in IoT). Cha et al. [53] proposed a Blockchain-based gateway that enables authentication for users when connecting to an IoT device. Through a Blockchain with Eris, a smart contract and an out-of-band authentication scheme, another group of authors protected an IoT system from exploitation by detecting malicious devices [54]. FairAccess [55] is a Blockchain-based access control framework for the IoT, which uses grant, get, and revoke access methods. Its decentralised nature, paired with its ability to enable transparency and security, makes it a suitable choice for diverse IoT applications.

All these properties put it in the lead as a suitably advanced technology to apply to our smart campus framework. Therefore, deploying Blockchain technology to our proposed smart campus framework will provide more benefits by managing the problems associated with a centralised IoT architecture, especially from a security perspective.

5 Design of the Proposed Smart Campus Framework

A novel smart campus framework architecture based on IoT and Blockchain has been designed to provide a comprehensive guiding framework [56]. The main goal will be the collection and aggregation of the data from various areas of the campus while increasing the data security and providing a better service to enhance the experience of the user. To accommodate the different areas of the campus, several devices and sensors will be used. These will be distributed around the campus and will work simultaneously with an aggregation of networks and software to generate heterogenous data, which will be released as reliable and valuable data from the IoT. Thus, there is a need for the system to support several network protocols to provide scalable and reliable communication across the networks. In addition, the system needs to handle the previously discussed problems related to current smart campus frameworks, especially security.

IoT architecture with six layers is applied for this framework. It employs Blockchain technology to eliminate the third party, decentralise the architecture and ensure data security. See Fig. 2 for the conceptual framework.

The following subsections describe the framework stack and Blockchain in details:

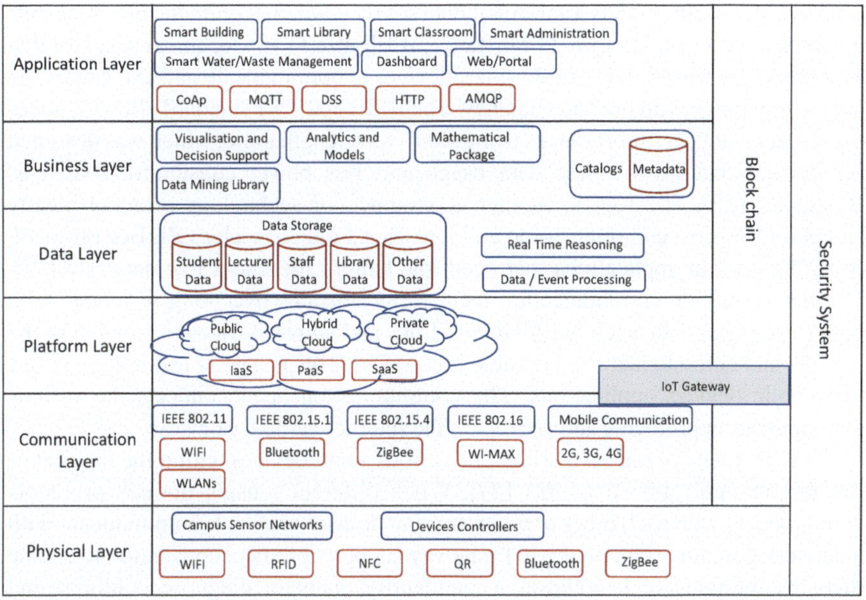

Fig. 2 A novel smart campus framework [56]

5.1 Physical Layer

On the lowest level of the proposed architectural framework, several sensors have been used to collect and share raw data with upper levels. There are two types of sensors in our system: deployed and wearable. The former is at fixed points in various areas of the campus such as RFID and GPS. Cameras are also installed at these locations to gather environmental data such as humidity, location, and temperature. The wearable type of sensor can be used by individuals daily in items such as tablets and smartphones. The layer also contains actuators that are responsible for the physical action of converting data signals coming from the upper layers. They have an opposite function to the sensors. Thus, smart devices are defined as any hardware component that can be connected to the internet and can operate with or without wires.

5.2 Communication Layer

Since IoT is about interoperability and connectivity, there is a need for communication between objects. Thus, the communication level is responsible for broadcasting the data across the layers in the network, the initial connection setup and dealing with data transmission errors. The use of communication infrastructures, including technologies and protocols, is necessary for the transmission of data. An application such as a smart campus has a variety of data sources, which require the use of diverse communication technologies. For instance, IEEE 802.11 is a technical standard that supports radio waves and permits smart devices to communicate and exchange data over a long range with power efficiency. The IEEE 802.15.1 standard allows devices to communicate over short ranges using short-wavelength radio, which was designed for devices that involve less data usage and less power consumption such as Bluetooth. IEEE 802.15.4 is another communication technology protocol ideally suited for low-cost communication and low digital radios, such as ZigBee protocol. It can be used in applications that need low battery life and a low data rate [57]. Cellular or mobile communication technology operates over long distances with high throughput data, such as 3G/4G/5G [58], and allows devices to connect to the internet and communicate at a reliable high speed. This is a good technology to use for mobile devices applications. The communication layer connects the various communication protocols and sensors to the internet via IoT gateways.

The IoT gateway acts as a bridge for various smart devices within the internet or public communication networks [59]. When different sensing domain protocols communicate with each other or when sensing domain protocols communicate with a network domain protocol, the IoT gateway supports protocol conversion to enable them to communicate. In addition, it can identify, maintain, diagnose, configure and control smart devices. In other words, the IoT gateway for different networks acts as a proxy and solves the integration issues.

Many have studied integrated Blockchain technology in this layer to provide security in communication. For example, Biswas and Muthukkumarasamy [47] proposed a smart city framework that integrated smart devices with Blockchain technology to support secure communication. The paper stated that the Blockchain needed to be integrated with the communication level, converting the transactions into blocks using Telehash to be sent to the network. This would provide privacy and security. Cha et al. [53] developed a gateway transmitting data connected to the Blockchain to mediate between IoT devices and users. The gateway also securely maintained and managed user preferences and prevented them from being altered while legacy smart devices were in use.

5.3 *Platform Layer*

A smart campus generates a significant amount of data from various sensors and resources that need to be processed and analysed in a short time. Traditional information systems are not effective for this as they are prone to data inconsistency and have no data sharing capacity [60]. Therefore, there is a need for advanced technology in the processing unit that can provide the resources needed and have the computational capability. Cloud computing technology is a shared pool of data information and computing resources that can analyse data and provide valuable results. Blockchain can be integrated with this layer to decentralise the system structure and can be used instead of cloud storage to store the data.

5.4 Data Layer

Campus systems have a massive amount of data called 'big data' that needs to be placed in a data layer. The layer can store and retrieve data and contains all the data that the campus needs in a variety of forms. Instead of storing campus data over the network, Blockchain technology can be used to decentralise data storage and use a data warehouse to add privacy and security.

5.5 Business Layer

Every organisation has specific objectives and goals that need to be accomplished. For this, they need to collect intelligence from analysed data and apply it to their business strategy and planning. This layer delivers an evaluation of the performance of existing services and applications and provides intelligence solutions. The success of an IoT system not only depends on the use of advanced technology but also depends on who the system delivers the services to, i.e. the users. The business layer is responsible for creating graphs, models, and flowcharts based on analysed data, as well as creating business rules, process management, and the process of decision making.

5.6 Application Layer

The application layer includes campus smart services such as learning, library, administration, building, and parking. The analytical data is collected, and the information is presented in a visual form. In addition, it contains a set of protocols that are responsible for transmitting messages from the application layer [61]. For example, Constrained Application Layer Protocol (CoAP) is a lightweight protocol, which is appropriate for deployment for machine-to-machine applications, particularly for communication in a local network and web services as it follows a request/response model which supports broadcast and multicast. CoAP is designed to limit fragmented messages by keeping the overhead small. Message Queue Telemetry (MQTT) is also a lightweight protocol. It is used in a publish-subscribe-model; therefore, it is appropriate for cloud/remote communication and for applications that involve short-range mobile communication, satellite links, and dial-up connections. MQTT is widely used for battery-run devices as it is efficient, has a compact size, consumes less power and delivers information to multiple receivers at the same time. Advanced Message Queuing Protocol (AMQP) is designed to support business applications and to transmit business messages on different platforms. Thus, merging these protocols increases the efficiency of device communication and enables interoperation in smart applications.

5.7 Blockchain Layer

Despite the benefits of IoT with connecting devices, it does have limitations in privacy and security as it relies on the centralisation of architecture, which leaves it open to attack. Blockchain technology recently merged with IoT technology. This has provided a decentralised ecosystem environment in several innovative cases beyond cryptocurrencies, such as healthcare, government, and learning, with enhanced security and reduced risk. The large volume of data in any ecosystem environment needs to be stored in such a way that it is immutable, available, and accessible. For example, student data in a higher education environment is created when a student enrols in a university. The data grows as their time at the university lengthens, and will include personal, courses, grades, finance and health information data. Student information needs to be accessible by different stockholders including administration, lecturers, and finance. However, the information should be immutable and private. Therefore, a structure of storing student records should be accessible and maintainable. Blockchain allows parties to share a student record in real time with privacy, integrity, and immutable features, which are all required in smart university systems.

Blockchain technology networks can be classified into two different models: permissionless and permission networks [62]. Permissionless or a public Blockchain network is a decentralised ledger opened to any peer to join and publish a block without authority or permission. This network is mostly open-source software and available to any user to download without permission and may be executed on millions of devices. In addition, any peer has the right to join the network, publish blocks, issue transactions, read the Blockchain and validate the transactions. Thus, any node in the permissionless Blockchain network can access the ledger to read or write without requiring the identity of a user. This type of Blockchain network applies consensus or agreement algorithms that run in the Blockchain's protocol to prevent malicious peers from accessing the ledgers. Proof of work is one of the well-known consensus algorithms; proof of stake is another one. Since every node in the network applies a consensus mechanism to verify the transactions to increase the security, this consumes a vast number of resources such as electricity and storage costs and has low transaction throughput. Examples of a permissionless or public Blockchain network are Bitcoin [63] and Ethereum [64]; they are wildly used in finance and cryptocurrency areas.

A permissioned Blockchain network is a network where only an authorised user can publish blocks. The network has restrictions as to who can access the Blockchain and issue transactions. Different restriction rules can be applied to a permissioned Blockchain network; it may allow any user to read the ledgers, it may be restricted to only authorised nodes, it may allow any user to issue transactions in the network or restrict it to permission users. A permissioned Blockchain network may have open or closed source software. A permissioned Blockchain network is suitable for use in organisations that are required to control their Blockchain, or where organisations working together do not fully trust each other. These organisations can establish a consensus algorithm based on the desired trust level. Transparency can be achieved in this type of network and may assist in improving business decisions. There are, therefore, two types of permissioned Blockchain networks: private and consortium.

The private Blockchain network is a network where only invited and selected nodes have permission to access the network. It is partially decentralised since a peer must control the permission to validate the transactions, change the rules and read and write the ledgers. Unlike the public Blockchain, a reduced number of involved nodes help in the efficient use of resources and transaction throughput. Nevertheless, a private Blockchain network affects security and trust because the trusted peer may be compromised with the reduction of transparency and the immutability of the network leading it to be highly centralised, although there are still more benefits than with a traditional database. Ripple [65] and Eris [66] are examples of this type of network.

The other type of Blockchain network is a consortium, which combines the advantages of both private and public Blockchain networks. As with a private Blockchain network, a selected number of peers have permission to join the network; however, all the invited nodes have the right to read and write to the ledger, as well as verify transactions. This is like the decentralised nature of a public Blockchain. In terms of transparency and immutability, it is midway between the structures of the two previous chain models. Its features lead to an enhanced efficiency compared to a public Blockchain network. In addition, nodes in the consortium Blockchain compute only their role either endorsing peers or committing peers, which helps to improve the network performance and throughput. Furthermore, this network is more trustworthy and less vulnerable to threats of security compared to a private network as the network is not controlled by a single node to validate, authorise, write, or read to the ledger. Hyperledger [67] and Corda [68] are examples of a consortium Blockchain network. In our view, using a consortium Blockchain network is the best fit for our smart campus framework.

To sum up, a smart campus includes applications, technologies, and smart features, which can be implemented in any area of the campus. There have been numerous studies done on ways to implement systems to develop a smart campus [29–31], however, no standard design has yet been developed. Our new framework provides a guiding architecture that may allow the integration of Blockchain technology with existing technologies as a base to develop various smart systems further. The framework will be tested in the next section for the higher education certification system for issuing authentic, verifiable, and sharable student credentials.

6 Blockchain Use Cases in Higher Education

This section discusses existing blockchain-based technology systems that have been developed in the field of higher education in various functionalities. According to Turkanović et al. [69], different higher education institutions (HEIs) worldwide are considering the adoption of blockchain technology as an aid to designing approaches and solutions for higher education. Several of these systems and solutions have adapted Bitcoin and Ethereum blockchain-based technology. Blockchain technology has been adopted in various domains and fields predominantly due to its accessibility, auditability, and distributed storage benefits.

6.1 Student Profile and Records

This section describes and discusses existing blockchain-based systems in the education field that handles student records and data.

MIT Media Lab (Blockcerts) Blockcerts, an open standard, was developed to enable the creation, publication, viewing, and validation of blockchain-based certificates [70]. Many digital records are registered on blockchains, which are cryptographically signed, tamper-proof, and shareable. The goal behind the innovation is that it poses the capacity of the individual and the sharing of the official records [70]. According to Jirgensons et al. [71], MIT Media Lab is the only institute in the United States that has established and developed a complete education credentialing system based on blockchain technology. The MIT team used the Bitcoin framework rather than Ethereum to develop its credentialing system, the rationale being that Bitcoin was considered a stronger technology to hold the transactions [71].

EduCTX The European Credit Transfer and Accumulation System (ECTS) is based on the global blockchain, and it was developed with the higher education credit platform known as EduCTX [69]. The planned system leverages the advantages of blockchains in terms of their decentralised architecture, anonymity, security, longevity, transparency, immutability, and global simplifications. In this regard, there is the creation of the international trusted credentialing system for higher education institutes [69].

Knowledge Media Institute, Open University (Badges) The UK-based Open University's Knowledge Media Institute (KMI) is one of the most innovative universities to employ blockchain technology [71]. The KMI created an Ethereum-based blockchain platform for academic applications known as OpenBlockchain in partnership with British Telecommunications (BT). The target of the KMI consists of students at UK universities, and the institute offers badges called "microcredentials" that are designed for the courses offered on the UK's massive open online courses (MOOCs) and open learning websites. Smart contracts control registrations for the earned microcredentials, and these provide details about the badge (e.g., the receiver, security assertion, issuer, evidence of accomplishment, and criteria applied) [71].

KAUST (Saudi Arabia) In November 2018, King Abdullah University of Science and Technology (KAUST) and Learning Machine Technology (LMT) announced that KAUST will start issuing blockchain-anchored credentials and blockers. The changes are adopted by KAUST among the most innovative institutions worldwide, including MIT, Stanford University, and the University of Melbourne [72].

6.2 Digital Credentialing Systems

In HEIs, credentialing systems are used to generate and manage certificates, degrees, and other achievements and rewards for students and alumni. These documents are necessary for university alumni to receive jobs that correspond to their degrees. With the many issues currently facing university credentialing systems, digitising this process is an ideal way to solve the issues and exploit promising opportunities. Existing credentialing systems use analog operations to manage certificate generation. Notably, these systems are slow and unreliable, and, in some cases, they may produce cultural and social tensions depending on the context of the education system. Therefore, creating a digital infrastructure for certificate-generating systems represents an important opportunity to leverage many promising new technologies, including blockchain. Nevertheless, given that credentialing systems deal with highly sensitive data and, moreover, represent a HEI's professional reputation, both need to be assured. When selecting an appropriate technology, decision makers should be fully aware of the technology, its design, and its characteristics, thereby promoting trust in the operation of such an important system.

In recent months, HEIs have become increasingly interested in using blockchain-based digital certificates [73]. The adoption of blockchain technology can help to establish a certification infrastructure that enables students to control the complete record of their achievements. Thus, such a system would enable students to have full access to their awards and certificates even in the event that the issuing HEI no longer exists [74]. Furthermore, students can share their credentials with other universities and prospective employers under the assurance that the information will be shared only among trusted parties [73]. Although blockchain adoption offers multiple opportunities to improve existing credentialing systems, using the technology does not offer a straightforward process to help overcome all the challenges facing credentialing.

6.3 Personalisation of Learning Courses

From a technology perspective, the use of internet-based platforms for higher education processes is not novel. For example, over the past decade, MOOCs have emerged as a core feature of the educational landscape [75]. The MOOC has been chosen as one of the existing developed learning creation systems designed for the higher education sector. The benefit of using MOOCs relates to their ability to provide an affordable and flexible method to acquire new skills and provide the learner with advanced career options. Therefore, MOOCs create a substantial opportunity for HEIs to deliver high-quality educational experiences at scale.

Approximately 7000 MOOCs are currently available for students across over 700 universities, and there are now around 58 million students enrolled in these programs [75]. As such, since their initial deployment in 2006, MOOCs have become established as a widely used aspect of the distance learning paradigm [76]. Regarding the exploration of online educational technology, MOOCs have had a significant impact on the educational landscape for nearly a decade, and they are considered one of the most distinguished ways to simplify the education sector.

6.4 Publishing

The field of scientific publication is another area that can leverage Blockchain technology. The main issue in this field is the system for submitting manuscripts. This is because, in most cases, the system provider is a third party that supports organisers of journals in terms of coordination and academic conferences in terms of the submission workflow, which starts from submitting the abstract and paper and moves onto arranging peer reviews and camera-ready submission. Although such systems have made the manuscript submission process more efficient, they have also raised concerns about potential dishonesty and technical weaknesses. For example, the naming scheme for all paper submissions made to Sheridan Printings, which is a software for conference management utilised by several conferences (e.g., ITS and TEI), was easily guessable for the 2004–2011 period. In particular, the naming scheme made it possible to retrieve all the documents submitted to a particular conference easily for anyone with the base URL [77]. Thus, in such cases, a researcher must trust the system when submitting their unpublished work, which leads to the possibility of data leakage or idea leakage. In addition, users must trust that anonymous peer reviewers and the program committee will not plagiarise or amend the results or findings.

Many research papers have proposed the use of Blockchain systems to resolve these issues and to decentralise manuscript submission systems. For example, Gipp et al. [77] developed a scientific manuscript system named CryptSubmit to provide data integrity and security for submissions, which relied on the use of Blockchain technology to verify timestamps as a tamper proof data point connected to each submission. Hepp et al. [78] proposed OriginStamp using Blockchain technology for intellectual property protection, offering a novel method for timestamping and storing documents. Pozi et al. [79] proposed a framework based on Blockchain technology that used proof of work to measure writer contributions and to enable the paper to be accessed transparently. In addition, Andi et al. [80] used Blockchain to design a plagiarism prevention model, along with digital signatures to protect the work and allow only reviews to access the paper without altering it. Therefore, thanks to Blockchain technology, particularly due to its features of decentralisation and immutability, it is possible to improve security and privacy in the field of education.

7 Developing a Framework for Adoption Blockchain in Higher Education Certification Process in Saudi Arabia: Case Study

Due to the establishment of new research and development programs, paired with the growth of new educational schemes, the Saudi Arabian educational system has improved dramatically in recent decades. A key aim of the Saudi Ministry of Education (MoE) is to establish an integrated services system that supports educational processes by improving performance, efficiency, and promoting the adoption of state-of-the-art technologies [81]. In 1999, there were only 10 public universities in Saudi Arabia, but this figure increased to 26 by 2017. This figure is anticipated to increase more in the future due to substantial population growth in Saudi Arabia, with estimates suggesting that the country's 2012 population of 29.2 million will increase to 35.9 million by 2020 [82]. In 2016, the number of students at university age in Saudi Arabia reached approximately 1.7 million, whereas this figure was 850,000 in 2009 and around 650,000 in 2006. With these considerations in mind, it is clear that the Saudi higher education system is undergoing a range of significant changes concerning its capacity, international connections, graduate outcomes, and research impacts [83].

7.1 Overview of the Proposed Framework

The purpose of the proposed framework is to fill a gap in the literature concerning the current state of research on blockchain adoption among higher education institutions (HEIs). Our motivation was to examine this topic and propose a DASC solution, thus contributing to the growing body of literature in this field. This chapter reviews previous studies that have examined blockchain adoption in HEIs, and it also explores differences between the proposed blockchain-based DASC and existing certificating or credentialing systems, focusing especially on the limitations of the latter. We argue that the proposed framework is an adequate one because it integrates all relevant actors, processes, and data storage units; it also accounts in a logical way for the principal obstacles in current systems, which include certificate fraud and dishonesty. The deployment of blockchains will make it possible to exploit the property of immutability for student records, which we believe will address critical challenges in existing HEI credentialing processes.

Currently, the design phase does not highlight any data storage issues that the system may encounter in a real-world setting, which means that system scalability cannot be analysed. Nevertheless, under a scenario where a blockchain is used as a database to store student records (e.g., containing information such as student ID, name, date of birth, department, courses, and badges achieved), the sheer number of data points would overwhelm the chain, which would include spreading all over the network's nodes. Over time, such a blockchain would encounter challenges regarding the maintenance and storage of these data for every student, thus potentially undermining system performance. In our future research, we intend to undertake experiments to test a prototype of the proposed system, which will involve investigating the impact of the framework's influential factors on the target user's feedback.

7.2 *Influential Factors*

The framework shown in Fig. 3 provides an overview of the main factors that influence the adoption of blockchain technology in the higher education sector. These influential factors will form the basis of the tests undertaken in the later parts of this research. Specifically, the purpose of this framework is to assess the acceptance level of blockchain technology adoption across four types of users in the certification process in HEIs. It represents four main categories of factors that reflect the key requirements for Blockchain-based system adoption.

Trust Given trust is impacted by a range of quantified and non-quantified properties, understanding the meaning of trust is complex. However, concerning the acceptance and use of novel technology, trust is a critical factor.

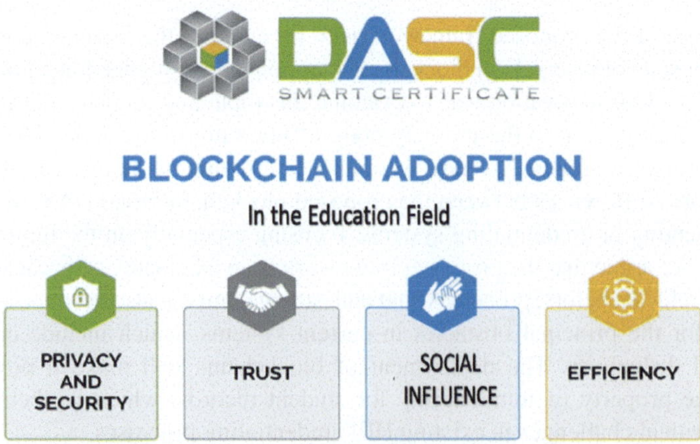

Fig. 3 Blockchain Adoption Framework Factors [84]

Security and Privacy It has been noted in several papers that a prominent reason for blockchain technology adoption is the technology's security properties. Nevertheless, some researchers have shown that a fundamental disadvantage associated with the adoption of blockchain technology is security. Ultimately, as noted by Halpin and Piekarska [85], the degree to which blockchains enable privacy and security must be studied in greater depth. According to Garcia-Font [86], a key area in current blockchain research is identity management. Presently, nearly every notable authentication system relies on usernames and passwords, which is inherently risky. Notably, the application of blockchain technology can account for these risks and disadvantages, increase security and privacy, and decentralise the storage of identifiers [86]. Additionally, blockchain technology can enable identity management without the need for usernames and passwords, which represents a critically important advantage [86].

Social Influence In the field of blockchain technology, social influence may expand to the user's perception of a service provided by the technology that is highly influenced by the perceptions of other fields and domains concerning technology adoption. According to the unified theory of acceptance and use of technology [87], social influence is one of the four primary factors that impact a user's decision to adopt a technology.

Efficiency Through the use of blockchain technology, it is possible to complete any transaction in an efficient way within a decentralised environment. Hence, this lowers the total cost and improves the efficiency of transactions. At the same time, blockchain technology reduces transaction fees and lowers the time needed to execute a transaction [14]. Due to blockchain's ability to reduce paperwork, mitigate process-associated administrative expenses, and elevate the efficiency of routinary processes involving several parties, the prospect of its adoption is promising in diverse areas [86]. Hence, we regard studying the effect of efficiency as a factor that affects the process of blockchain adoption and higher education.

Having outlined these four influential factors, it is important to note that these are not the only relevant considerations when investigating the acceptance of novel technology in a sensitive field such as higher education.

7.3 The Design of the Study

A questionnaire was disseminated online to students enrolled in higher education and prospective employers. The rationale for including prospective employers was that these stakeholders are pivotally concerned with validating student credentials, as a result of which they may impose pressure on HEIs to adopt blockchain-based certificating systems. The target population of students consisted of all students enrolled in Saudi Arabian universities, particularly Riyadh's universities (i.e.,

Table 1 Demographic characteristics of the case study

Characteristics		Student (*n = 405*)	%	Employer (*n = 22*)	%
Age	18–25	244	60.24	0	0
	26–35	88	21.72	11	50.0
	36–45	49	12.09	9	40.9
	46–60	21	5.18	2	9.1
	+60	3	0.7	0	0
Total		**405**	**100**	**22**	**100**
Gender	Female	285	70.4	6	27.3
	Male	119	29.4	16	72.7
	Prefer not to say	1	0.2	0	0
Total		**405**	**100**	**22**	**100**
Education level	High school	25	6.17	1	4.5
	Undergrad	220	54.32	0	0
	Bachelor's degree	88	21.72	8	36.4
	Postgraduate or higher	72	17.77	13	59.1
Total		**405**	**100%**	**22**	**100**
Field domain	Science, Technology, and Engineering	208	51.35	12	54.6
	Business and Economics	47	11.60	5	22.7
	Humanities and Art	49	12.20	4	18.2
	Other	101	24.94	1	4.5
Total		**405**	**100%**	**22**	**100**

universities in the country's capital). To maximise the response rate, the questionnaires were disseminated in Arabic and English. The types of data obtained via the questionnaire were the following: demographic information, technology awareness, current process situation, and factors influencing the adoption of blockchain in the certificating process in Saudi Arabia.

Four hundred twenty-six responses were received from students to the questionnaire, 405 of which are included in our analysis because the students provided full consent. The total number of prospective employers who responded to the survey was 34, but 4 participants provided incomplete responses and a further 8 did not consent. For this reason, the total number of prospective employers who are included in our analysis is 22.

Table 1 provides an overview of the participants' demographic characteristics. For the students, as the table indicates, 70.4% of the participants were female while 29.4% were male. A further 0.2% did not specify their gender. The largest age group represented in the sample was 18–25 years (60.2%), and in terms of the highest educational qualification, most participants were undergraduates (54.3%), which constituted the target population of the research. A further 21.7 of the participants held bachelor's degrees. Additionally, most of the participants were studying a field in Science, Technology, and Engineering (54%), while 12.2% were studying Humanities and Arts fields.

For the prospective employers, most were aged between 26 and 35 years (50%), followed by the age group from 36 to 45 years (40%). Representing a contrast with the student participants, most employers were male (72.7%), and in terms of the highest educational qualification, 59% were holders of postgraduate degrees or higher. Most of the employers had studied a field in the domain of Science, Technology, and Engineering, followed by Business and Economics, and then by Humanities and Arts. A total of 33.3% of the prospective employers reported that they were moderately aware of blockchain technology.

7.4 Analysis of the Results

This section provides an in-depth discussion of the questionnaire results and maps the results to the four influential factors. Questionnaire data were checked and pre-processed for statistical tests, and Qualtrics[1] was used to generate the report containing all survey data. These data were imported into Excel for quality evaluation. In turn, once the data cleaning and pre-processing phases were finished, the Statistical Package for the Social Sciences (SPSS) was applied for quantitative data analysis.

In this study, to measure the reliability of the data, Cronbach's alpha was used as a measure of internal consistency (where the value ranges from 0 to 1), which is one of the most common techniques applied in the literature [88]. Opinions differ about how to interpret Cronbach's alpha values, but a commonly used schema views acceptable alpha values as those in the range of 0.70–0.95. Table 2 includes all the alpha values for each factor in the framework that was analysed to establish each factor's reliability. Cronbach's alpha values in the 'student' category ranged from 0.784 to 0.969, whereas in the 'employer' category, they ranged from 0.752 to 0.876. Since these values exceed the threshold, they are indicative of good internal consistency and reliability. To measure the validity of the study's data, the Average Variance Extracted (AVE) was selected as the most appropriate approach to validate the framework. In Table 3, the AVE values are greater than the 0.6 threshold, except the trust factor, which indicates an acceptable convergent validity. Thus, the results indicate the model is valid, and the framework factors are related.

The analysis involves investigating the relationships between the factors in the proposed framework using Pearson's correlation test, as shown in Table 3. The results indicate a significant linear correlation between the proposed factors: trust, privacy and security, efficiency, and social influence. Trust in blockchain technology in the certification process is significantly related to users' view of blockchain's level of enhanced efficiency and the reduced associated costs, which was the strongest correlation among all the factors. Also, the relationship between the users' perception of the impact of user awareness in a social context and the security

[1] www.qualtrics.com

Table 2 Cronbach's Alpha test values and analysis

	Student		Employer	
Factor	α value	Analysis	α value	Analysis
Trust (*T*)	0.957	Excellent	0.951	Excellent
Functionality and transparency	0.955	Excellent	0.876	Very good
Knowledge and familiarity	0.846	Very good	0.752	Good
Easy to access and share	0.955	Excellent	0.846	Very good
Applicants credentials authenticity (employer)			0.864	Very good
Social influence (SI)	0.936	Excellent	0.835	Very good
User awareness (student)	0.879	Very good		
Privacy and security (PS)	0.937	Excellent	0.919	Excellent
Immutability and reliability	0.957	Excellent	0.859	Very good
Perceived risk	0.784	Good	0.864	Very good
Fraud and dishonesty			0.847	Very good
Efficiency (E)	0.969	Excellent	0.873	Very good
Efficient smart certificate	0.964	Excellent	0.795	Good
Cost reduction	0.930	Excellent	0.817	Very good

Table 3 Means, AVE, standard deviation, reliabilities, and correlations

Factor	Mean	AVE	SD	Trust (*T*)	Social influence (SI)	Privacy and security (PS)	EFFICIENCY (*E*)
Trust (*T*)	2.60	0.55	0.67	1			
Social influence (SI)	2.54	0.7	0.68	0.802**	1		
Privacy and security (PS)	2.81	0.6	0.67	0.739**	0.805**	1	
EFFICIENCY (*E*)	2.44	0.8	0.74	0.812**	0.872**	0.786**	1

Note: ** Correlation is significant at the 0.01 level

services provided by blockchain is positively related to users' increased intention to adopt blockchain.

Overall, the relationships between all the framework factors were positive and significant, which encouraged the authors to move to the next phase of this research. Moreover, the relationships between some factors and other variables of the study were analysed and are presented in the following section. This analysis helps the researchers to detect any influencing factors and recognise more opportunities for blockchain to be adopted by higher education sector.

7.4.1 Relationship Between Gender and Trust Factor

In this study, the majority of the sample is female, which is an important matter to study in developing countries like Saudi Arabia. Thus, the authors are motivated to evaluate the female participants' intention to trust blockchain technology as a new

Table 4 Relationship between gender and trust

Constructs	Correlation coefficient	Sig.	Analysis
Gender—trust	0.053	0.286	Weak positive
Gender—functionality and transparency	0.038	0.450	Weak positive
Gender—knowledge and familiarity	−0.012	0.809	Weak negative
Gender—Ease of use (success and share)	0.050	0.312	Weak positive

technological base for certification systems in the higher education sector. Spearman's correlation coefficient was used to determine the degree of relationship and influence between gender and all the items listed under the trust factor in this study, including functionality and transparency, knowledge and familiarity, and ease of use. In Spearman's correlation, statistical significance is set at ≤ 0.05. As shown in Table 4, there was a weak positive correlation between the female gender and trust in blockchain technology in the certification process. Moreover, this relationship indicates a significant negative association between female gender and level of knowledge and familiarity with blockchain technology.

These findings could be interpreted as a lack of experience with blockchain technology since the majority of participants were between 18 and 25 years old. Also, blockchain is considered an innovative technology that is mostly related to cryptocurrencies. Even though the results do not show a strong correlation between female gender and trust in blockchain technology in the certification process, the study is important because the majority of the participants were female, and the study gave participants an opportunity to report what factors enhanced their intention to adopt this new technology.

7.4.2 Relationship Between User Awareness and Privacy and Security Factors

The survey results were analysed to verify the relationship between users' awareness of blockchain technology and users' perceptions about the level of privacy and security provided by blockchain technology. We investigated these associations by using Spearmen's rho statistical test and then evaluating the significance of the investigation. Table 5 indicates a strong, positive relationship between user awareness about the blockchain features and their perceptions about how secure the blockchain is to handle student's data and transactions.

Based on the results, it is clear that user awareness had a significant impact on how users perceived the risk associated with adopting blockchain technology. Finally, we examine the accuracy of the assumption we stated in the first phase of the research [84] that users' awareness positively influences their intention to adopt blockchain technology in the certification process.

Table 5 Relationship between user awareness and security Aspect of blockchain

Constructs	Correlation coefficient	Sig.	Analysis
User awareness—security and privacy	0.782	0	Strong positive
User awareness—perceived risk	0.636	0	Strong positive

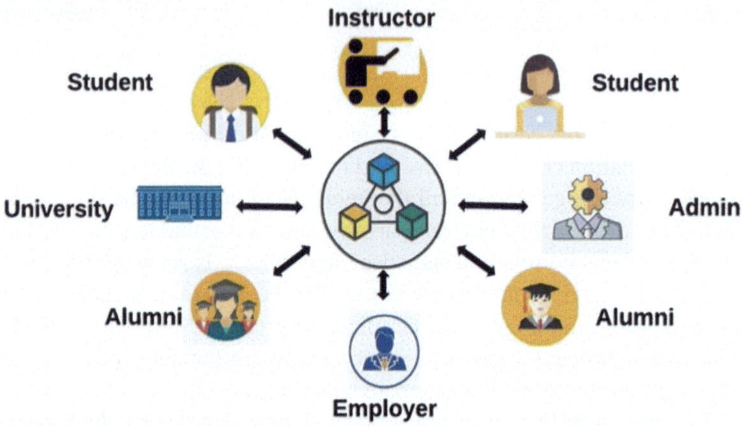

Fig. 4 Actors in DASC [84]

7.5 Proposed System DASC

The use of blockchain technology enables the elimination of third-party intermediaries and improves interactions between all related participants. To resolve the identified problems and challenges in existing certificating systems in higher education, we proposed the structure and functionality of DASC. As Fig. 4 shows, there are five principal actors in the system: alumni, students, instructors, administrators, and prospective employers.

To ensure that the aims and goals of the research are achieved, the DASC should provide solutions to the following questions:

- How is blockchain technology beneficial in resolving ongoing problems in the higher education sector when generating student certificates and accreditations?
- How can blockchain-based systems enhance the degree to which it is efficient to generate certificates in the higher education sector?

DASC aims to maintain a log of student data, including credits, skills, badges, and course registrations. The system should have the ability to enable student data sharing with authorised stakeholders, including prospective employers, university staff, and university administrators. The high level of transparency afforded by the system should enable HEIs to design and implement distinctive and personalised teaching methods for each student. The DASC should function as a standard information repository that integrates students' information—including transcripts, achievements, and digital certificates—from different HEIs. Therefore, it will be possible for students to maintain authentic records of all of their certificates for use as a long-term e-portfolio, along with a complete log of their grades, courses, and achievements. Given that prospective employers will be allowed to use the proposed system to verify the authenticity of a candidate's qualifications and transcript, it will eliminate certificate fraud and dishonesty.

The survey results were used to improve the initially proposed model. Quantitative analysis indicated that the four influential factors a substantial impact on students' and prospective employers' acceptance of blockchain technology. This indicates the validity of the model structure and is ready to be implemented and test the user feedback afterward.

7.6 *High-Level Conceptual Infrastructure*

Figure 5 illustrates the high-level conceptual infrastructure of the DASC, which represents the blockchain as the left dashed box (on-chain transactions), connected directly to the front-end system and centralised database systems (off-chain transactions). On-chain transactions are transactions that occur directly on the distributed ledger network, while off-chain transactions are external transactions that occur outside the distributed ledger [89]. The proposed DASC system enables students to view their credentials with a high level of integrity in a single location, and each student can decide whether to share this view with other stakeholders. As shown in the conceptual infrastructure, the DASC enables direct interactions between prospective employers and front-end systems that are overseen by system administrators allocating the correct permissions.

DASC provides students with a single view of their credentials data with a guarantee of data integrity. Such a view can be shared with external parties with the student's permission. As represented in the conceptual infrastructure, the DASC allows direct interactions between prospective employers and front-end systems, which are controlled by system administrators giving the appropriate permissions.

7.7 *Scenarios and Use Cases for the System Prototype DASC*

This section presents the main cases and scenarios the DASC system's users experience while using the system. The system deals with four types of users, which are student, admin, instructor, and prospective employers. Mainly, we discuss three main case scenarios of the systems' user. They are about the main transactions obtained by different user types while using the system. Moreover, this section shows that the use case scenario can represent a various range of potential transactions with indications to all the efficient functionality provided by the system. The

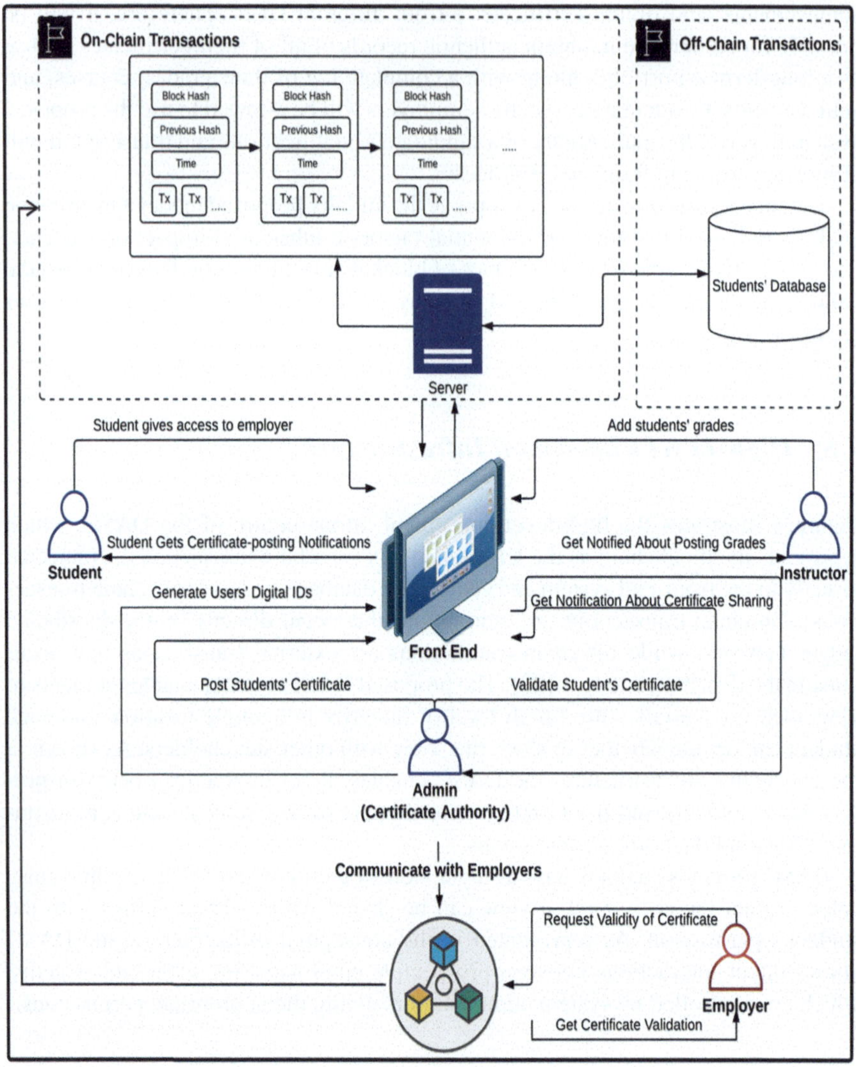

Fig. 5 High-level conceptual infrastructure of DASC [84]

user experience scenario illustrates the common transactions faced by the user in both a graphic format and descriptive text. The purpose behind this is to show how Blockchain technology is deployed to remedy these situations. These scenarios are real processes provided by higher education institutes which are used as models in a design process for future implementation of DASC (Fig. 6).

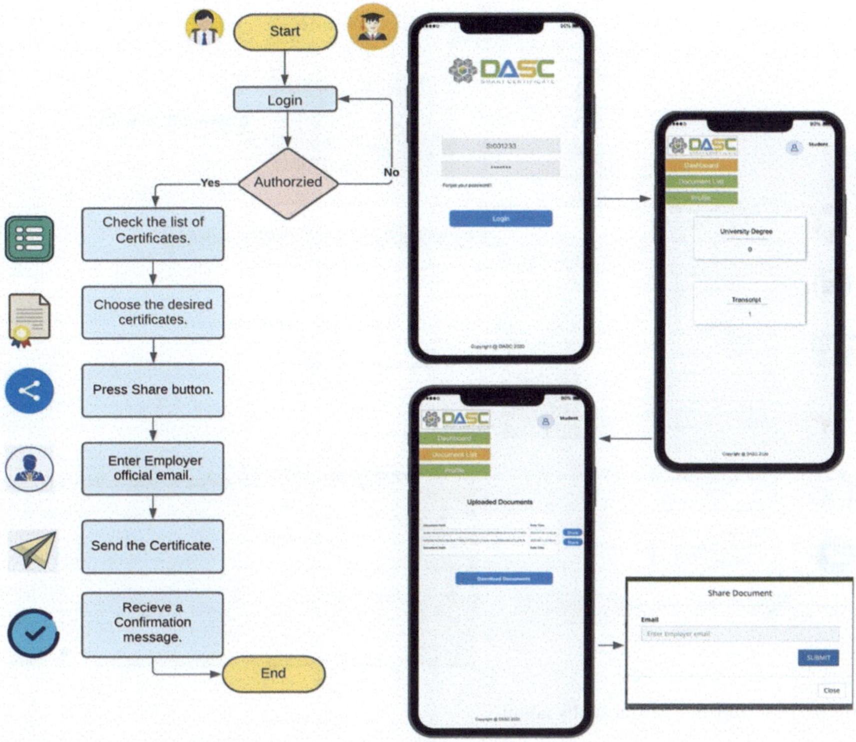

Fig. 6 Sharing student's certificate with prospective employer

7.7.1 Sharing Certificate with Prospective Employer

The process emphasises the idea of sharing students or alumni certificates. It comes as one of the main benefits the Blockchain can offer as new technology in this field. The sharing process in the case of Blockchain-based system is performed by the student which will be given by the system to share the student's own credentials with any other party. Sharing certificate feature allows the students to have the right to present the earned qualifications with others in any point of time with no extra effort or permissions. Figure 7 shows the flow chart of the process that the student will follow to share any posted certificate with external party.

7.7.2 Post Student's Certificate

As stated previously, the main benefit of using Blockchain in generating student's qualifications is leveraging the properties which can be represented in this process. The process starts once the student earns a new certificate and the university's admin wants to publish the certificate to the chain so the student can maintain the full

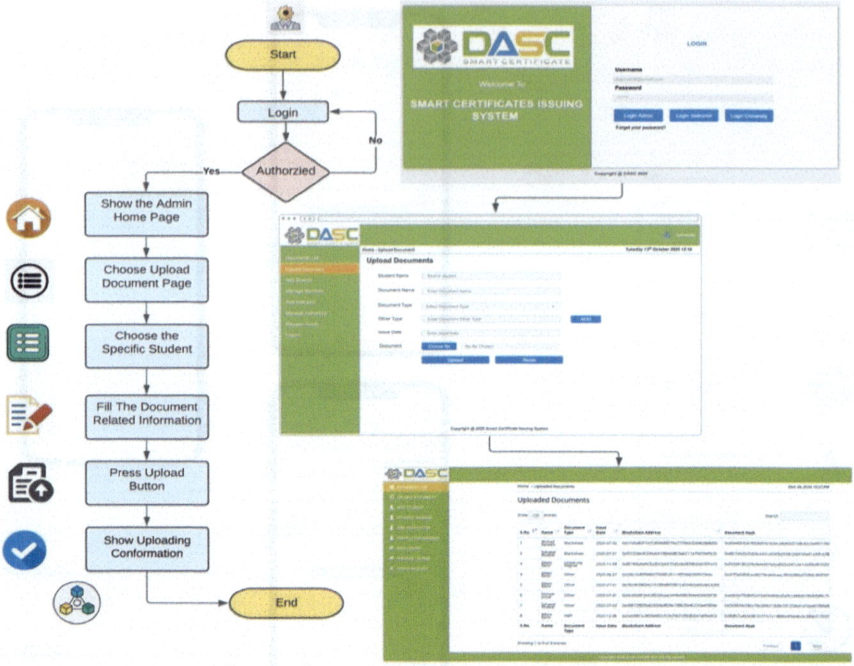

Fig. 7 Post student's certificate by admin

record of earned qualifications with no count effort. Moreover, the student can share this certificate with other parties such as prospective employer. Figure 8 represents the flow of the steps involved in this process along with the screenshots from the actual prototype of DASC.

7.7.3 Verify the Student Certificate by DASC

This process starts with the desire from the employer to verify any received qualifications or credentials from prospective candidates. As the proposed system is based on the architecture of Blockchain technology, this process will be carried on with so much trust from the employer side. The verification process guarantees that this applicant holds all the assumed qualifications form the authorised institutions. As shown in Fig. 8, the employer has two options to get access to the desire student's certificate; either by receiving sharable link via email sent by the applicant or by using the DASC system to search about the certificate either by student's details or university name.

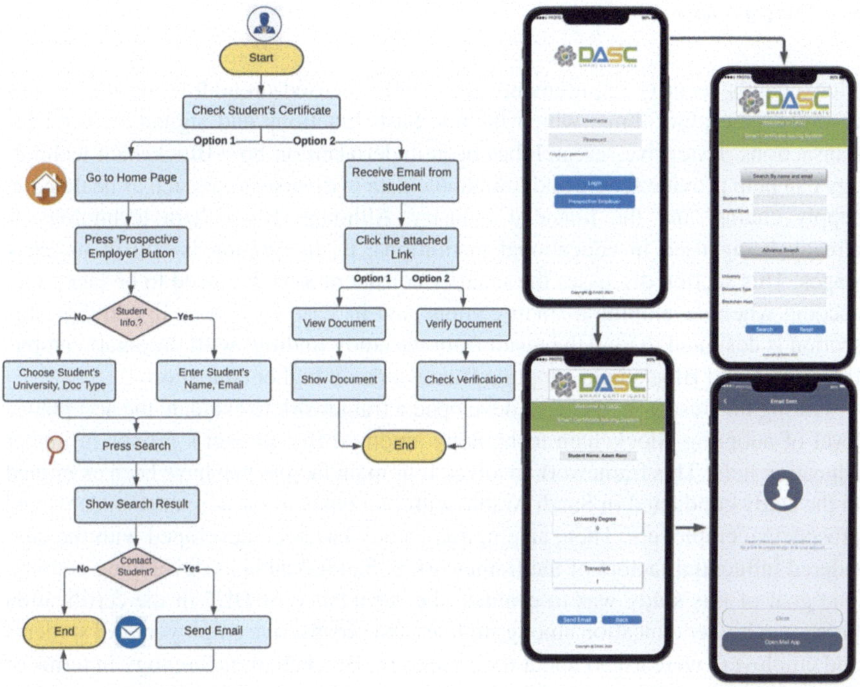

Fig. 8 Verify the certificate by DASC

7.8 Summary

To summarise, this case study provides empirical evidence showing that the four examined factors positively influence students' and employers' acceptance of Blockchain-based systems in the context of certification processes in higher education. Additionally, this study offers several contributions to theory and practice in higher education, especially in the certification process used in developing countries such as Saudi Arabia, which served as our case study. The researchers created the suggested Blockchain technology adoption model to examine the factors that the literature indicates are likely to influence the intention of higher education institutions to adopt smart certification. In turn, this provided a model that was tested on two types of participants: students and employers. Descriptive analysis of the collected data indicated the substantial influence of all proposed factors. The next step of the study will evaluate and validate the applicability of the proposed DASC. Therefore, the system should be tested according to specific criteria. The team will be tasted the prototype usability and efficiency on different user groups in order to test the applicability of using Blockchain as a decentralised technology in the higher education sector. Note that this is an ongoing research project, and its progress will be reported after we have completed the next phases.

8 Discussion

Blockchain's security requirements are mainly focused on addressing the various cryptographic algorithms, which include hash functions and signatures used for transactions. Extensive research has been undertaken on how Blockchain technology can help provide secure and immutable data in diverse areas such as healthcare, supply chains, and the financial industry. Although Blockchain technology is already being used in educational institutions, its integration is still in its early stages. This section discusses the security considerations that need to be taken into account when developing a smart campus and its case scenarios. In addition, this section is designed to map the result from the study findings with the smart campus framework and Blockchain adoption frameworks in higher education.

During this research, the team developed a framework to evaluate the acceptance level of adopting Blockchain technology among different user's groups in higher education field. This framework involved four main factors that have been examined in the study conducted in Saudi Arabia with two types of participants: students and prospective employers. Then, an empirical study has been developed with the considered influential factors of the framework in Saudi Arabia as developing country. The goal of this study was to evaluate the acceptance of DLT in the certification process in higher education among students and prospective employers. All students and employers were asked about their views on Blockchain technology in terms of the factors of trust, social influence, privacy and security, and efficiency, with detailed items associated with each factor. We regarded prospective employers as the prime driver of the research according to the benefits that stand to be attained if Blockchain technology is adopted in the certification process in higher education.

One of the main advantages associated with the use of Blockchain technology is tamper resistance and, in particular, the technology's ability to provide secure data. With this feature, it is almost impossible to alter data due to the simultaneous control of all the devices that contain the distributed data. In education, preserving the privacy of students' sensitive information plays an important role, especially when sharing personal and confidential data with others [90]. Due to the definition and features of Blockchain technology mentioned above, it is possible to ensure security and privacy. In our study conducted in Saudi Arabia, in the privacy and security factor, the employers agreed on the general importance of security-related benefits in integrating Blockchain technology in higher education. Meanwhile, most of the participants were neutral about believing in the ability of Blockchain to maintain a high level of security, including data protection, integrity, and privacy, which leads to an influence on the trust towards it. In the matter of the reliability of Blockchain technology, most of the participants agreed that Blockchain enhances the reliability and transparency of prospective employees' certificates by establishing secure connections between all included parties and easing their interactions. The study examines the user perception of low risk associated with the use of Blockchain technology that would positively influence their intention to adopt Blockchain technology in the certification process.

Data integrity is an important aspect of storage security. It can help users identify the causes and consequences of data changes and prevent such changes from happening. An accepted transaction in Blockchain network is stored by a distributed ledger in a block and then each block must reference its previous block to build a chain, any published block that did not refer to its preceding block would be rejected by other nodes. Therefore, applying Blockchain technology to our framework can provide data integrity, which leads to increase users trust in the system, as in the next section. In the perspective of the reliability and immutability of blockchain technology, the majority of the participants in the conducted study agreed that blockchain enhances the reliability of prospective employees' certificates by establishing secure connections between all included parties and easing their interactions. Another advantage that Blockchain technology has is transparency since all transactions are validated before they are added to the blocks. If there was an invalid transaction made by a malicious peer, other nodes would detect and reject the transaction. Unlike a centralised model, a peer must trust that the transaction received is validated by the owner of the ledger. When we asked the employers, who participated in the study if the transparency composes a good motivation behind adopting Blockchain technology, the result shows that most of the employers agreed on the effect of the transparency feature on making it a suitable option for managing educational certificates.

Trust has been investigated in studies on the adoption of technologies that involve handling, storing, or processing sensitive information [87]. Blockchain technology ensures data persistence since it is stored in a distributed manner, which means many copies of the same ledger are shared, updated and synced among nodes. When a new peer joins the blockchain system, it can access a whole copy of the ledger, but it is more difficult to lose or destroy the ledger. Unlike cartelised systems that own the ledger, blockchain provides data persistence by backing up the system, meaning that users have no choice but to trust the system. Providing employers and students with a trustworthy and transparent system such as a blockchain-based system would drive the adoption process among higher education institutions. The case study's results clearly revealed that students were overwhelmed by the existing certification system, which strongly shaped their perception of new technology adoption. In the trust factor items, the majority of students agreed on the effect of blockchain's transparency and its immutable features, which make it a trustworthy technology for managing educational certificates. Additionally, most students believed that blockchain can handle all forms of academic credentials, transcripts and certificates. The participants indicated that implementing a system that eliminates the control of third parties would increase their level of trust in blockchain technology.

One of the most critical security aspects is user authentication. It is essential to maintain secure communications between devices and systems to prevent unauthorised access. Thus, authorisation is the process of validating a peer's identity to authenticate them and enable communication with each other. The use of a single factor or a combination of multiple factors is not sufficient to provide an adequate level of protection. Thus, the use of Blockchain technology in our research can prevent the leakage of private information and enhance the authorisation level.

Moreover, the result of conducted study shows that most employers agreed that Blockchains are useful for authenticating student identities and smart certificates, and they also suggested that the use of the technology can reduce the risk of duplication and fraud.

Data in a Blockchain is provenanced when the transaction or data stored in a distributed ledger needs to be processed through cryptographic mechanisms such as hash functions and digital signatures. This process ensures tamper resistance, and the authenticity of transactions. While in a centralised system a user must trust the owner of the ledger and accept that the transactions are not being altered. From the results collected in the case study, most of higher education's students believed in the ability of blockchain technology to facilitate the sharing of their official documents directly with external parties, thereby embodying the learning outcomes and enhancing their chances in appropriate career opportunities. However, a low number of the responses from the students' indicated having a good level of awareness about the benefits provided by DLT in the general aspects. Thus, higher education institutes have to embrace new strategies to enhance the user awareness regarding Blockchain technology as a step to ease the adoption in the future.

As discussed before, scalability is considered one of the challenges facing blockchain adoption in various fields. Thus, in the DASC, off-chain transactions have been applied to students' data that could be maintained and monitored outside the blockchain ledger. The results of the case study show the feasibility of applying the frameworks to various processes within higher education, via either a smart campus or blockchain. To sum up, deploying Blockchain technology to our proposed smart campus framework will provide more benefits by managing the problems associated with a centralised IoT architecture, especially from a security perspective.

9 Conclusion

The several aspects of smart environments have been the subject of various studies. The security and privacy concerns have become more prevalent due to the increasing number of connected devices. This chapter analysed the various components of a smart campus, including architectures, platforms, and technologies with their limitations. Then we proposed a new smart campus architecture that combines the advantages of both the Internet of Things and Blockchain technology. The framework considered as guid base to develop various campus services such as smart learning, smart building, and smart parking. In addition, this work presented an overview of Blockchain technology and discussed the challenges and problems in current higher education systems. Moreover, this chapter includes a case study about the acceptance of Blockchain in the certification process in the higher education sector in developing countries. This study was conducted to investigating the target users' intention to the adoption process by examining four influential factors which are trust, privacy and security, social influence, and efficiency. Then we proposed the DASC, a solution to the abovementioned problems that uses a

Blockchain-based credentialing system to generate and maintain student credentials. In the next phase of the research, an experimental approach will be used to test the system's validity, and the research team will also evaluate and validate the applicability of the implemented solution. It is necessary to verify the efficiency of Blockchain as a decentralised technology in the higher education sector. Importantly, this is an ongoing research project, and its progress will be reported after we have completed the next phases and future work.

References

1. W. Villegas-Ch, X. Palacios-Pacheco, S. Luján-Mora, Application of a smart city model to a traditional university campus with a big data architecture: A sustainable smart campus. Sustainability **11**(10), 2857 (2019)
2. A. Alam, S. Ullah, Adaptive 3D-virtual learning environments: From students' learning perspective, in *Proceedings—14th International Conference on Frontiers of Information Technology, FIT 2016* (2017)
3. A. Abuarqoub et al., A survey on internet of thing enabled smart campus applications, in *Proceedings of the International Conference on Future Networks and Distributed Systems—ICFNDS'17* (2017), pp. 1–7
4. R. Mei, J. Guan, B. Li, University course timetable system design and implementation based on mathematical model, in *2010 The 2nd International Conference on Computer and Automation Engineering, ICCAE 2010* (2010)
5. Y. Atif, S.S. Mathew, A. Lakas, Building a smart campus to support ubiquitous learning. J. Ambient. Intell. Humaniz. Comput. **6**(2), 223–238 (2015)
6. S. Li, L. Da Xu, S. Zhao, The internet of things: A survey. Inf. Syst. Front. **17**(2), 243–259 (2015)
7. N. Mishra, P. Singhal, S. Kundu, Application of IoT products in smart cities of India, in *Proceedings of the 2020 9th International Conference on System Modeling and Advancement in Research Trends, SMART 2020* (2020), pp. 155–157
8. A. Nayyar, V. Puri, D.-N. Le, Internet of nano things (IoNT): Next evolutionary step in nanotechnology. Nanosci. Nanotechnol. **7**(1), 4–8 (2017)
9. A. Al-Fuqaha, M. Guizani, M. Mohammadi, M. Aledhari, and M. Ayyash, "Internet of things: A survey on enabling technologies, protocols, and applications," IEEE Commun. Surv. Tutorials, 2015 17 (4): 2347-2376
10. K. Zhao, L. Ge, A survey on the internet of things security, in *Proceedings—9th International Conference on Computational Intelligence and Security, CIS 2013* (2013), pp. 663–667
11. A. Reyna, C. Martín, J. Chen, E. Soler, M. Díaz, On Blockchain and its integration with IoT. Challeng. Opportun. **88**, 173–190 (2018)
12. Z. Zheng, S. Xie, H. Dai, X. Chen, H. Wang, An overview of Blockchain technology: Architecture, consensus, and future trends, in *Proceedings*—2017 *IEEE 6th International Congress on Big Data, BigData Congress 2017* (2017), pp. 557–564
13. S. Nakamoto, Bitcoin: A peer-to-peer electronic cash system
14. M. Conoscenti, D. Torino, A. Vetr, D. Torino, J.C. De Martin, Blockchain for the internet of things: A systematic literature review (2016)
15. D. Yaga, P. Mell, N. Roby, K. Scarfone, Blockchain technology overview, arXiv (2019)
16. G. Chen, B. Xu, M. Lu, N.-S. Chen, Exploring Blockchain technology and its potential applications for education. Smart Learn. Environ. **5**(1), 1 (2018)
17. S. Huckle, M. White, Fake news: A technological approach to proving the origins of content, using Blockchains. Big Data **5**(4), 356–371 (2017)

18. R. Bhattacharya, M. White, N. Beloff, A Blockchain based peer-to-peer framework for exchanging leftover foreign currency, in *Proceedings of the Computer Conference 2017*, vol. 2018 (2018), pp. 1431–1435
19. S.M. Idrees, M. Nowostawski, R. Jameel, A.K. Mourya, Security aspects of Blockchain technology intended for industrial applications. Electronics **10**(8), 951 (2021)
20. S. Huckle, R. Bhattacharya, M. White, N. Beloff, Internet of things, Blockchain and shared economy applications. Procedia Comput. Sci. **58**, 461–466 (2016)
21. S. Huckle, M. White, R. Bhattacharya, Towards a post-cash society: An application to convert fiat money into a cryptocurrency by. First Monday **22**(3) (2017)
22. S. Huckle, M. White, Socialism and the Blockchain. Futur. Internet **8**(4), 49 (2016)
23. J. Golosova, A. Romanovs, Overview of the Blockchain technology cases, in *2018 59th International Scientific Conference on Information Technology and Management Science of Riga Technical University* (2018), pp. 1–6
24. Idrees, S. M., Agarwal, P., & Alam, M. A. (Eds.). (2021). Blockchain for Healthcare Systems: Challenges, Privacy, and Securing of Data. CRC Press.
25. S.M. Idrees, M. Nowostawski, R. Jameel, Blockchain-based digital contact tracing apps for COVID-19 pandemic management: Issues, challenges, solutions, and future directions. JMIR Med. Inform. **9**(2), e25245 (2021)
26. ARMONK and TOKYO, Sony and Sony Global Education develop a new system to manage students' learning data, built on IBM Blockchain (2017) [Online], https://www-03.ibm.com/press/us/en/pressrelease/52970.wss. Accessed 6 Dec 2018
27. F. Vidal, F. Gouveia, C. Soares, Analysis of Blockchain technology for higher education, in *Proceedings—2019 International Conference on Cyber-Enabled Distributed Computing and Knowledge Discovery CyberC 2019* (2019), pp. 28–33
28. K. Panetta, Gartner, *5 Trends Emerge in the Gartner Hype Cycle for Emerging Technologies, 2018—Smarter with Gartner* (Gartner, Stanford, 2018)
29. T. Narendrakumar, A.S. Pillai, Smart connected campus, in *2017 International Conference on Intelligent Computing, Instrumentation and Control Technologies, ICICICT 2017*, vol. 2018 (2018), pp. 1591–1596
30. I. Hossain, D. Das, M.G. Rashed, Internet of things based model for smart campus: Challenges and limitations, in *5th International Conference on Computer, Communication, Chemical, Materials and Electronic Engineering, IC4ME2* 2019 (2019)
31. H. Zhe, W. Xiaojun, W. Qiang, Research on campus information service platform based on locating system of campus wireless LAN, in *Proceedings—2017 International Conference on Smart Grid and Electrical Automation ICSGEA 2017*, vol. 2017 (2017), pp. 623–626
32. H. Yan, H. Hu, A study on association algorithm of smart campus mining platform based on big data, in *Proceedings—2016 International Conference on Intelligent Transportation, Big Data and Smart City, ICITBS 2016* (2017), pp. 172–175
33. H. Hu, H. Yan, A study on discovery method of hot topics based on smart campus big data platform, in *Proceedings—2016 International Conference on Intelligent Transportation, Big Data and Smart City, ICITBS 2016* (2017), pp. 176–179
34. V. Agate, F. Concone, P. Ferraro, WiP: Smart services for an augmented campus, in *Proceedings—2018 IEEE International Conference on Smart Computing, SMARTCOMP 2018* (2018), pp. 276–278
35. J. Enqing, R. Peixiang, W. Huanjin, S. Yanping, Discussion on construction method of smart campus basic platform based on 3D geographic information technology, in *Proceedings—2017 Chinese Automation Congress, CAC 2017*, vol. 2017 (2017), pp. 7790–7794
36. W. Lihong, Research on the construction of smart campus social platform based on Hadoop, in *Proceedings—2020 International Conference on Computer Engineering and Application, ICCEA 2020* (2020), pp. 214–217
37. P. Agarwal, G.V.V. Ravi Kumar, P. Agarwal, IoT based framework for smart campus: COVID-19 readiness, in *Proceedings of the World Conference on Smart Trends in Systems, Security and Sustainability, WS4 2020* (2020), pp. 539–542

38. O. Debauche, R.A. Abdelouahid, S. Mahmoudi, Y. Moussaoui, A. Marzak, P. Manneback, RevoCampus: A distributed open source and low-cost smart campus, in *3rd International Conference on Advanced Communication Technologies and Networking, CommNet 2020* (2020)
39. M.A. Khan, A survey of security issues for cloud computing. J. Netw. Comput. Appl. **71**, 11–29 (2016)
40. H.F. Atlam, G.B. Wills, A. Alenezi, M.O. Alassafi, Blockchain with internet of things: Benefits, challenges, and future directions. *Int. J. Intell. Syst. Appl.* (2018)
41. H.F. Atlam, E. El-Din Hemdan, A. Alenezi, M.O. Alassafi, G.B. Wills, Internet of things forensics: A review. Internet Things **11**, 100220 (2020)
42. C. Wang, S. Chen, Z. Feng, Y. Jiang, X. Xue, Blockchain-based data audit and access control mechanism in service collaboration, in *Proceedings—2019 IEEE International Conference on Web Services, ICWS 2019—Part of the 2019 IEEE World Congress on Services* (2019), pp. 214–218
43. M. Chowdhury, S. Ferdous, K. Biswas, Blockchain platforms for IoT use-cases (2018), pp. 3–4
44. N. Kshetri, Blockchain's roles in strengthening cybersecurity and protecting privacy. Telecomm. Policy **41**(10), 1027–1038 (2017)
45. P. Kianmajd, J. Rowe, K. Levitt, Privacy-preserving coordination for smart communities, in *Proceedings—IEEE INFOCOM* (2016)
46. G. Zyskind, O. Nathan, A.S. Pentland, Decentralizing privacy: Using Blockchain to protect personal data, in *Proceedings—2015 IEEE Security and Privacy Workshops, SPW 2015* (2015)
47. K. Biswas, V. Muthukkumarasamy, Securing smart cities using Blockchain technology, in *Proceedings—18th IEEE International Conference on High Performance Computing and Communications, 14th IEEE International Conference on Smart City and 2nd IEEE International Conference on Data Science and Systems, HPCC/SmartCity/DSS 2016* (2017)
48. Y. Chen, L. Li, Z. Chen, An approach to verifying data integrity for cloud storage, in *Proceedings—13th International Conference on Computational Intelligence and Security, CIS 2017*, vol. 2018 (2018), pp. 582–585
49. B. Liu, X.L. Yu, S. Chen, X. Xu, L. Zhu, Blockchain based data integrity service framework for IoT data, in *Proceedings—2017 IEEE 24th International Conference on Web Services, ICWS 2017* (2017)
50. D. Yue, R. Li, Y. Zhang, W. Tian, C. Peng, Blockchain based data integrity verification in P2P cloud storage, in *Proceedings of the International Conference on Parallel and Distributed Systems—ICPADS*, vol. 2018 (2019)
51. V. Yatskiv, N. Yatskiv, O. Bandrivskyi, Proof of video integrity based on Blockchain, in *2019 9th International Conference on Advanced Computer Information Technologies, ACIT 2019—Proceedings* (2019), pp. 431–434
52. N. Fabiano, Internet of things and Blockchain: Legal issues and privacy. The challenge for a privacy standard, in *Proceedings—2017 IEEE International Conference on Internet of Things, IEEE Green Computing and Communications, IEEE Cyber, Physical and Social Computing, IEEE Smart Data, iThings-GreenCom-CPSCom-SmartData 2017*, vol. 2018 (2018), pp. 727–734
53. S.C. Cha, J.F. Chen, C. Su, K.H. Yeh, A Blockchain connected gateway for BLE-based devices in the internet of things. IEEE Access **6**, 24639–24649 (2018)
54. L. Wu, X. Du, W. Wang, B. Lin, An out-of-band authentication scheme for internet of things using Blockchain technology, in *2018 International Conference on Computing, Networking and Communications, ICNC 2018* (2018), pp. 769–773
55. A. Ouaddah, A. Abou Elkalam, A. Ait Ouahman, FairAccess: A new Blockchain-based access control framework for the internet of things. Secur. Commun. Netw. **9**(18), 5943–5964 (2016)
56. M. Alkhammash, N. Beloff, M. White, An internet of things and Blockchain based smart campus architecture, in *Intelligent Computing*, (Springer, Berlin, 2020), pp. 467–486
57. K. Ponnusamy, N. Rajagopalan, Internet of things: A survey on IoT protocol standards. Adv. Intell. Syst. Comput. **564**, 651–663 (2018)

58. F. Samie, L. Bauer, J. Henkel, IoT technologies for embedded computing: A survey, in 2016 *International Conference on Hardware/Software Codesign and System Synthesis, CODES+ISSS 2016* (2016)
59. B. Kang, D. Kim, H. Choo, Internet of everything: A large-scale autonomic IoT gateway. IEEE Trans. Multi-Scale Comput. Syst. **3**(3), 206–214 (2017)
60. I.K. Azeemi, M. Lewis, T. Tryfonas, Migrating to the cloud: Lessons and limitations of 'traditional' is success models. Procedia Comput. Sci. **16**, 737–746 (2013)
61. C. Sharma, N.K. Gondhi, Communication protocol stack for constrained IoT systems, in *Proceedings—2018 3rd International Conference on Internet of Things: Smart Innovation and Usages, IoT-SIU 2018* (2018)
62. A. Shanker, Public vs. private Blockchains, PC Magazine (2017)
63. R. Grinberg, Bitcoin: An innovative alternative digital currency. Hast. Sci. Technol. Law J. **159**, 197 (2011)
64. V. Buterin, Ethereum white paper, Etherum (2014)
65. J. Schneider et al., Blockchain—Putting theory into practice (2016)
66. J. King, ERIS, in *ERIS* (2020)
67. Hyperledger, Hyperledger whitepaper. Publ. (2016), https://wiki.hyperledger.org/groups/whitepaper/whitepaper-wg
68. M. Hearn, Corda: A distributed ledger. Whitepaper (2016)
69. M. Turkanović, M. Hölbl, K. Košič, M. Heričko, A. Kamišalić, EduCTX: A Blockchain-based higher education credit platform, arXiv, vol. 6 (2017)
70. Blockcerts, 2019 [Online], https://www.blockcerts.org/. Accessed 23 Feb 2019
71. M. Jirgensons, J. Kapenieks, Blockchain and the future of digital learning credential. Assess. Manag. **20**(1), 145–156 (2018)
72. KAUST, KAUST set to issue digital Blockchain credentials using Blockcerts open standard | King Abdullah University
73. S. Apurv, Digital Certificates on Hyperledger Fabric. Coinmonks (2018)
74. A. Curmi, F. Inguanez, Blockchain based certificate verification platform. Lect. Notes Bus. Inf. Process. **339**, 211–216 (2019)
75. S. Joksimović, C. Mills, S. Dawson, A.C. Graesser, C. Brooks, How Do We Model Learning at Scale? A Systematic Review of Research on MOOCs, vol. 88, no. 1 (2018)
76. J. Xi, Y. Chen, G. Wang, Design of a personalized massive open online course platform. Int. J. Emerg. Technol. Learn. **13**(4), 58–70 (2018)
77. B. Gipp, C. Breitinger, N. Meuschke, J. Beel, CryptSubmit: Introducing securely timestamped manuscript submission and peer review feedback using the Blockchain, in *Proceedings of the ACM/IEEE Joint Conference on Digital Libraries* (2017)
78. T. Hepp, A. Schoenhals, C. Gondek, B. Gipp, OriginStamp: A Blockchain-backed system for decentralized trusted timestamping. Inf. Technol. **60**(5–6), 273–281 (2018)
79. M.S.M. Pozi, G. Muruti, A.A. Bakar, A. Jatowt, Y. Kawai, Preserving author editing history using Blockchain technology, in *Proceedings of the ACM/IEEE Joint Conference on Digital Libraries* (2018), pp. 165–168
80. A.R. Purba, R. Yunis, Application of Blockchain technology to prevent the potential of plagiarism in scientific publication, in *Proceedings of 2019 4th International Conference on Informatics and Computing, ICIC 2019* (2019)
81. MOE, Education and vision 2030, (2017) [Online], https://www.moe.gov.sa/en/Pages/vision2030.aspx
82. Statista, Saudi Arabia: Total population from 2012 to 2022 (2018) [Online], https://www.statista.com/statistics/262467/total-population-of-saudi-arabia/. Accessed 6 Dec 2018
83. ICEF, Saudi Arabia's expanding higher education capacity (2018) [Online], http://monitor.icef.com/2018/07/saudi-arabias-expanding-higher-education-capacity/. Accessed 10 Dec 2018
84. M. Alshahrani, N. Beloff, M. White, Revolutionising higher education by adopting Blockchain technology in the certification process, in *2020 International Conference on Innovation and Intelligence for Informatics, Computing and Technologies, 3ICT 2020* (2020)

85. H. Halpin, M. Piekarska, Introduction to security and privacy on the Blockchain, in *Proceedings—IEEE European Symposium on Security and Privacy Work, EuroS PW 2017* (2017)
86. V. Garcia-Font, Blockchain: Opportunities and challenges in the educational context. Lect. Notes Data Eng. Commun. Technol. **34**, 133–157 (2020)
87. V. Venkatesh, J.Y.L. Thong, X. Xu, A I S Association for information systems unified theory of acceptance and use of technology: A synthesis and the road ahead. Journal (2016)
88. M. Tavakol, R. Dennick, Making sense of Cronbach's alpha. Int. J. Med. Educ. **2**, 53–55 (2011)
89. B. Ramamurthy, *Blockchain in Action* (Manning Publications, Shelter Island, 2020)
90. M.J.M. Chowdhury, A. Colman, M.A. Kabir, J. Han, P. Sarda, Blockchain as a notarization service for data sharing with personal data store, in *Proceedings—17th IEEE International Conference on Trust, Security and Privacy in Computing and Communications, 12th IEEE International Conference on Big Data Science and Engineering Trust* 2018, (2018), pp. 1330–1335

Transforming Education Through Blockchain Technology

Ravi Verma

1 Evolution of Blockchain Technology Towards Applications

In the early 1990s the very big invention of blockchain technology was introduced into the world of computer science and technology. That is why we knew that blockchain technology is the biggest invention of the twenty-first century because it considers various areas of applications around the world such as finance, medical, cryptocurrency, as well as in education.

The popularity of blockchain technology comes from a few years back; therefore, in this section we would like to explore the journey of blockchain technology from 1990 to 2021 toward improving the finance, health, and education systems.

Today's digitalization of services is continuing, people's trust in the system is in all directions, but the integration and involvement of advanced technology such as blockchain can make such digitalization more trustworthy. In order to improve the trust of people in internet services, blockchain technology brings a platform that transforms the existing area of operation, with more reliability, security, and privacy as in Hawlitschek et al. [1]. Figure 1 describes the evolution and invention of involvement of blockchain technology year by years in various service areas.

Figure 1 describes the progressive direction of involvement of blockchain technology in a different timeline frame; the historical journey of blockchain technology defines blockchain technology covers different aspect of digital platform? So that one can imagine the future of blockchain technology with different areas of application for the consideration of security, privacy, and safety of confidential matters. Table 1 defines the ongoing activity process system with blockchain technology in different timeframes, to take place for the advancement of traditional service platforms to advance a blockchain-based system.

R. Verma (✉)
Department of CSE, Radharaman Institute of Technology and Science, Bhopal, India

S. M. Idrees, M. Nowostawski (eds.), *Transformations Through Blockchain Technology*, https://doi.org/10.1007/978-3-030-93344-9_2

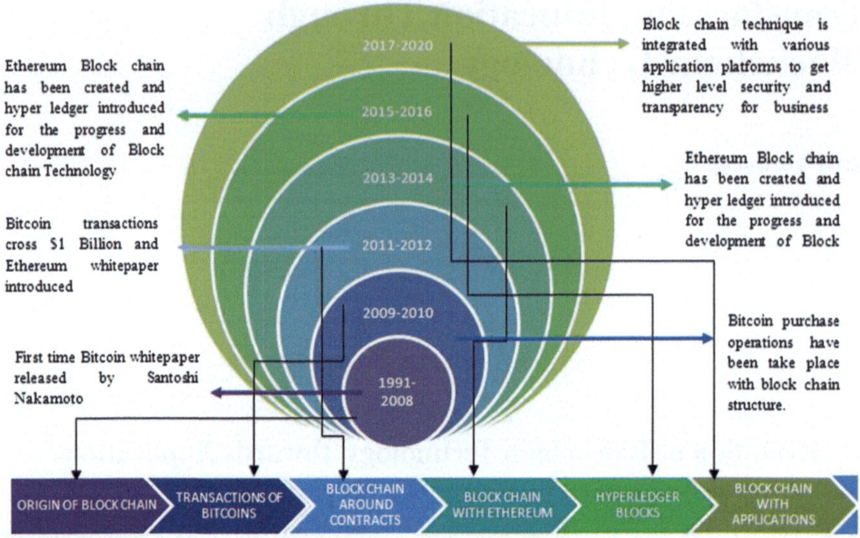

Fig. 1 Evolution of blockchain technology

2 Introduction to Blockchain Technology

Education is the foundation of every country in the world; therefore, one needs to protect the values of the education sector in the form of degree certificates, copyright documents, university policies, and grading systems so that no one devalue these things, because as digitalization is proceeding, educational institutions such universities move fast through e-learning or remote learning platforms, this makes our educational systems more transparent and meaningful in all ways, although it is very challenging to keep student records, copyrighted documents, and confidential certificates as safe as possible. Nowadays crime in the education sector is very common; many people are doing business in providing fake and duplicate certificates to students, and this situation is very dangerous for people themselves, for nations, and maintaining the value of education for life. Blockchain technology can make this possible, as it provides a magical mechanism to protect the original values of confidential certificates and other documents [2].

Blockchain technology did great things in the last century via transactions using cryptocurrency. At that time blockchain technology brought the concept of implementing a decentralized virtual exchange of currency, which was accepted by many business people for the purposes of investment. Also, the blockchain mechanism utilized bitcoin for their effective and efficient use all around the world, which has been reflected in many day-to-day social activities. Blockchain has already provided various effective solutions for finance and health care systems and further improvements are continuously being worked on [3].

Table 1 Blockchain revolution in different time frame

Timeframe	Objectives	Technological development
1991–2008 (origin of blockchain technology)	Invention of block chain technology for the purpose of invention of cryptographical chains of blocks to be invented for improving the security of confidential credentials	In early 1991, the first time that block chain technology comes into the knowledge of people, which was invented by Stuart Haber and W.S. Stornetta and in later stages Satoshi Nakamoto introduced the first block chain digital technology project on Bitcoin for the management of ledger applications in blocks [1]
2008–2013 (transactions)	To improve security on transactions with digital currency	Block chain technology is dedicatedly used for transactions on cryptocurrencies such as bit coins; therefore, bitcoin is the very first application of blockchain [1]
2014–2015 (block chain around Ethereum)	Bit coins have a limitation, which was overcome by Ethereum in 2013 with an additional functionality. Ethereum is difference from bitcoin blockchain in the manner that it allows contracts and other information to be recorded as well	Block chain technology has many limitations as it used the peer-to-peer system; therefore, Vitalik Buterin introduced Ethereum as a latest public blockchain along with additional functionality [1, 2]
2016–2018 (NEO applications)	Advancement of blockchain with different applications such as NEO	During this period block chain is not limited to Ethereum and bitcoin currency but now it is adopting the NEO application platform of China for various customer services [2]
2019–2021 (blockchain and Internet of things)	Integration of blockchain technology with internet of things and other service platforms	Now are the days blockchain technology has cropped up with many services platforms for different verification processes [1]

The services provided by the blockchain network deal with more technological advancement for various other service sectors such as education, which can be expanded with more additional functionality to provide quality education with confidentiality, security, validity, and privacy.

Now, blockchain technology helps us to solve a variety of problems associated with the education sector, where it will be beneficial for educators as well as for students regarding different learning outcomes, because blockchain technology is effectively useful for securely storing and maintaining tamper-proof credentials using a peer-to-peer network policy. In this way, all relevant confidential data such as certificates can be managed with a tamper-proof system that can provide validation and verification of the same thing at the same time, via the blockchain network. Our education system can witness a full flashed secure environment for a higher level of security against fraud, as well as reducing the chances of confidential documents such as degree certificates being tampered with.

Ethereum blockchain technology makes our traditional education system more impressive by providing a way of managing smart contracts using Ethereum blockchain techniques to verify and store all relevant data of a specific record with a high level of security and transparency. The invention of Ethereum in the blockchain network extends the utility and application of traditional blockchain technology, the invention of designing and deploying additional data properties as an asset makes blockchain technology more scalable and useful for a large area of application. Looking forward to the present and future of blockchain technology, it may be applicable for various private and public sectors according to the need and scope of the applications and the scalability issues of the blockchain technique [4]. As in Idrees et al. [4], taking into consideration various areas of application of the education system, educational organizations are facing many problems related to the online and distance learning paradigm. The verifying authorities take a long time over the period to provide validated and authenticated certificates and documents, but at the same time, blockchain technology resolves time consumption and server failure issues in a more effective, secure, and proper manner. Blockchain technology works on the concept of a distributed computing environment instead of a centralized management system; therefore, it creates different leaders of blocks in the form of chains in a different place for the purposes of highly secure validation and verification. As a result, this process takes less time for the entire task.

In Dwyer [5], defines different types of methodology for achieving the operation with a proof of work system such as an algorithm of a cryptographic puzzle, using the blockchain technique. In Saberi et al. [2] the way in which blockchain techniques manage transactions in the form of electronic cash is defined, where blockchain considers only two associated parties of transactions using a peer-to-peer communication policy without involving any other outsiders or mediators during the course of the operation. Tapscott and Tapscott [3] also define t how the integration of methods of digital signature and digital certification can help centralized universities and educational institutes using distributed data storage techniques to maintain high security, safety, privacy, and confidentiality of important documents compared with a centralized data storage system.

Blockchain networks are also prone to hacking attacks. In order to hack a blockchain network, one needs to get access to at least 50% of the systems of the entire blockchain network, which is practically impossible; therefore, a blockchain network is completely protected against such types of attacks. Similarly, to provide the safest way for digital payment of cryptocurrency, blockchain works on the concept of maintaining knowledge of relevant data in the form of a cryptographic data set that has proceeded throughout the blocks of chain on every peer node of the blockchain network, as defined in Dwyer [5]. To make such payment successful the blockchain network uses the concept of the Ethereum blockchain, which manages all the associated relevant sets of data, as described in Peters and Efstathios [6], and is implemented for the secure payment of cryptocurrency.

3 Key Elements of Blockchain Technology

3.1 Digital Transactions

Blockchain always follows the policy of proceeding with digital transactions in a distributed manner, as the back bone of blockchain technology is the peer-to-peer network; therefore, a chain of transactions is followed by each and every node of the peer network. For this reason, whenever a single change is made on this, every peer node has to agree with this change, otherwise changes made by any transaction will not be accepted. Thus, for the purposes of security, a blockchain network is digitally signed, verified, and maintains every transactional request uniquely using a cryptographical approach.

3.2 Shared Ledgers

Shared ledgers have been distributed throughout the blockchain process. They are also known as distributed ledgers. In the transaction process, once a ledger proceeds with its unique value, then it must be distributed to all the participants equally so that each participant has the corresponding copy of the ledger. When any participant updates any ledger, a copy of this updated ledger is also received by each node of the blockchain network [7].

3.3 Encrypted Token

To transmit the cryptocurrency blockchain technology uses a cryptographical token, which is basically a string of knowledge. It consists of the relevant information on currencies such as bitcoin as a knowledge base. In addition to this information, it also has some algorithmic encryption code called a secret code, which should be transmitted to all the associated corresponding nodes of the network.

3.4 Payment Channel

To manage the payment methods of cryptocurrency at every level with security constraints blockchain technology needs to use Ether as a fuel in the blockchain distributed network. Ether technology enables smooth and qualitative services for many applications, which may also make the network healthy and accurate.

3.5 Ethereum Network

Ethereum network is a type of distributed decentralized network that manages strings of knowledge as smart contract properties, which may also be used for the purposes of validation and verification of associate requested, which has spread to entire blocks of the blockchain network. Various platforms and programming languages are available for designing and implementing smart contracts in the form of strings of knowledge that can be deployed in the blockchain network.

3.6 Gas Limit

This is an essential element of the blockchain network that provides a way to process the transaction in a given gas limit so that whenever a transaction is created it will have an associated gas limit value. This value defines how much work has been done. If the out of gas limit function is triggered, it means that system needs to revert back all the corresponding transactions at that moment.

3.7 Solidity

Solidity is a scripting language tool that follows the concept of object-oriented programming. In blockchain technology solidity is used for the development of smart contracts. For this purpose, mostly the Remix development environment has been used, which makes an active and useful environment for the development of solidity programmers with the help of an inbuilt tool. Figure 2 describes the essential elements of blockchain technology.

4 How Does Blockchain Work?

Definition Blockchain consists of distributed peer-to-peer communication networks of blocks that maintain cryptographical hash values from peer-to-peer node in order to bring security, immutability, and updatability at every stage of the process. The future of blockchain technology is very bright because it brings highly secured infrastructure to various platforms such as education, health, finance, and many customer services platforms [8].

Blockchain transactions are very secure compared with other options because of its tremendous working style where it continuously spreads the list of leaders (records) to each and every interconnected linked node of the peer network; therefore, this silent feature of the blockchain network brings a high level of transparency

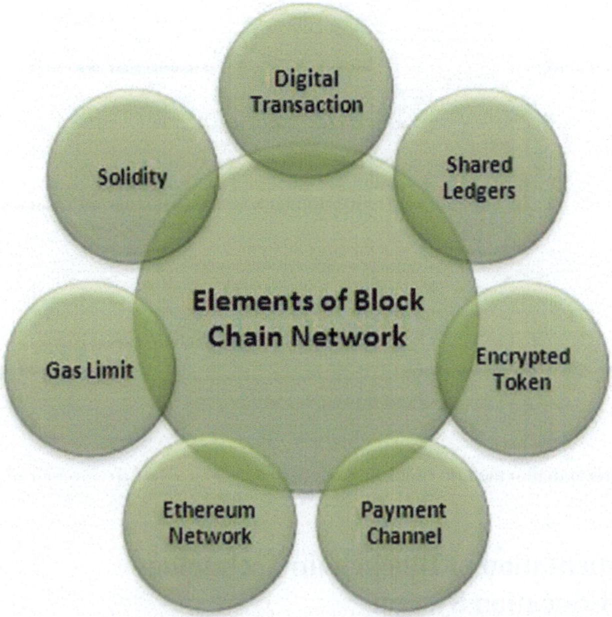

Fig. 2 Elements of blockchain technology

and vulnerability at every stage of the transaction. Apart from this, every block of the network is uniquely connected with its followers and previous blocks of leader through a digitally signed mechanism so that the network could be protected against unauthorized changes made during the operation. The key invention of this technology is that if a participant wants to transfer any leader across the network, then they can do so because they need not take permission from a third party as in centralized systems. Since the very beginning of internet, there has always been big issues with virtual currency transactions, because in virtual currency transactions the problem of double spending arises, which needs a third party to solve the problem, but blockchain realize the thing and resolves the problem of double spending in a superior manner because of its most secure and independent infrastructure where it eliminates the need to have a mediator or third party to allow the transactions to proceed with trust. Similarly, blockchain technology is also expanding its services, in the same way as for other noncommercial areas of application such as smart health, education industry. Figure 3 describes the story of transaction processing in blockchain technology.

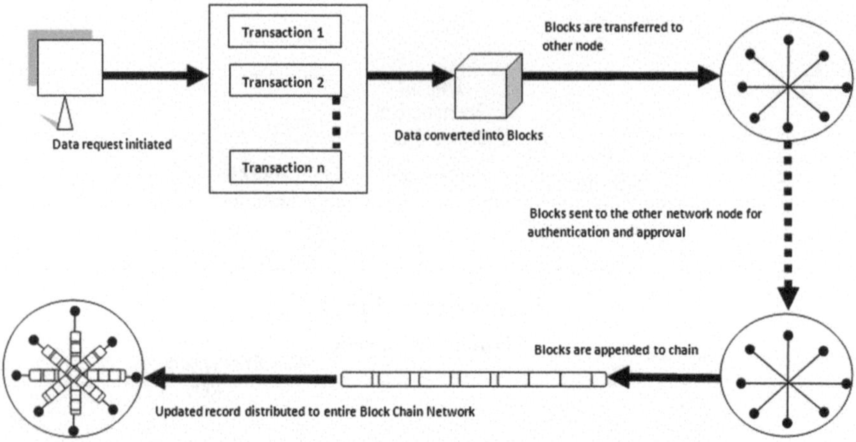

Fig. 3 Transaction deployment in blockchain network

5 Implementation of Blockchain Technology in the Education System

The blockchain technique undergoes the process of validation and verification every time a change has been made to records. In order to process the verification and validation of certificates blockchain first identifies the relevant information regarding digital certificates and credentials that will be issued to the relevant learner and educator. At the end, these certificates should be verified via educators and learners, which is mostly provided in paper form along with a secret digital value for the purposes of security. During this phase, the maintenance of consistency, integrity, and immutability is necessary, which has been taken care of by the blockchain mechanism itself.Blockchain finishes all these process in a very small amount of time without the need for additional resources compared with the centralized certificate management system, as defined in Sompolinsky and Zohar [9], and the university of Nicosia only issuing degree certificates digitally using blockchain technology. Figure 4 describes the conceptual procedure of implementation of blockchain technology in the education system.

6 Role of Blockchain Technology in the Education System

In the education system certificates are the symbol of special achievements, which is why education has always been represented in the form of certificates since the beginning. The quality of the learner and educator is always considered in the form of text materials such as academic grades, as mentioned on particular certificates;

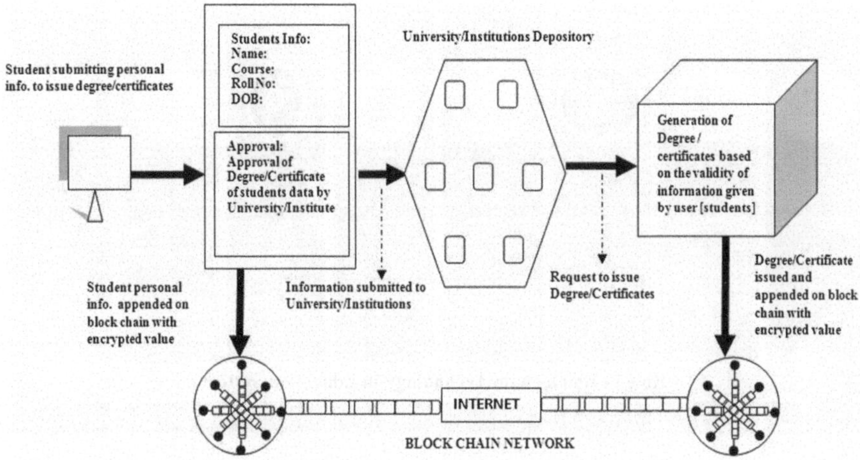

Fig. 4 Implementation of blockchain technology in the education system

therefore, certificates are important assets for achievement. In the education industry various activities can be mentioned in the form of physical papers, paper certificates, and documents are prone to tampering. Anyone can easily make a duplicate copy or they could be stolen for personal use.

On the other hand, the management of certificates and documents in digital format via blockchain technology are very easy to handle and are highly secure to unauthorized access. Therefore, it is tamper-proof every time. In blockchain network certificates are stored and maintained via a distributed peer-to-peer network, which is decentralized in nature; thus, records can be managed without interference of a centralized authority or third party [10].

In order to manage the importance of certificates, credentials, and degrees, blockchain technology plays an important role. The following are the major characteristics of the blockchain technique in the education system. Figure 5 describes the role of blockchain technology in the advancement of blockchain network.

6.1 Resilient Education System

Blockchain technology uses the concept of the sharing of strings of tokens as ledgers across the network and business platform, which allows appendage only and is guaranteed to eliminate a single point of failure to make the education system resilient.

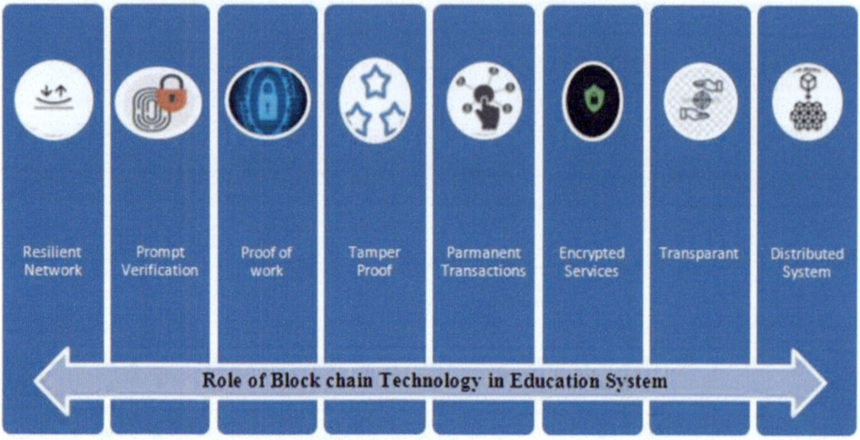

Fig. 5 Role of blockchain technology in the education system

6.2 *Prompt Verification*

Verification of every appendage is necessary to make the education system more reliable; therefore, the blockchain technique allows only those particular requests that are agreed by the entire network node to be accepted. Such a process is called a network-verified process and it may make the education system more efficient and consistent.

6.3 *Proof of Work*

As blockchain technology manages the complete set of records via blocks of chain so that blockchain always has proof of work to prove the latest request based on history. Often, when someone has lost their certificates or credentials, the latest request created for the generation of new degree certificates can be analyzed and matched with the proof of work system in a much smaller amount of time, which makes the education system more impressive and effective.

6.4 *Tamper-Proof Storage and Management*

It is always considered a big challenge in the education system to prevent all confidential certificates and credentials records to be managed in a tamper-proof manner. It is important, but the blockchain technique has a guaranteed solution to this problem. The technique enables a committed record-keeping system that is a completely tamper-proof and reliable mechanism to ensure a high level of security with the tamper-proofing of certificates using shared ledgers.

6.5 Permanent Transactions

In order to ensure a high level of security, confidentiality, privacy, and safety of credentials we need a highly protective and preventive educational storage system that enables a trustworthy environment for managing such important and valuable documents, as the traditional education system is facing challenges during the issuing of degree certificates. Crimes of issuing fake certificates and similar fraud has been increasing in past years. Maintaining highly sensitive data in open environment is always difficult to, on the other hand. The blockchain network is completely immune and protected from such challenges because of its permanent record management system, which stipulates that once a record has been placed it can never revert back to such permanent transactions, making our education system highly integrated, consistent, and reliable [8, 9].

6.6 Encrypted Security

The blockchain system undergoes a digitalized encryption-based security mechanism that integrates a secret code called a hash value, which is uniquely created and generated by the blockchain system itself during operational mode. This encrypted secret code has been associated with every shared leader, which makes the safest operational environment for various financial and nonfinancial applications such as the education system.

6.7 Transparent

Transparency is one of the most desired features of the education system. If one wants to get their credentials any time or if someone got a job opportunity, then we need a transparent system that can provide necessary and accurate information on a real-time basis. Blockchain technology offers a completely transparent, accurate, and reliably trustworthy system on the basis of its distributed infrastructure. Here, one can obtain the desired information on time and with accuracy.

6.8 *Distributed (Decentralized) System for Education*

Blockchain technology deals with multiple blocks in which the shared leader of information has been distributed over the entire blockchain network so that if any other user want to make changes, these changes can be tracked or monitored by other participants of blockchain networks. In this way the distributed database has spread over the entire network so that it helps to avoid degree fraud or tampering issues, because once a leader has been deployed on a network it cannot be altered by anyone, which is why the distributed system can maintain only verified and valid credentials [10].

7 Blockchain Layered Framework for Educational Industries

At present, blockchain technology is not very popular in teaching, learning, and education-oriented business applications, but since 2009 it has been very famous from the deployment of blockchain for bitcoin cryptocurrency transactions, Most of the research defines that a large number of the financial business platforms have agreed to work with blockchain. Blockchain takes into consideration the maximum expectations of the financial business world by providing a trustworthy transactional environment across the internet, but in the latest research report it has been specified that now other nonfinancial areas are also thinking of adopting blockchain services for the betterment of their business. For example, in educational applications, the blockchain technique is always fruitful for learners, students, teachers, and the management authorities. The basic features of the blockchain technique are defined that through blockchain deployment we can make our education system more productive, efficient, secure, reliable, and trustworthy. Similarly, blockchain is also good for retail business platforms and the health care system. The integration of blockchain with the internet of things, machine learning technology, and data science strategies provides a more impressive impact on the quality and productivity of different business platforms. As the popularity of blockchain is increasing in day-to-day business life we can imagine that in the twenty-first century all such nonfinancial services platforms did a great job and provided smart business solutions all around the world [11]. Deployment of blockchain technology in teaching and education defines that blocks of chains are integrated with large data sets. Along with the conceptual deployment of other technology such as artificial intelligence, the internet of things, and cloud computing network, blockchain technology constitutes a mixed (partially distributed and centralized) model [6]. This model consists of various layers of application services, as shown in Table 2, such as network management, services, and applications to meet the desired objectives of the education system.

Table 2 Layered service architecture of blockchain technology

Service architecture of Blockchain network	
Application layer	Blockchain transcript
	Learning certificates via smart contacts
	Effective individual learning
	Shared ledger
	Teaching resource management
	Monitoring and tracing of resources
Service layer	Authentications
	Distribution of accessing rights
	Audit services
	Tracing and monitoring of records
Management layer	Encryption-based computation
	Hash value management
	Validation and verification of data
	Digitalized certification
	Smart contract management
Network layer	Distributed computing based on peer-to-peer networking
	Blockchain cloud network

In order to provide systematic services for different business platforms the blockchain network works with the following layers.

7.1 Application Layer

The blockchain application layer provides different levels of application services in a seamless manner, such as issuing certificates, validating credentials, and for improving quality in education. It helps to provide transparent direction to evaluate and monitor the quality parameters needed for quality education. The application layer, consisting of the following different functions, bring application services in consistent and integrated manner.

7.1.1 Blockchain Transcript

A blockchain transcript works as a string of information. This function consists of the details of different credits, all credentials and skill sets all around proof of work for every individual that he/she has achieved throughout his/her life. This simple thing has been considered as a blockchain transcript [8]; through transcription any employee can get the valid credits from anywhere and anytime. In this way transcript brings special verification and validation services for each business platform [9].

7.1.2 Learning Certificates

The blockchain application layer provides the facility to obtain issues and certifies to one's special learning achievements through online or offline educational courses across the country. This makes talent frequent and meaningful to make this possible. Blockchain consists of smart contacts that play an important role in obtaining certification on the basis of one's talent.

7.1.3 Effective Individual Learning

At present, many online learning platforms are increasing, especially owing to the popularity of mobile devices. One can learn lots of new things anytime and anywhere. All such records of learning have been associated through a smart contract on the blockchain network so that it always help institutes or centers to know the learning level of individuals along with their background. This practice makes individual learning more impactful and effective.

7.1.4 Shared Ledger

The distributed infrastructure of the blockchain network can make sense for all the educational institutions in the way that via blockchain technology one can manage the permanent records of learning using a shared ledger. Therefore, as defined in Xu et al. [11], the institute of the future has put forward a proposal for the shared ledger technique. This proposal describes shared ledger as having all the relevant information of learners achievements in education and certification so that such ledgers make blockchain technology more impressive for the future of educational institutions and organizations.

7.1.5 Teaching Resource Management

The globalization of teaching and learning through different online platforms requires customization of the teaching learning resources so that various learning materials such as text books, question bank, reference books, syllabus requirements can be maintained in the form of records over a blockchain network. In this way, only authenticated teaching material is used by learners. Such resources are available as chain of blocks so that anyone can get them on demand. They can also put in their request for new resources, which may be appended with existing blocks at a later stage.

7.1.6 Monitoring and Tracing

Evaluation is one of the important and necessary parts of every learner's journey and educational institutions can manage and maintain this through the blockchain technique in more proper way. Blockchain technology plays an important role in reforming our traditional education system. It helps to maintain the overall learning material as well as the progress of individual students along with keeping the entire record of feedback so that in a transparent way everyone can monitor and track the individual learning records and progress of the organization [12].

7.2 Service Layer

The service layer of blockchain technology is responsible for the management and maintenance of different blocks of data in encapsulated form for different individuals. In addition, this service layer provides the function of authentication, access control, audit, and monitoring services, which plays an important role in smoothing the operations in the blockchain network.

7.3 Management Layer

The management layer is responsible for the entire management of operational services such as using smart contracts, data encryption, and digital certification techniques during operation. The management layer ensures the process of validation and verification before, during, and after every operation in the blockchain network.

7.4 Network Layer

The back bone of blockchain technology is the distributed peer-to-peer network, which provides the facility to use peer-to-peer network services for the entire operational domain of blockchain technology. A distributed network uses the concept of shared ledger and uses smart contracts to get avoid tampering, and fraud for commercial and noncommercial applications [13].

7.5 *Blockchain Cloud Network*

The blockchain cloud provides an appropriate environment to bring blockchain services across the network. It ensures the availability of blockchain resources such as software, hardware, and other application requirements and security concerns.

8 Essential Condition for Blockchain as a Solution

Blockchain technology looks ahead to the great future of the digital computing platform because it brings various secure services to its participants along with the trust of users; therefore, nowadays many business applications are looking forward to the blockchain infrastructure, which is very suitable for commercial and noncommercial applications. Blockchain is excellent for their services, but it is also true that it is not a single solution for all problems; there are many essential conditions that need to be kept in mind during the use of blockchain as a solution. The following are the various conditional properties that are defined as prerequisites for the implementation of blockchain over different areas of application [14].

8.1 *Shared Data Should Be Updated by Multiple Participants*

To use blockchain as a solution, it is essential to get updates on all the shared data from all participants so that the remaining blockchain process can proceed in a pipelined manner to take the advantage of technology.

8.2 *Verification as an Essential Process*

The process of verification of recorded data, which is shared through multiple participants, needs to be verified so that blockchain technology can gain the trust of all its participants. This condition is only possible when the system meets the first condition as defined above.

8.3 *Need to Reduce Complexity*

In order to reduce the complexity of the entire system it is necessary to eliminate unwanted parties so that the dependency of the data over multiple participants can be reduced. The involvement of additional intermediaries always adds complexity to the operation.

8.4 Need to Reduce Delay

The priority of business is to complete all the tasks on time and reduce delays; therefore, it is essential to eliminate factors such as interaction among the participants so that business objectives can be met on time.

8.5 Process Interaction

Since the blockchain technique provides a solution for transactions, the presence of multiple transactions creates interaction and dependency.

Figure 6 describes the essential prerequisites for the smooth working of blockchain technology. If the first condition is met, it means that the remaining three conditions are also met in the blockchain environment. In this case blockchain technology can be considered the best solution for service platforms.

9 How Digital Credentials Are Processed in Blockchain for Educational Applications

The management of degree certificates, and student credits in an open place through an centralized distributed online storage system is always a challenging task in the face of online data base handlers and certificate distributors. The blockchain

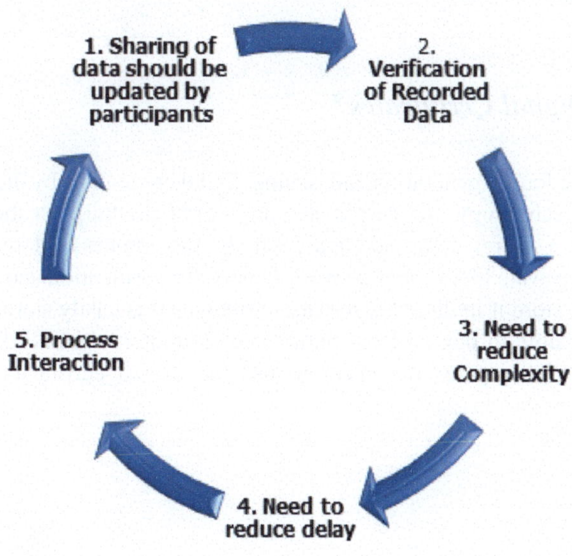

Fig. 6 Essential prerequisites for blockchain solutions

technique is one of the best solutions for maintaining the confidential records of student credits in a very secure and superior manner, as blockchain technology manages the credentials using a digital encryption technique in which recorded data have been maintained using a secret code called a hash value, Blockchain continuously generates this value and appends it to every block of the blockchain network. The traditional encryption technology has been encapsulated in blockchain to operate the entire blockchain network with digitally computed values in this way. The blockchain technique provides a high level of security to issue digital degree certificates, which eliminates the issues of fraud and tampering with confidential data and certificates [15].

In order to define the life cycle of the generation, production, and distribution of digital certificates, the following terminologies are involved in blockchain networks.

9.1 Digital Certification and Distribution

In order to issue degree certificates a student provides his details as initial inputs into the system along with some unique ID such as roll number. This information is recorded at the student's blockchain and is later appended as a block with the network and then for verification purposes, this information is forwarded to the verifying authority such as the university or institute. Once the verification process has been completed successfully, a token is created, which transfers over to students and it is also kept and appended as a block over to the university's or institute's blockchain network itself. Figure 7 describes this process for understanding the role of digital certificates in blockchain networks.

9.2 Why Digital Certificates?

As per the procedure of generating and issuing digital credentials by the deployment of encryption technology, the certificates are stored digitally on the blockchain network using a secret code called a hash. In this process all the authorities have their own private blockchain network to perform verification and validation of certificates and store it in digital certificate format, as it is easily storable, manageable, and accessible through a large number of computers digitally from anytime and anywhere. It is quite convenient to use the digital certification technique [13, 14].

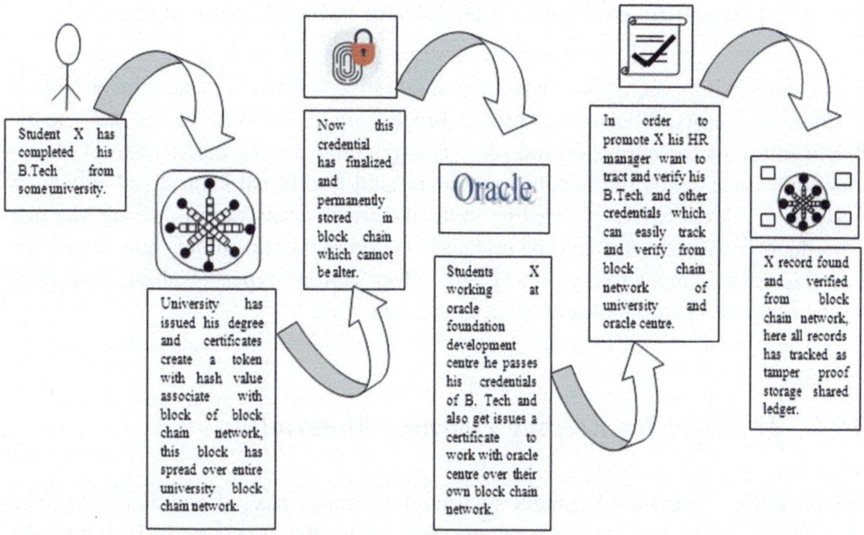

Fig. 7 Management of digital credentials in blockchain

9.3 *Digital Signature*

In order to make the recorded data tamper-proof one needs to get these recorded data signed using a digital signature technique in which all the confidential documents and credentials need to be integrated with a locking system. To deploy this feature of security, every intermediate and participant authority needs to agree on the deployment of encryption and decryption techniques during the generation and distribution of digital certificates. Once a document has been signed digitally, it means that the document is digitally locked for further processing and it allows only that user having that specific signature is able to unlock the signed document. The digital signature completely brings our documents into highly protective modes, which avoids the chances of unauthorized access and tampering with the certificates for someone to commit a crime for personal benefit. The blockchain follows the process of the encryption and decryption techniques, as provided by traditional networking concepts, in which every user has two keys: one is called a private key, used for digitally signing over the documents, and the second one is called a public key, which has been announced or shared with a specific valid recipient of the document to get it unsigned. This technique is very secure for maintaining the confidentiality and security of the documents compared with other techniques.

10 Applications of Blockchain in the Educational Sector

Blockchain brings such a wonderful and qualitative service to education, as well as to different sectors of business. Through blockchain technology we can imagine the bright future of the education industry. The dynamic service mechanism of blockchain always extends the benefits of manual and traditional centralized education systems to a great extent. According to the different survey reports we can say that our education system needs to be updated via deployment of blockchain networks. For this advancement today, blockchain offers various types of application platforms for the education system, such as the following.

10.1 Manager for Keeping Learning Materials

Traditionally, educational centers such as universities have the authority to keep students' learning records of credentials. This means that it is very difficult for a job provider to get access to and to verify the originality and validity of the records from official storage, which maintains the entire record of learning including course title, category, and the exact credentials of any student gained during completion of the course. In this case such official data are managed through a single and unique authority. In this case verification of official data is very hard to obtain and analyze for further decision making [15].

Blockchain technology is specially invented and designed for this kind of problem. The structure of the blockchain is completely appropriate for storing and keeping the learning records of students in a very efficient and effective manner. Pursuing a higher level of security measurements makes this possible. Blockchain introduces the concept of a smart contract, which ensures management of the entire credentials of any student. Through blockchain, any employer can access these records for verification of students' data to make decisions regarding employment. As defined in Gatteschi et al. [16], blockchain is a novel approach to keeping learning materials, in which learning credentials are considered as virtual coins managed as blocks, and then, at a later stage, appended to blockchains by the teacher. In this way, all the credentials are now available for accessing, verifying, and validating by employers or any third party. In addition to verification services, keeping the learning record also helps in observing and evaluating a learner's professional, mental, physical, conceptual, and technical abilities, which helps the employee to get the job and the employer to select someone with complete convenience.

10.2 Authority to Issue Degree Certificates

The magical and impressive service network of blockchain brings the features of transparency and immutability, which help one to get a higher level of security to maintain overall credentials and the degree certificates of learners. These credentials proved that particular learners had obtained this award and professional certificates during their schooling or academic studies [17]. In this manner, through the blockchain network a learner or employer can validate the needed information anytime and anywhere. The blockchain network works in a transparent way to manage records permanently around all the academic and business requirements by eliminating the distance and data accessing permission factor to save time and avoid burdensome situations.

The traditional education system is facing many challenges in managing and maintaining certificates and credentials with consideration of security, but technically, various institutes and universities still face problems of fraud and misuse of credentials by unauthorized users, as blockchain offers the best solution to this problem. In 2015, an MIT lab initiated and started to distribute digital certificates to their learners so that they can access their credentials and control them themselves. Similarly it can also be verified by any third party authority without wasting time or interacting with many people to obtain validated credentials [18].

As in Aitzhan and Svetinovic [19], another user of blockchain technology is the University of Nicosia, which has deployed it to obtain verified and distributed degree certificates in a digitalized manner. Similarly, a new concept of permission based on blockchain technology has been proposed, called "Credence Ledger," which is especially used to obtain verified academic credentials in a decentralized manner, deploying the services of the management of digital certificates through a ledger, to obtain verified records by the stakeholder itself or if required so that any third party intermediates can also access them [14]. As defined in Wang et al. [20] EduCTX is another process mode provided from the Ark blockchain technological platform to gain access, control, and manage the credits and origin of students in a most superior manner for their students and third-party intermediates.

10.3 Managing Education through Decentralized Database

Managing the education system through a decentralized database is the back bone of blockchain technology. Decentralized management of data brings transparency among all the parties such as learners, teachers, and intermediaries, which means that each participant of the blockchain network can have a copy of the ledger to get validated and update the required changes. This features brings a collaborative environment to all its users and participants [21].

The decentralization in the education industry brings great efficiency to institutes, colleges, and schools as defined in Bore et al. [22], with Sony global education system adopting the latest framework of the blockchain network instead of using a traditional process to evaluate the results and learning activities, which makes the education system more reliable, fair, and efficient. In Wang et al. [20] the authors describe an application of blockchain called "school information hub" used in Kenya to manage and validate the entire learning records of school students. This practice helps management in decision making for individual learners' performance and improvement in academia.

10.4 Tool for Effective Communication

The application of blockchain has also served as a tool for effective communication for the benefit of the teaching/learning process since 2019. During the COVID-19 pandemic the E-learning platform played an important role for the entire learning platforms, which require effective communication between teachers and learners for better learning and credits. In this approach learning material exchange should be handled through blockchain network topology to definitely encourage teachers and learners to communicate efficiently with a high level of security, privacy, and transparency benefits [23].

Similarly, in Li et al. [24], the authors proposed a novel blockchain application of E-learning that supports the provision of additional awards in the form of incentives as a virtual currency to the learners who secure a top rank. In this way we can encourage learners to engage with E-learning platform services.

10.5 Security Keeper of Credentials

Blockchain technology works on the concept of encapsulating the recorded data into blocks called ledgers, which consist of digitally encrypted values throughout the sharing of blocks with each and every participant; therefore, the process of accessing and updating shared data is always controlled by the digital signature technique, which allays the security concerns of every participant. That is why blockchain technology is the best network for providing services as a secure keeper of credentials.

10.6 Tool for Managing Financial Issues

Blockchain technology also addresses a solution for financial issues raised in educational institutions with regard to taking fees and managing scholarship records. In many institutes or schools we can see that learners have raised financial matters and problems very frequently. Many academic organizations have faced this challenge for a long time, but the deployment of blockchain technology is the innovative invention in the educational industry, and now all the financial matters can be handled by the blockchain network as easily as it handles all the credits and results of students because blockchain techniques avoid the chances of fraud and crimes related to fees and other financial factors, completely eliminating the opportunity to claim against financial things. In addition, through blockchain technology, every educational institution can gain the trust of its students before, during, and after their learning courses. The application of blockchains in finance brightens the image of the education industry for past, present, and future learners [25].

11 Issues and Challenges in the Education Industry

We knew the fact that behind every successful technology there is always a big chain of challenges and issues. Similarly, blockchain technology is also facing various challenges in today's business world as many educational organizations have believed in the traditional approach to learning and deployed centralized management for issuing degree certificates. By comparison, blockchain technology is in the experimental phase of its beginning. Its brings many good things that are favorable for teaching and learning, although there are many challenges that blockchain technology still needs to face. To understand these challenges we divide it into two major categories of application, that is, technological and nontechnological challenges.

11.1 Technological Challenges

To integrate or deploy a new advancement over an existing platform, we need to think about the various challenges raised before, during, and after the deployment of new technology. At present we have found that many technical challenges have been faced by blockchain networks because they are in the beginning phase. People in the business domain are not very familiar with blockchain technology; thus, before deployment of blockchain technology over an existing one, we need to rethink following technical challenges [26, 27].

11.1.1 Scalability

According to the latest reports released on blockchain technology in Fairley and Bore et al. [14, 22], scalability is one of the major challenges faced by the educational industry because according to the nature of the data, the need for storage on a blockchain network at every stage of communication requires a massive demand for storage and the propagation speed should be less in a network. As per the technological requirement of the blockchain network each participant's node needs to store transactional data and manage relevant information to pursue synchronous operation to make it possible for every participant to have to maintain sufficient storage. Meanwhile, this practice leads to additional overheads on nodes and the network, which degrades the overall performance of the network. As a result, it reduces the efficiency and response time of the node significantly [28].

11.1.2 Security

Security is another major challenge faced by blockchain technology in education, even though we knew that blockchain technology has a sufficient solution to ensure security, but as fact defines that if an attacker were to hack more than 51% of the nodes of the network, information and recorded data could also be tampered with and leaked; however, the cryptographic mathematical model proves that the chances are very low to make it possible, but the model doesn't guarantee that the blockchain network is completely immune from attackers and that it will not be cracked by unauthorized access in future. This results in the leakage of confidential information from academic institutions [29]. In addition, as blockchain technology manages all its records and transactions in an open environment with transparency, this transaction can be traced or predicted by anyone for their individual benefit. In this case blockchain does not insure the services against privacy and safety benefits.

11.2 Nontechnical Factors

As we have discussed in previous sections, similarly, nontechnical factors also need to be considered a big challenge for blockchain technology. When one talks about the advancement of the existing platform they need to think about both technical and nontechnical challenges. Every business needs to keep in mind that before the deployment of new technology one should prepare in advance to face both technical and nontechnical challenges and need to address their solution for the betterment of their business future. In the following, the major nontechnical challenges are described that have been faced by blockchain technology in its deployment replacing an existing platform.

11.2.1 Pool of Skilled Engineers or Trainers

Blockchain applications are considering a large service domain, In the past few years many business platforms have been looking to deploy and integrate services with a blockchain platform as per the demand and new business policies. It might be possible that in a very short period of time blockchain networks will be accepted by business platforms such as education, health, finance, and noncommercial and commercial platforms, as block chain experts and well-trained engineers will be very scarce if the blockchain technology continues to grow. In that case we need a large amount of skilled and trained manpower that have the qualities to manage and solve the complex technical and computation problems as fast as possible. Because blockchain ledgers and smart contracts follow complex mathematical models, unskilled engineers are not able to understand and solve the problems [30, 31].

11.2.2 Existing Traditional Policy of Organization

Many organizations have bound their work and services under the norms of an existing policy and rules that do not allow a service and operational platform to be switched from the existing one to an advanced one. Moreover, every organization has different levels of authority for flow rights and for accessing permission regarding operation mode protocols; therefore, often we can see that the variations in business decisions do not meet the requirements of transferring the existing business structure and operational mode to another platform. This seems to be a very complex problem with many of today's business sectors.

11.2.3 Budget

The deployment of technology from old to new demands a high budget and strong financial background so that financial factors are also raised as a big challenge in the face of blockchain technology, owing to insufficient funds, meaning that many business service platforms cannot switch their business over to a new development because the technological and infrastructure-based investment needs an adequate budget [32].

12 Future of Blockchain Technology in Educational Applications

Blockchain is one of the fastest growing technologies of the future digital platform. As the business grows it promotes online transaction services with safety and security constraints, but in 2009 the financial crisis caused the commercial platforms to crash. Since then, they have lost their clients' trust, which affects the entire business world and digital operations. According to the requirements of digital services through blockchain technology, one can perform transactions based on different operation in different areas of application such as health, smart cities, digital payment, the stock market, communication, education, and agriculture. Blockchain brings the most secure, efficient, and effective service architecture for future business service. It addressed the solution of transactions based on cryptocurrency such as bit coin. Blockchain offers secure services for a large area of application, which means that blockchain technology will be a great supporter and part of multiservice platforms in the future. According to the various reports in the literature, the future of blockchain technology is very bright, but in this chapter we ignore the other areas of application and concentrate on the future of the educational application of blockchain technology [33]. Blockchain offers the facility to process transactions using digital currency, which allows learners to get it into a wallet and motivates learners to gain admission to different educational places or join learning activities at national and international levels using online learning platforms. Since 2019, from the time of the COVID-19 pandemic, e-learning platforms have been promoted for providing online learning on different courses. In the future blockchain technology will make this system more proper and trustworthy in every direction and in a similar fashion it also motivates learners to learn. The following are major future possibilities for considering blockchain technology as a solution for different educational applications [34, 35].

The education system needs to track and manage all the records of learning activities in the most superior manner so that it can help someone to know the direction of improvement in a specific learner's profile. In future, blockchain can be a role model for the management of formative assessment in a systematic direction [36].

Blockchain technology can be a great helping hand in today's education system and in the near future because it uses smart contracts to eliminate the need for involvement of a third party. The concept of smart contract management is also helping to manage the entire record of teachers and learners in a more systematic way, which solves a large number of other problems raised during academic and non-academic activities [37].

13 Conclusion

In this section we conclude the importance and applications of blockchain technology in educational systems. With the help of different sections of this chapter we have tried to give a systematic description of the role of blockchain in educational organizations and how it is useful for the future of this area, as blockchain works on the basis of the integration of different encryption techniques and consensus algorithms and brings the benefits of tamper-proofing, validation, efficient performance, and an immutable environment to make education systems more effective. Finally, we can say that we have met the objective of this chapter by providing the maximum fundamentals of blockchain technology in respect of education systems and future possibilities with the blockchain network. The proposed consensus-based algorithm improving the services of traditional blockchain networks along with the integration of the OLSR protocol provides better network services in addition to security and privacy. Meanwhile, in a very short space of time we can see the future aspects and responsibilities of the education industry will definitely adopt a blockchain-based education system to ensure that education can be pursued with confidentiality, privacy, and security. I hope that this chapter will be very helpful to its readers for understanding the basics and fundamentals of blockchain technology and its applications in education systems. We aim to consider all categories of learners; thus, this material should be productive for beginners and those experienced for further research and development of blockchain technology.

References

1. F. Hawlitschek, B. Notheisen, T. Teubner, The limits of trust-free systems: a literature review on block-chain technology and trust in the sharing economy. Electron. Commer. Res. Appl. **29**, 50–63 (2018)
2. S. Saberi, M. Kouhizadeh, J. Sarkis, Block-chain technology: a panacea or pariah for resources conservation and recycling? Resour. Conserv. Recycl. **130**, 80–81 (2018)
3. D. Tapscott, A. Tapscott, *The Blockchain Revolution: How the Technology Behind Bitcoin is Changing Money, Business, and the World* (Penguin Publishing Group, London, 2016), p. 72
4. S.M. Idrees, I. Aijaz, R. Jameel, M. Nowostawski, *Exploring the blockchain technology: issues, applications and research potential.* iJOE **17**(07), 49 (2021)
5. G.P. Dwyer, The economics of Bitcoin and similar private digital currencies. J. Financ. Stab. **17**, 81–91 (2015)
6. G.W. Peters, P. Efstathios, Understanding modern banking ledgers through block-chain technologies: Future of transaction processing and smart contracts on the internet of money, in *Banking Beyond Banks and Money*, (Springer, Cham, 2016), pp. 239–278
7. J.P. Shim, M. Warkentin, J.F. Courtney, D.J. Power, R. Sharda, C. Carlsson, Past, present, and future of decision support technology. Decis. Support. Syst. **33**(2), 111–126 (2002)
8. G.W. Peters, E. Panayi, Understanding modern banking ledgers through blockchain technologies: Future of transaction processing and smart contracts on the internet of money, in *Banking Beyond Banks and Money*, (Springer, Cham, 2016), pp. 239–278
9. Y. Sompolinsky, A. Zohar, Secure high-rate transaction processing in bitcoin, in *International Conference on Financial Cryptography and Data Security*, (Springer, Berlin, 2015), pp. 507–527
10. N. Kshetri, Block-chain's roles in meeting key supply chain management objectives. Int. J. Inf. Manag. **39**, 80–89 (2018)
11. Q. Xu, K.M. Aung, Y. Zhu, K.L. Yong, A block-chain-based storage system for data analytics in the internet of things, in *New Advances in the Internet of Things*, (Springer, Cham, 2018), pp. 119–138
12. B.A. Scriber, A framework for determining blockchain applicability. IEEE Softw. **35**(4), 70–77 (2018)
13. Idrees, S. M., Agarwal, P., & Alam, M. A. (Eds.). (2021). Blockchain for Healthcare Systems: Challenges, Privacy, and Securing of Data. CRC Press
14. P. Fairley, Blockchain world – feeding the blockchain beast if bitcoin ever does go mainstream, the electricity needed to sustain it will be enormous. IEEE Spectr. **54**(10), 36–59 (2017)
15. T.T.A. Dinh, R. Liu, M. Zhang, G. Chen, B.C. Ooi, J. Wang, Untangling blockchain: A data processing view of blockchain systems. IEEE Trans. Knowl. Data Eng. **30**(7), 1366–1385 (2018)
16. V. Gatteschi, F. Lamberti, C. Demartini, C. Pranteda, V. Santamaría, To blockchain or not to blockchain: that is the question. IT Prof. **20**(2), 62–74 (2018)
17. G. Pîrlea, I. Sergey, Mechanising blockchain consensus, in *Proceedings of the 7th ACM SIGPLAN International Conference on Certified Programs and Proofs*, (ACM, New York, 2018), pp. 78–90

18. M. Turkanović, M. Hölbl, K. Košič, M. Heričko, A. Kamišalić, EduCTX: a blockchain-based higher education credit platform. IEEE Access **6**, 5112–5127 (2018)
19. N.Z. Aitzhan, D. Svetinovic, Security and privacy in decentralized energy trading through multi-signatures, blockchain and anonymous messaging streams. IEEE Trans. Depend. Secure Comput. **15**(5), 840–852 (2018)
20. J. Wang, M. Li, Y. He, H. Li, K. Xiao, C. Wang, A blockchain based privacy-preserving incentive mechanism in crowd sensing applications. IEEE Access **6**, 17545–17556 (2018)
21. S. Kim, J. Kim, POSTER: mining with proof-of-probability in blockchain, in *Proceedings of the 2018 on Asia Conference on Computer and Communications Security*, (ACM, New York, 2018), pp. 841–843
22. N. Bore, S. Karumba, J. Mutahi, S.S. Darnell, C. Wayua, K. Weldemariam, Towards blockchain-enabled school information hub, in *Proceedings of the Ninth International Conference on Information and Communication Technologies and Development*, (ACM, New York, 2017), p. 19
23. Y. Yuan, F.-Y. Wang, Blockchain and crypto currencies: model, techniques, and applications. IEEE Trans. Syst. Man Cybernet. Syst. **48**(9), 1421–1428 (2018)
24. Z. Li, J. Kang, Y. Rong, D. Ye, Q. Deng, Y. Zhang, Consortium blockchain for secure energy trading in industrial internet of things. IEEE Trans. Ind. Inf. **14**(8), 3690–3700 (2018)
25. C.-W. Chiang, EberBetanzos, and Saiph Savage., Exploring blockchain for trustful collaborations between immigrants and governments, in *Extended Abstracts of the 2018 CHI Conference on Human Factors in Computing Systems*, (ACM, New York, 2018), p. 531
26. S.M. Idrees, M. Nowostawski, R. Jameel, A.K. Mourya, Security aspects of blockchain technology intended for industrial applications. Electronics **10**(8), 951 (2021)
27. L. Luu, D.H. Chu, H. Olickel, P. Saxena, A. Hobor, Making smart contracts smarter, in *ACM Conference on Computer and Communications Security*, (ACM, New York, 2016), pp. 254–269
28. V. Buterin, Ethereum: a next generation smart contract and decentralized application platform (2013), https://github.com/ethereum/wiki/wiki/White-Paper. Accessed 1 December 2020
29. K. Zhang, H. Jacobsen, Towards dependable, scalable, and pervasive distributed ledgers with blockchains, in *2018 IEEE 38th International Conference on Distributed Computing Systems (ICDCS), Vienna*, (IEEE, Piscataway, 2018), pp. 1337–1346
30. F. Hofmann, S. Wurster, E. Ron, M. Böhmecke-Schwafert, The immutability concept of blockchains and benefits of early standardization, 2017 ITU Kaleidoscope: challenges for a data-driven society (ITU K). Nanjing **2017**, 1–8 (2017)
31. M. Benchoufi, R. Porcher, P. Ravaud, Blockchain protocols in clinical trials: Transparency and traceability of consent. F1000Research **6**, 66 (2017). https://doi.org/10.12688/1000research.10531.5
32. C. Dannen, Solidity programming, in *Introducing Ethereum and solidity*, ed. by C. Dannen, (Springer, Berlin, 2017), pp. 69–88
33. C.D. Clack, V.A. Bakshi, L. Braine, Smart contract templates: foundations, design landscape and research directions. CoRR abs/1608.00771 (2016)
34. F. Tschorsch, B. Scheuermann, Bitcoin and beyond: a technical survey on decentralized digital currencies. Cryptology ePrint Archive, Report 2015/464 (2015)
35. S.M. Idrees, M. Nowostawski, R. Jameel, Blockchain-based digital contact tracing apps for COVID-19 pandemic management: Issues, challenges, solutions, and future directions. JMIR Med. Inform. **9**(2), e25245 (2021)
36. Svein, A. Jansen, Block-chain technology as a support infrastructure in e-government, in *International Conference on Electronic Government*, (Springer, Cham, 2017), pp. 215–227
37. W. Li, A. Sforzin, S. Fedorov, G.O. Karame, Towards scalable and private industrial blockchains, in *Proceedings of the ACM Workshop on Blockchain, Cryptocurrencies and Contracts*, (ACM, New York, 2017), pp. 9–14
38. W. Yin, Q. Wen, W. Li, H. Zhang, Z. Jin, An anti-quantum transaction authentication approach in Blockchain. IEEE Access **6**, 5393–5401 (2018)

39. R. Henry, A. Herzberg, A. Kate, Blockchain access privacy: challenges and directions. IEEE Secur. Privacy **16**(4), 38–45 (2018)
40. K. Kotobi, S.G. Bilen, Secure blockchains for dynamic spectrum access: A decentralized database in moving cognitive radio networks enhances security and user access. IEEE Veh. Technol. Mag. **13**(1), 32–39 (2018)
41. M. Wang, MeijiaoDuan, and Jianming Zhu., Research on the security criteria of hash functions in the blockchain, in *Proceedings of the 2nd ACM Workshop on Blockchains, Cryptocurrencies, and Contracts*, (ACM, New York, 2018), pp. 47–55
42. N.T. Courtois, L. Bahack, On subversive miner strategies and block withholding attack in Bitcoin digital currency. CoRR, vol. abs/1402.1718 (2014), https://arxiv.org/abs/1402.1718
43. C. Natoli, V. Gramoli, The blockchain anomaly, in *15th International Symposium on Network Computing and Applications (NCA)*, (IEEE, Piscataway, 2016), pp. 310–317

Development of a Blockchain-Based Survival Game for Blockchain Education

Chibuzor Udokwu, Alex Norta, and Roxana Voicu-Dorobantu

1 Introduction

Blockchain is a recent technology that seeks to address some of the organizations' important needs such as trust and transparency when dealing with each other [1]. Research has shown that the main reason organizations plan to adopt blockchain technology is to increase the effectiveness of their inter-organization collaborations by solving trust and transparency problems [2]. Understanding new technology is one of the first steps in increasing adoption, and thus, blockchain-technology education has largely lagged behind due to the blockchain complexity which is difficult for common users to comprehend. Blockchain comprises several technologies such as public-key identity management systems, digital signatures, consensus and smart contracts programming. Presenting these sub-technologies and their relationship is complex and difficult to conceptualize for most users that lack technical skills. As a result, most education materials in the blockchain domain are mostly unpopular and unused. Some educational materials already exist to help individuals to understand and apply blockchain technology in their organizations. The study [3] shows a flowchart system that outlines a series of steps in determining if a blockchain solution can be adopted in solving a given problem. Furthermore, other research works explore the integration of blockchain educational material in existing study curricula [4–6]. Research shows that gamification has the potential to simplify complex concepts [7]. The same gamification concepts can also be applied in explaining various aspects of blockchain technology to a non-technical audience. Still, only a few

C. Udokwu (✉) · A. Norta
Department of Software Science, Tallinn University of Technology, Tallinn, Estonia
e-mail: alexander.norta@taltech.ee

R. Voicu-Dorobantu
The Bucharest University of Economic Studies, Bucharest, Romania
e-mail: roxana.voicu@rei.ase.ro

S. M. Idrees, M. Nowostawski (eds.), *Transformations Through Blockchain Technology*, https://doi.org/10.1007/978-3-030-93344-9_3

studies explore the application of gamification in describing blockchain. The study [8] explores potential opportunities in gamifying blockchain concepts for educational purposes. The study [9] applies gamification in explaining blockchain concepts such as the block-creation process in PoS consensus mechanisms.

Employing the right gamification concepts to explain blockchain technology must take into account the challenging issues that mitigate the adoption and use of blockchain. Blockchain involves the concept of monetary economics, consensus governance and risk rewards [10]. Grasping the concept of money and money economics is necessary for understanding blockchain cryptocurrencies [11]. The concept of governance, especially decentralized governance, is necessary for understanding various blockchain-consensus systems [12]. The concept of risks and rewards provide the necessary background for users to understand blockchain tokenization of physical assets and the exchange of such assets [13].

This chapter proposes the gamification of blockchain concepts to enable non-technical users in understanding the various technologies that constitute a blockchain. We design a survival game of people (game players) left on an island without external human interaction that captures the essential concepts of blockchain technology. Blockchain technology provides the possibility for the island residents to record their daily activities for transparency and immutability of records (each player's activity), while the resources (assets) generated by island residents are valued and exchanged using blockchain tokens. The rewards for engaging in meaningful activities by island residents are saved as blockchain tokens as blockchain consensus systems provide the island residents with the possibility of establishing their rules (governance) such as acceptable island activities, reward amounts and a consensus system for recording their activities.

Therefore, the research question addressed in this study is how to apply gamification concepts in simplifying the conveyance of complex aspects of blockchain technology? The rest of this chapter is structured as follows. Section 2 provides the main background of the study outlining blockchain concepts and the scope of the survival game. Section 3 provides a high-level description of the proposed game outlining the main game elements such as the player activity records, block hash records and the phases of players locations within the game. Section 4 evaluates existing online resources available for the simulation of the game. Section 5 provides discussions of this paper results and related studies. Finally, Sect. 6 provides the conclusion, limitations and future work.

2 Background

This section contains background on blockchain concepts and the scope of the survival game. The blockchain concepts in Sect. 2.1 describe the main components that comprise a blockchain and the survival-game scope in Sect. 2.2 outlines the various phases that describe the simulated game.

2.1 Blockchain Concepts

A blockchain is a decentralized network that provides the participants with the possibility to record data on a distributed ledger. For example, cryptography ensures that the recorded information remains immutable. The main elements of a blockchain network include digital signature, consensus mechanism and a decentralized storage ledger [10].

The public key cryptography adopted in a blockchain ensures that all participants are duly identified by their digital signatures. Each user is associated with two keys, public- and private keys. The public key is used for identifying the members of the network and the private key for signing data before they are updated on a blockchain [14]. The consensus mechanism of a blockchain network defines the condition for adding new records of data. Transactions outlining the activities of the participants are gathered in blocks and updated on the ledger [12]. The distributed storage ensures that the blocks of data recorded on this network are updated and synchronized across all the ledgers of the participants, thereby ensuring the immutability of the stored records [15].

2.2 Survival-Game Scope

The proposed game shows the phases of life for island inhabitants that are disconnected from external human interactions. The phases are as follows, the resources-extraction phase, investment phase, disruption phase and retirement phase.

In the *resource-extraction phase*, the players have to extract resources from the ground or water for their survival. Such resource extractions are conducted by engaging in farming, fishing or hunting. The *investment phase* involves players investing their resources in assets such as food possession, real estate and crypto tokens. The investment phase is followed by the *disruption phase* where the players experience several disruptions such as accidents, or losses from disasters. The player investments and savings from the previous phase will determine if they proceed to the next phase.

The last phase is the *retirement phase*. If the players navigate luckily and invest wisely in the previous phases, they can proceed to final retirement and win the game that can be simulated and played on a web platform, or as a paper-board game. For simplification and explanation of the game concepts, the paper-board game is presented.

3 High-Level Game Description

We show in Fig. 1 a diagrammatic representation of the blockchain survival game outlining the main concepts and high-level requirements. The board game can be divided into two, player location as they move towards retirement and a scratch-note section for recording player activities, blocks generated and dice rolling.

3.1 Scratch-Note Section

This section consists of each player's activity records, block-hash records and the dice roll.

Dice Roll A six-sided dice is used in generating probability results of different game activities and also for selecting the next user who gathers transactions (activities) and creates a block where each dice roll represents a transaction. A complete dice roll for all the game players represents a set of transactions that can be used in creating a block. Another dice is rolled to determine the user who creates the block. The probability of selection is the same for all users. Still, users who invest in tokens have higher chances of selection in the proof of stake. Every player activity (dice roll) is powered by token savings and the user who creates the block is rewarded with tokens. The reward for the block creator is the summation of the transaction fee (token charges) for each player activity.

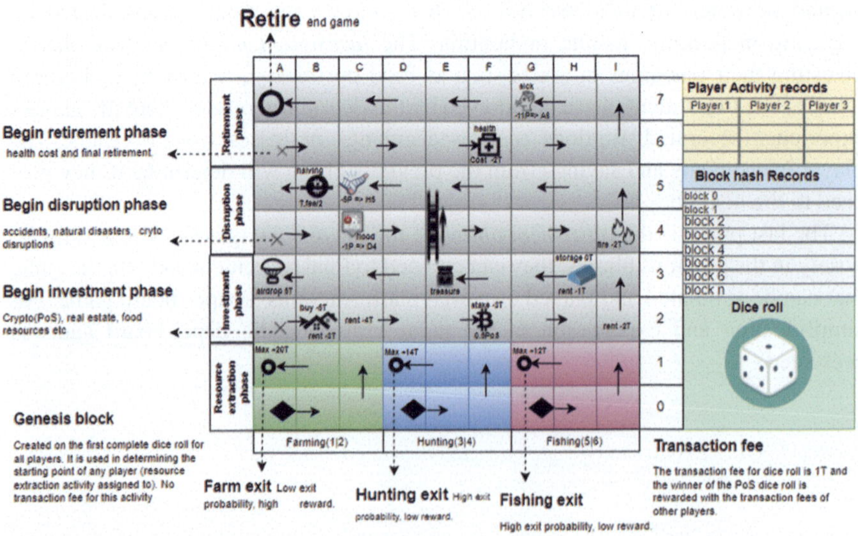

Fig. 1 Diagrammatic representation of the blockchain survival game concept

Player-Activity Records The section records the activity of the players as they move in the square boxes of the game. The player movement starts from either A0 or D0 or G0 towards A6. The player positions after each dice roll are recorded on the player-activity records. All the obstacles and successes of each player are recorded on the activity records as they move towards retirement. Therefore, two types of data can be recorded: player positions and player earnings. The positions are generated by combining the row- and column labels of the board.

The player earnings (or losses) are recorded as token gains, or losses depending on the successes and obstacles experienced as they journey towards retirement (G0). For instance, given a player with an id: Uch1, with the current position on the board as A2 and token balance of 31 tokens. The player rolls the dice that results in a face of 4. The player's new position is now E2. Assuming an obstacle that results in a loss of 5 tokens is associated with the position, the new player activity record is as follows: *ID_U ch*1:*P_E*3:*T*_26.

Block-Hash Records The block-hash record shows the hash of a combination of all transactions for a complete dice roll among all players. Given three players with the new records *ID_Uch*1:*P_E*3:*T*_26*; ID_Kkl*:*P_C*4:*T*_41 *and ID_Iju*:*P_C*5:*T*_12, using a common hash function such as sha256 and assuming the previous block hash is the genesis block with a hash of *5feceb66ffc86f38d952786c6d696c 79c2dbc239dd4e91b46729d73a27fb57e9*, the new block hash is now the overall hash of:

- the previous block hash,
- the combination of transactions for the new block.

The result is the hash of *5feceb66ffc86f38d952786c6d696c79c2dbc239dd4e91b-46729d73a27f b57e9* + *ID_Uch*1:*P_E*3:*T* _26 + *ID_Kkl*:*P_C*4: *T*_41 + *ID_ Iju*: *P_C*5:*T*_12.

The plus sign and white spaces are removed before calculating the hash. The new block hash is now:

ae5fe9d69ea848b9f5175585b32fed38eae4acd6cd69802608d13efa93ade059.

Transactions in the block are listed as follows:

1. ID_U ch1: E3:T_26.
2. ID_Kkl:P_C4:T_41.
3. ID_Iju: P_C5:T_12.

Block header: The timestamp and the block number are recorded as the block header, e.g., *Block*1 14:04:2020 13:48.

3.2 Player Locations

The player locations are represented as squares on the left side of Fig. 1. Each square shows where each player can be located at any point in the game. The locations are structured according to their phases of life while surviving the desolated island. The details of the phases and the possibilities of proceeding to the subsequent phases are presented as follows.

Resource Extraction Phase This phase begins by creating the genesis block as each player proceeds to a specific resource-extraction activity that is either farming, hunting or fishing. There is an equal probability of selecting any of these activities, still, the probability of exit and amount of token gain for each extraction activity varies. The probability of any ground activity is $1 = 3$ For all the three ground activities: $(1 = 3) * 3 = 1$.

The dice face 1&2 are allocated to farming, 3&4 are allocated to hunting and 5&6 are allocated to fishing. Therefore, for any number of players, there are equal chances of selecting one ground activity. Assuming three players with the ids *ID_ Uch*1, *ID_Kkl* and *ID_Iju* with the following dice face after the face roll 1, 5 and 6. The player *ID U ch*1 is be assigned farming while *ID_Kkl* and *ID_Iju* are assigned fishing. Since in a real-life scenario, the reward for farming is usually expected in a couple of months, this is unlike the other ground extraction activities that have instant rewards. Still, farming produces higher rewards compared to other ground activities, and thus, the following maximum rewards and exit probabilities are assigned to the ground activities.

- **Farming**: the maximum available reward is 20 tokens and the probability of exit is 1/6 (e.g., a face of 6 on the dice).
- **Hunting**: the maximum reward is 10 tokens and the probability of exit is 1/2 (e.g., a face of 4, 5 or 6 on a dice).
- **Fishing**: the maximum reward is 8 tokens and the probability of exit is 2/3. (e.g., a face of 3, 4, 5 or 6 on a dice).

A fixed transaction fee is applied to every activity in the game. This implies while every player remains in the ground extraction phase, an amount of 1 token is extracted from their maximum possible rewards every time they fail to exit the ground activity phase.

Investment Phase Each player has the opportunity to invest in property by purchasing a house or acquiring warehouse storage for free. The first player to arrive on this property can decide to own and collect rent from other players when they arrive at the location. The players can also invest in crypto by staking for PoS, or by getting lucky in winning the airdrop tokens. For the first player to invest in PoS, the chance of winning the selection of block creation is 50%. For instance, with four players in the game, the player who invests in PoS will create the next block if the block creation dice results in 1, 2 or 3 while the other three players evenly share the remaining dice face. Finally, the players can also discover treasures on the island, which speeds up their success towards retirement by moving 22 positions forward.

Disruption Phase The players experience several disruptions that slow their movement or results in the loss of tokens. Any player who owns a property and arrives at a flooded location, the player ceases to collect rent (tokens) for any property they own. Any player who arrives at a fire location loses specified tokens. Injured players move backwards to the specified location. During crypto-associated disruptions such as halving, the player with the PoS tokens, the staking advantage is eliminated and the probability of selection remains the same for every player unless another player arrives at the PoS location.

Retirement Phase Now, the players may invest in their healthcare when they arrive at the health location. If a player has not invested in personal health and arrives at the sickness location, the player moves backwards to the specified location. The first player who reaches the final exit A7 wins the game. If a player exhausts their token balance, the player cannot proceed further in the game and consequently, the position is moved outside the box to restart again. In that case, the next dice roll for the player is an extraction-activity selection.

4 Evaluation of Available Game-Simulation Resources

The blockchain-based survival game described above can easily be simulated and played using online tools such as Hash Calculators[1] and virtual Dice Rollers.[2] The hash calculators may be used for generating hashes for blocks of data generated from player activities. The virtual dice roller generates random numbers represented by the dice faces. The players' positions on the physical paper board are trackable using a distinguishable physical tiny object to indicate their locations. Furthermore, a notepad is needed by each player to record the activities during the game. Still, to design and implement a web platform of the game that potentially integrates actual smart-contract enabled blockchains, further research is needed.

[1] Hash calculators: https://xorbin.com/tools/sha256-hash-calculator

[2] Dice Roller: https://www.calculator.net/dice-roller.html

4.1 Simulation Results

We show in Table 1 the simulation data generated by executing the described game using the online resources presented in the initial part of this section. The simulation involves three players with the following identification *Ama*1, *Bor*2 and *Cud*3. The governance rules for the game simulation are initialized as follows:

- ***Genesis block***: In allocating the ground activity for the genesis block, a player with a face of 1 or 2 is initialized to farming activity. A face of 3 or 4 is initialized to hunting activity. A face of 5 or 6 initializes fishing activity.
- ***Ground-activity exit***: When a player reaches the end of the resource extraction activity, represented by locations *A*2, *D*2 and *G*2 on the board, to exit, the player needs any of the following dice faces. For farming, a face or 1 or 2. For hunting, a face of 1, 2, 3 or 4. For fishing, a face 1, 2, 3, 4 or 5. These represent the difficulty in exiting each activity.
- ***PoS initialization***: For the PoS dice roll, a face of 1 or 2, shows the next block is created by *Ama*1. A face of 3 or 4, shows the next block is created by *Bor*2. A face of 5 or 6 shows the next block is created by *Cud*3.

The first row of Table 1 shows the initialization of the players with null values for player positions and current token values. Therefore, the players' IDs are as follows *ID Ama*1: *P* 0: *T* 0, *ID Bor*2: *P* 0: *T* 0 and *ID Cud*3: *P* 0: *T* 0. The rest of the rows represent dice rolls for player activities and block creations. The even rows (*Roll0, Roll2, Roll4....*) represent player activities showing the player movements on the board. The odd rows (*Roll1, Roll3, Roll3....*) represent the blocks created by the user who wins the PoS dice roll. The second, third and fourth columns represent the players' positions on the board and the *PoS* column shows the player who creates the next block. The *block hash* is generated by combining player positions with the previous block hash, while the *header* column shows the title for each block created. Finally, the *dice* column shows the face of the dice.

Some noteworthy events that occurred during the game simulation are presented as follows. *Roll1* represents the creation of the genesis block. After *Roll4*, the player Ama1 remains in the farming stage, the player Bor2 bought a house property −5T and received the rent of +4T and the player Cud3 rented a property for −4T. In *Roll6*, Cud3 pays rent of −2T. In *Roll8*, Bor2 gains 5T from airdrop Cud3 gets a treasure and move 18 steps ahead. In *Roll12*, Ama1 wins 5T worth of token airdrop. After *Roll20*, Cud1 wins the game by moving up to retirement in position A7, Bor2 moves steps backwards to position H4 after injury and Ama1 came second in position D5.

Table 1 Blockchain-educational survival-game simulation data

Initialize	ID_ Ama1:P_0:T_0	ID_ Bor2:P_0:T_0	ID_ Cud3:P_0:T_0	PoS	Block hash	Header	Dice
Roll0	ID_Ama1:P_ B0:T_20	ID_Bor2:P_ E0:T_14	ID_Cud3:P_ D0:T_14				2,4,2
Roll1	ID_Ama1:P_ B0:T_20	ID_Bor2:P_ E0:T_15	ID_Cud3:P_ D0:T_14	PoS:Bor	1c45a892219b980e07c498461a9d 310c5e01c2bf0690accfb076a6bfee 943344	Block0_23.08.2020_11:39	3
Roll2	ID_Ama1:P_ A1:T_19	ID_Bor2:P_ D1:T_14	ID_Cud3:P_ D1:T_13				4,6,5
Roll3	ID_Ama1:P_ A1:T_19	ID_Bor2:P_ D1:T_14	ID_Cud3:P_ D1:T_14	PoS:Cud	0fb3c3854ebf6f977474098bb8a1c 973c2c26e98037b33aeeab1240493 be93ec	Block1_23.08.2020_11:55	6
Roll4	ID_Ama1:P_ A1:T_18	ID_Bor2:P_ B2:T_13	ID_Cud3:P_ C2:T_10				6,2,3
Roll5	ID_Ama1:P_ A1:T_18	ID_Bor2:P_ B2:T_13	ID_Cud3:P_ C2:T_11	PoS:Cud	eef8f5966e3f51224dde1a705f0c8c 3eb14278a03ff0976238c05cc 7c06b6e94	Block2_23.08.2020_12:12	5
Roll6	ID_Ama1:P_ A2:T_17	ID_Bor2:P_ E2:T_12	ID_Cud3:P_ I2:T_8				1,3,6
Roll7	ID_Ama1:P_ A2:T_17	ID_Bor2:P_ E2:T_13	ID_Cud3:P_ I2:T_8	PoS:Bor	fc886ae6ed218579aad5bb5746924 fd838fbc1cb31cbe8af3ca6240 60192bced	Block3_23.08.2020_12:23	2
Roll8	ID_Ama1:P_ G2:T_16	ID_Bor2:P_ E2:T_17	ID_Cud3:P_ E5:T_7				6,5,5
Roll9	ID_Ama1:P_ G2:T_16	ID_Bor2:P_ E2:T_17	ID_Cud3:P_ E5:T_8	PoS:Cud	728d85014045febb502c7320507a2546 cb41a5fdab4e3e3648d6136c5 fcbbec9	Block4_23.08.2020_12:39	6

(continued)

Table 1 (continued)

Initialize	ID_Ama1:P_0:T_0	ID_Bor2:P_0:T_0	ID_Cud3:P_0:T_0	PoS	Block hash	Header	Dice
Roll10	ID_Ama1:P_F3:T_15	ID_Bor2:P_F4:T_16	ID_Cud3:P_A6:T_7				6,6,5
Roll11	ID_Ama1:P_F3:T_15	ID_Bor2:P_F4:T_17	ID_Cud3:P_A6:T_7	PoS:Bor	a29fd9852fb155d4be9416ca8d7e2adf9d444f57000f46729a6586b5aee1ab15	Block5_23.08.2020_12:52	3
Roll12	ID_Ama1:P_A3:T_19	ID_Bor2:P_G4:T_16	ID_Cud3:P_B6:T_6				5,1,1
Roll13	ID_Ama1:P_A3:T_19	ID_Bor2:P_G4:T_17	ID_Cud3:P_B6:T_6	PoS:Cud	35b37d8550b5d1d0c6c7e9c45966d2c9f68863c51972e9b2843dafcc502943e0	Block6_23.08.2020_13:06	3
Roll14	ID_Ama1:P_D4:T_18	ID_Bor2:P_F5:T_16	ID_Cud3:P_H6:T_5				4,6,6
Roll15	ID_Ama1:P_D4:T_18	ID_Bor2:P_F5:T_16	ID_Cud3:P_H6:T_6	PoS:Cud	f54c5316bb882f07c35477fc6f4900cf4a298d57c8c4bf7ab224f6f7d04f12d6	Block7_23.08.2020_14:01	6
Roll16	ID_Ama1:P_F4:T_17	ID_Bor2:P_E5:T_15	ID_Cud3:P_F7:T_5				2,1,5
Roll17	ID_Ama1:P_F4:T_17	ID_Bor2:P_E5:T_16	ID_Cud3:P_F7:T_5	PoS:Bor	26276f510cf8cb266709c7b15209b05abb1ebdaf6c4df5e4d5809a1a46c8bc4f	Block8_23.08.2020_14:14	4
Roll18	ID_Ama1:P_I5:T_16	ID_Bor2:P_D5:T_15	ID_Cud3:P_B7:T_4				4,1,4
Roll19	ID_Ama1:P_I5:T_17	ID_Bor2:P_D5:T_15	ID_Cud3:P_B7:T_4	PoS:Ama	0d080666538fe74b0eefbac7303653791bf5cd71c61bb2df1280b7f346e18df9	Block9_23.08.2020_14:23	2
Roll20	ID_Ama1:P_D5:T_16	ID_Bor2:P_H4:T_14	ID_Cud3:P_A7:T_3				5,1,2
Roll21	ID_Ama1:P_D5:T_16	ID_Bor2:P_H4:T_15	ID_Cud3:P_A7:T_3	PoS:Bor	3fc85fae2a166e50eddf49662b83bcad2d10d0105a745879bded8946b7251b64	Block10_23.08.2020_14:31	4

4.2 *Evaluation of the Players' Understanding of Blockchain Concepts*

The header column of Table 1 containing the game simulation data shows that the game lasts for ca. 3 h, including a short break time. After the game, the players are individually interviewed to show their understanding of blockchain concepts such as blockchain governance, consensus, block creation, verifiability of records, transparency of records and the immutability of blockchain data. The players show a clear understanding of these concepts by using events within the game to explain them.

In explaining PoS consensus, one of the players mentioned, "*when I won the dice roll for block creation, I did not pay a transaction fee of 1T for the activity*". In this case, the PoS reward is used in offsetting the transaction fee cost for the player. In explaining immutability, one of the players said as follows: "*When we paused the game for a short break, the physical objects for identifying the positions of the players were mistakenly removed. But when we returned, we were able to place each player's positions correctly by the blockchain data in each of our notebooks. Interestingly, we all had the same results*". This statement shows that the players are able to independently verify data recorded that demonstrates the understanding of verifiability and immutability of data recorded on a blockchain. Finally, on the question about the transparency of a blockchain, one of the players responds as follows "*we were able to individually see the game-winner and runner-up on our notepads without depending on anyone in particular*". The statement prooves that blockchain does not require any central authority in decision making since everyone has the same copy of the game data on their notepads. This, therefore, shows the understanding of transparency and the decentralized nature of blockchains.

5 Discussions

To highlight the novelty of the contributions of this study, we discuss the results of this study with similar studies in blockchain game development and education. The study [9] presents a conceptual framework for developing serious games that focuses on blockchain education. Some of the points considered in the framework are target of game, data flow, scenarios, and challenges. These main points that make up the game development framework for blockchain education are well covered in the blockchain survival game developed in this study.

The main targets for this game are higher education students that seek to understand the main concepts of blockchain technology and business organizations that wish to explore blockchain application in their organizational processes. The data flow in the survival game is represented by the players' activities generated during the game. These data flows are recorded in a transparent ledger available to all the players. The scenario of the survival game is represented by a closed system where players generate, exchange assets and record their activities to get through in such a closed environment. The players also face challenges in making specific decisions considering the amount of their token balance and assets they own since their decisions also affect their survival in the game. Also, they face a challenge in correctly calculating the block hash since only correct calculations ensure the verifiability of transactions/activities generated during the game.

The study [16] focuses on identifying the application of blockchain technology in solving educational challenges. The study shows that blockchain can be applied in verifying and validating the authenticity of certificates from higher educations. In the game scenario presented in this paper, we showed how blockchain can be used in verifying the ownership of assets generated in a closed system. Since the players' activities are recorded in a transparent ledger visible to all players, each player can independently verify the activities of other players and the assets they own.

6 Conclusions and Future Work

The focus of this study is to provide gamification concepts that improve the understanding of blockchain technology by simplifying complex blockchain concepts. To achieve this goal, this chapter provides blockchain educational materialby using the gaming concepts outlined in a survival game. The phases of the game cover the activities of the residents of a given island that are physically cut off from civilization. Blockchain, therefore, provides this set of people a system for recording their daily activities and a monetary system for exchanging their assets in a transparent and trustless manner.

This study also shows the availability of online resources such as hash generators, virtual dice rollers and notepads for simulating the proposed educational game. Evaluating the results from the game simulation by interviewing the players shows a clear understanding of complex blockchain concepts such as mining, data verifiability and immutability of records. Besides, the players also show an understanding of concepts such as centralized- versus distributed authority and transparency of records.

One of the main limitations of the study is that simulation carried out in evaluating the developed game is filled with manual processes. Further research is needed to automate these processes by developing a web platform for the game simulation. Furthermore, the evaluation of this game is based on the experiences of only three players. A proper evaluation that is based on multiple numbers of players is needed to capture and analyze empirical evaluation data. In this case, a base is set by evaluating users' understanding of blockchain technology after training with standard educational presentations and compare the results with users trained with the blockchain survival game. This is considered as future research for this study.

References

1. J. Biggs, S.R. Hinish, M.A. Natale, M. Patronick, *Blockchain: Revolutionizing the Global Supply Chain by Building Trust and Transparency* (Rutgers University, Camden, NJ, 2017)
2. C. Udokwu, A. Kormiltsyn, K. Thangalimodzi, A. Norta, The state of the art for blockchain-enabled smart-contract applications in the organization, in *2018 Ivannikov Ispras Open Conference (ISPRAS)*, (IEEE, Piscataway, NJ, 2018), pp. 137–144
3. M.E. Peck, Blockchain world-do you need a blockchain? This chart will tell you if the technology can solve your problem. IEEE Spectr. **54**(10), 38–60 (2017)
4. Z. Chen, Re-examination of a bioinformatics database course: Engaging blockchain technology. Proc. Comput. Sci. **162**, 368–374 (2019)
5. S.R. Kursh, N.A. Gold, Adding fintech and blockchain to your curriculum. Bus. Educ. Innov. J. **8**(2), 6–12 (2016)
6. A. Ravishankar Rao, R. Dave, Developing hands-on laboratory exercises for teaching stem students the internet-of-things, cloud computing and blockchain applications, in *2019 IEEE Integrated STEM Education Conference (ISEC)*, (IEEE, Piscataway, NJ, 2019), pp. 191–198
7. A.C.Y. Hung, Gamification as design thinking. Int. J. Teach. Learn. Higher Educ. **30**(3), 549–559 (2018)
8. S.S. Oyelere, L. Tomczyk, N. Bouali, F.J. Agbo, Blockchain technology and gamification-conditions and opportunities for education, in *Adult Education 2018-Transformation in the Era of Digitization and Artificial Intelligence*, (2019)
9. Y.E. Oktian, I.K. Singgih, F.N. Ferdinand, Serious game for blockchain education purposes (using proof-of-work consensus of bitcoin), in *2019 5th International Conference on New Media Studies (CONMEDIA)*, (IEEE, Piscataway, NJ, 2019), pp. 177–183
10. Idrees, S. M., Nowostawski, M., & Jameel, R. (2021). Blockchain-based digital contact tracing apps for COVID-19 pandemic management: Issues, challenges, solutions, and future directions. JMIR medical informatics, 9(2), e25245.
11. D.L. Kuo, L.G. Chuen, Y. Wang, Cryptocurrency: A new investment opportunity? J. Altern. Invest. **20**(3), 16–40 (2017)
12. Idrees, S. M., Agarwal, P., & Alam, M. A. (Eds.). (2021). Blockchain for Healthcare Systems: Challenges, Privacy, and Securing of Data. CRC Press.
13. D. Mazzei, G. Baldi, G. Fantoni, G. Montelisciani, A. Pitasi, L. Ricci, L. Rizzello, A blockchain tokenizer for industrial IOT trustless applications. Futur. Gener. Comput. Syst. **105**, 432–445 (2020)
14. M. Pilkington, Blockchain technology: Principles and applications, in *Research Handbook on Digital Transformations*, (Edward Elgar, Cheltenham, 2016)
15. H. Natarajan, S. Krause, H. Gradstein, *Distributed Ledger Technology and Block Chain* (World Bank, Washington, DC, 2017)
16. G. Chen et al., Exploring blockchain technology and its potential applications for education. Smart Learn. Environ. **5**(1), 1–10 (2018)

Integrating Blockchain with Education: Proposed Model, Prospects and Challenges

Sarishma Dangi, Amogh Aggarwal, and Pranjal Rastogi

1 Introduction

Blockchain has recently emerged as a shining light leading us all towards a decentralized world. With applications ranging from vehicular ad hoc networks [1], distributed energy via smart grids [2], smart cities [3], healthcare [4], internet of things [5], etc. to the most popular ones, i.e., cryptocurrencies [6], blockchain is paving the path to decentralized governance and control in different facets of working of our modern world. Education serves as the most important asset of the twenty-first century which is leading the deep dive into technology and its applications. The field of education needs to evolve constantly in order to aspire for a better future for everyone involved in it. However, it often gets stuck with old technology and a slower shifting as well as adoption of new technologies since the sheer number of people as well as cost involved is far too high for everyone to upgrade often on time.

Blockchain is based on decentralization of trust via a distributed ledger. It offers both permission-less and permission-based application for different use cases. In this work, we would be working towards integration of blockchain technology with the education sector. We would be focusing on India and how blockchain could revolutionize the Indian educational landscape but we believe that most of this information would be valid for almost all countries and institutions present globally. With more than 60% of India's population belonging to rural areas, quality education has always been a challenge [7].

Our aim for this work is to show how blockchain can be used to reduce the irritants in delivery as well as validity of quality education. The idea is to give trust to various kind of new courses, degrees and small courses, which have recently sprung up worldwide so that no student has to ever again try and prove that she/he actually studied and knows the courses in her/his resume.

S. Dangi (✉) · A. Aggarwal · P. Rastogi
Graphic Era Deemed to be University, Dehradun, Uttarakhand, India

S. M. Idrees, M. Nowostawski (eds.), *Transformations Through Blockchain Technology*, https://doi.org/10.1007/978-3-030-93344-9_4

MIT Media Lab together with Learning Machine is an open source ecosystem already implementing and working to certify, share, and verify educational certificates [8]. We believe that this gives a slight assurance that the same could be implemented on a country-wide scale where anyone could easily verify the originality of the certificates given by students and employees and also see exactly how valuable those certificates are.

Another important implementation of this technology could be to bring smaller institutes and their students closer to international institutes and make it easy for them to apply at small companies, startups which might not have the resources to normally check the caliber of every applicant, and analyze how valuable and reputable their degrees actually are. This method would also be extremely useful to debunk the aura surrounding online educational courses and would serve to bring to light the actual effectiveness of players like Byju's, Vedantu, and White Hat Junior.

Currently, with heavy marketing campaigns and testimony videos of students from all age groups, these institutes have created a reputation of being a saving grace for all students with all kinds of calibers but this is not backed by any institution or government. It works more like an autonomous offering of education which is surviving in the economy due to its very need. However, no standard testing, validation or performance comparison is ever done on the quality of education imparted by these institutions. Blockchain could be an answer to this and show everyone exactly how much each course actually contributes to a person's portfolio. We offer a solution to move towards this direction.

It is the belief of the writers of this chapter that since India has one of the highest number of graduates joining the workforce each year and because all the major colleges and universities are already linked to the government through AICTE; India would be the perfect landscape to start the large-scale revolution of pushing the education system to be decentralized and open to all. The key contribution of our work can be summarized using six main goals as outlined below:

1. To propose a blockchain-based certificate management and validation system with the underpinnings of introducing a knowledge currency.
2. To change the dependence of employers on paper certificates and degrees by offering a universal means to validate and test the authenticity of a user's documents. To decentralize the process of obtaining education and make it possible for students to focus on the field they love and help build a society with more people who are masters of their domain.
3. Hold institutes liable for the graduates they pump out every year if they underperform in the future and reward them with better rankings when their graduates end up helping the society. Blockchain can offer accountability for this for every entity involved including students, parents, institutes, and recruiters.
4. Create a bullet-proof certification system which cannot be faked or tampered with.
5. Turn the process of making a curriculum vitae or resume "digital" which would help everyone to maintain a similar pattern making it easy for interviewers and

institutes to read, understand, and update anyone's resume instead of making it a burden on the student.

6. Make quality education accessible to everyone by recognizing the completion of all kinds of courses; be it an MOOC or an instructor-led seminar held in a huge auditorium with hundreds in attendance.

The rest of the chapter is organized as follows: Section 2 reviews the existing work done in the integrated area of blockchain and education. Section 3 presents a discussion on the integration of education with blockchain. Section 4 presents the necessary technical background for understanding the rest of the work. If you are familiar with blockchain basics, you can safely skip this section. Section 5 presents the proposed layered blockchain-based model along with a working model, flow chart and smart contracts. Section 6 points out the future research directions followed by the conclusion of the work.

2 Related Work

Blockchain finds application in numerous areas where education has more or less received lesser attention as compared to the development in areas such as supply chain management and cryptocurrencies [9]. In terms of education, students worldwide face a plethora of challenges while managing their certificates and proving their credibility as well as authenticity to the recruiting agencies. Several institutes have already taken initiative for developing a framework for a blockchain-based verification of academic achievements or to issue diploma certificates like the National University of La Plata, Argentinian College CESYT, Parisian Leonardo da

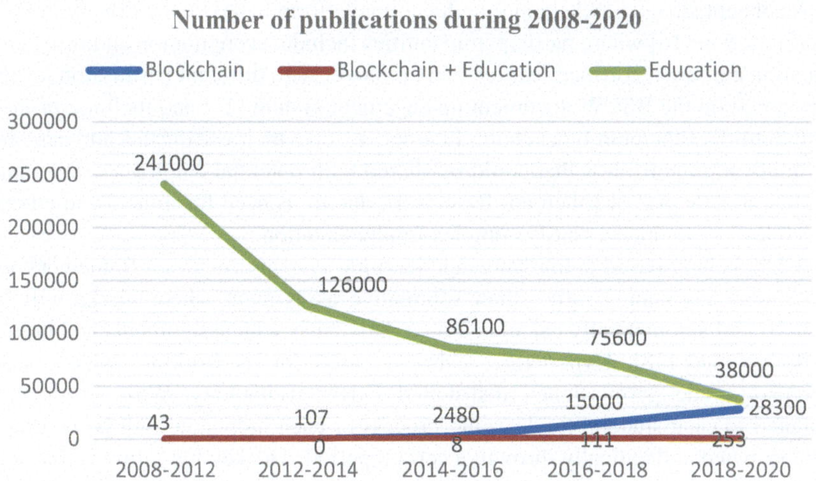

Fig. 1 Yearly distribution of number of publications for the presence of keywords in the title

Vinci Engineering School (ESILV), and the Holberton School (San francisco). In this section, we review the existing work in the area of blockchain and education and review it critically. The distribution of number of publications over the last decade is presented in Fig. 1. It is evident that more contribution is needed in the area of decentralized education paradigms in future.

Authors in [1, 10, 11] provided a basic technical background for using blockchain. However, the previous work does not provide any implementation and working details. Instead, the focus is on conceptual underpinnings of the work in a primal manner. Three partial implementation approaches are presented in the literature for integrating blockchain with education, namely Blockcert [8] by MIT and Oxford, Edublock [12] for IFTF, and the Edgecoin project [13]. Blockcert is an open standard proposed by MIT Media Labs for issuing, verifying, and sharing of digital certificates using the blockchain [8]. Blockcert is also integrated with the certification management system currently implemented at the University of Fernando Pessao and University of Birmingham [14].

EduCTX is a blockchain-based higher education credit platform [15]. It proposes a global higher education credit platform, named as EduCTX which is based upon the European Credit Transfer System (ECTS). ECTS assesses and compares students' academic performance and achievements and enables them to transfer credits effortlessly from one higher education institution to another. ECTS credits reflect learning that is based on pre-defined learning outcomes and their associated workload. In this system, generally each ECTS token represents 25–30 work/study hours of workload directly or an estimated number based upon the exam or the certification at hand. The transactions of these tokens are executed using blockchain. DLT used for the implementation is ARK technology with consensus protocol as Delegated Proof of Stake. Since the work proposes to store only ECTS tokens using an API for the employers to check the number of tokens, the employer would not be able to verify the authenticity of the certificates or the courses that the student claims.

A conceptualized blockchain for education platform based on the Ethereum DLT is proposed in [16] where participating entities include: accreditation authority, certification authority, certifier, learner, and employer with different permission levels. It is stored on the BSCW document management system [17] and its fingerprint is written to the Ethereum blockchain. Learners receive their certificates and can create application portfolios that could be shared with potential employers. The main limitation is the use of Ethereum framework due to its need for miners and ethers, which lead to an unnecessarily complex implementation.

A blockchain-based online language learning system provides a framework for continuous assessment rather than summative assessment where marks will be updated and automatically calculated on a daily basis with assignments and homework included [18]. Hyperledger Fabric is used for its working, which is an enterprise-grade permissioned distributed ledger technology. With an identity assigned to each student, the learning process of each individual will be recorded and evaluated authentically through a certain period. The key limitation is that it is narrowed to only one organization or institute instead of scaling it up to a blockchain that can expand across different institutes in order to store all the academic credentials and transcripts that a student earned throughout their life.

A novel blockchain-based education records verification solution is proposed to store all the education certificates, projects, and experience-based learning records onto the blockchain and to allow students to keep as well as maintain their learning history across different institutions and organizations [19]. Only certified institutions are allowed to access and update the data in the system under the restricted rules to maintain the authenticity of the blockchain. Using smart contracts employers could directly receive the latest update if there is an employee belonging to a particular educational program. Since the framework that they are using needs miners, they need bounty for those miners. This will lead to a need for huge financial resources for mining rewards and computation power when this blockchain scales up.

Application of blockchain technology in online education provides the concept of storing all the educational certificates onto a blockchain [20]. The student's educational data including learning time, course files, and test results can be recorded on the blockchain in chronological order with each data record marked with a T. The chapter does not provide the framework and working implementation of the concepts. Also, one problem with MOOCs is that many of those certificates can be claimed without working for them and without passing any qualifying exams for the same. This will impact the authenticity and validity of the online certification courses in general.

A perspective of using blockchain for higher education institutes is provided by [21] in the light of three key challenges, i.e., automation, democratization, and the lag of technological adoption. Another work proposes a blockchain-based approach to create a model of trust in open and ubiquitous higher education by implementing their concept to expand on MIT's Blockcerts by deploying Ethereum [12]. They aim to make it easier for students to have a degree or course that can be verified by any employer and also make it easy for the employer to validate the competency of the interviewee. However, the use of Ethereum means the requirement for miners which would make this system expensive in terms of resource usage and the security vulnerabilities.

In this work, we aim to propose a model that overcomes the limitations existing in the above-mentioned works. Use of open-source technologies and relying on a balance of decentralization and centralization, we provide a model which takes into account the Indian administrative structure which gives power to the majority of institutions across the country. We also aim to bring up the informal learning solutions which have recently sprung up. The coming decade will witness a migration of learning towards these new solutions. Using a blockchain-based decentralized alternative for maintaining certificates as well as learning credit will serve as a step into the future along with the handling of key irritants existing in education today. In the next section, we provide a discussion on some key parameters with regard to this futuristic integration.

3 Integrating Blockchain and Education

While working around this area, we found a few points which need immediate attention. These changes could greatly help in charting the future course of the educational scenario worldwide. The problems are deep rooted and they are born as repercussions of long-standing practices. These are further elaborated under the following subheadings.

3.1 The Current State of Education, Certification, and Hiring

To get a respectable job and have a career with a reputable future, one needs to have a strong resume, which includes proof of the claimed certificates and degrees. This is generally a mix of small courses and degrees along with some certificates of team activities and topic-specific MOOCs. Currently, most corporate jobs have a clear hierarchy system where most of the new recruits start with grunt work and move on towards management as they gain more experience. With the introduction of blockchain and having clear reports on every student and employee, it is possible to turn the current situation around. With clear certification, people can be allotted the work they are best at and trained in. The following model proposes the same.

3.2 Changing Dependence

Companies are currently dependent on paper certificates which leads to the submission of hundreds of resumes of potential interviewees even for a couple of positions that open up. The introduction of blockchain could lead to the companies calling for a specific type of person to interview for the job as all the certificates and caliber of the interviewee would be clearly visible to the company as soon as they apply. This could lead to saving of resources and time for both the parties involved.

3.3 Decentralization of Education

To get an education with value in the corporate world, one needs to go through a long and gruesome procedure of getting a degree and studying a range of subjects. This could be remodeled using blockchain and its integration with education as people could use the internet to take a quality course that is a perfect match for what they wish to learn and then use it to get a valued certificate that can be easily stored and verified when needed by any employer or institute. MIT and Oxford already have a form of this implemented in some of their courses where they are awarding digital certificates to their students for some of the webinars and degrees. These are stored on the blockchain and can be easily verified by anyone.

3.4 Validation of Academic Credentials and Transcripts

Academic Scores serve the all-important role of telling any employer or institution of higher education that the student has studied and is proficient in the basics of the field they are applying to advance in. However, verification of the originality of these scores and certificates has led to the creation of a whole new market with companies helping to verify and certify the authenticity of the student's certificates by jumping through the required hoops for the student or employer for a fee. Adding these certificates to a blockchain could bring the verification to the hands of everyone as the userID would be enough for a company to verify everything they need and lead to faster turnaround times, benefiting everyone involved.

3.5 Accountability of Parent Institutions

Currently, most colleges are not accountable for their students after they graduate. Even if the student is not able to get a stable job or a job that does not pay them handsomely for the amount and type of work they do, the student is solely blamed for not doing well. But if a student is placed well, the institutes boast with banners and full-page ads in the newspaper in a manner which says that the college was solely responsible for the student's success. Introduction of blockchain with the student's full information can easily tell everyone if the student bagged a nice job solely on the college degree or needed to take courses from elsewhere and had to study on their own to get their dream job. This could easily lift the veil that colleges hide behind and charge exorbitant fees from students who see the banners and parents who believe that even their son or daughter could bag those jobs and change the living conditions of the family.

3.6 Vision of Education as a Currency

In today's fast-moving world, education and knowledge are powerful tools for anyone to stay ahead and face the world head-on. This chapter looks at how, with the use of blockchain, education could be monitored and monetized helping people get the maximum value out of their education and skills. It is the goal of the writers that the tokens proposed in this chapter could help anyone prove and show their exact level of proficiency in a field and get jobs best suited for them and get paid accordingly. Decentralization of the field of education would put power in the hands of the people even in the corporate world.

4 Technical Background

In this section, we present the necessary technical background relating to blockchain and its working. If you are familiar with these concepts, we advise to move on to the next section. We introduce blockchain, consensus mechanism, role of timestamps, and hashing verification in this section.

4.1 Blockchain

To understand the following paper, it is important for the reader to first understand the concept of blockchain. Although the early idea for blockchain came about in 1991, it did not start gaining traction until 2008 [22] when an individual or a group of people under the pseudonym Satoshi Nakamoto presented a white paper on Bitcoin; a peer-to-peer electronic cash system which developed on the work done by Stuart Haber and W. Scott Stornetta on a cryptographically secured chain of blocks [23]. Haber's work was limited to creation of tamper-proof documents by linking them to each other in a chain, whereby changing one document would mean a change to the whole chain. Satoshi Nakamoto evolved the system to record transaction histories and make the whole system fully decentralized by the introduction of independent miners and anonymous users. Since the whole system is based on hash encoding with strong computational problems, the highest level of security is guaranteed.

As the name suggests, blockchain comprises of a chain of blocks where the first block is called as *genesis block*. These blocks are connected using hashes of previous block which all work synchronously with the concept of *Merkle tree root*. The underlying encryption technology ensures that even a small change to any one of the blocks would render the whole chain useless and thus it is almost impossible to manipulate or falsify data in a blockchain. This is the reason why blockchain could be the future of information storage and lead the change from centralized information storage to decentralized mode of storage [24, 25].

A blockchain is a ledger of information which works like any other ledger but every person involved in the chain has information about every entry added to the ledger. The most widely used application of this technology is *Bitcoin* where a ledger of who owns whom, who owns what, and every transaction is noted in the blockchain using the transactions which ultimately form a block. The same system can also be used to store other data as well.

4.2 Smart Contract

The term *smart contract* was first proposed in the year 1994 by Nick Szabo, an American computer scientist who referred to it as "*a set of promises, specified in digital form, including protocols within which the parties perform on these promises*". Szabo defined smart contracts as computerized transaction protocols that execute terms of a contract. Smart contracts are self-executing contracts with the terms of the agreement among all the parties involved being directly written into lines of code. This code executes the actions when the agreed upon conditions have been met and verified. The blockchain is then updated when the transaction is completed. Since they are automated and the conditions are predetermined, they provide a more trustful and secure way of executing actions on an "if/else" basis.

The use of smart contracts scales down the need of trusted intermediators, arbitrations, enforcement costs, fraud losses, etc. Generally, they cannot be revoked if after the agreement one party tries to back out; this makes them more reliable.

4.3 Consensus Mechanism

Proof of Work is the requirement of most cryptocurrency chains today, wherein any miner in order to verify the block that they are adding has to solve mathematical problems to get the right hash function required by the chain to prove that all the transactions in the block are legitimate [22]. This requires immense amount of computational energy and thus leads to the use of vast amounts of electricity, making it unsustainable in long run. Thus, the model we are proposing is a hybrid model where the initial addition to the chain can only be done by verified institutes and the authenticity of the addition would be guaranteed by the government and the people maintaining the chain.

Afterwards, the added element can be inspected by anyone with the specific users' public key. This will prevent the increasing harm done by miners in today's world and also utilize the power of blockchain in the field of awarding and authenticating degrees as well as certificates for users in a decentralized manner. The reader also needs to understand that any block in a chain can be designed to hold almost any kind of data and be linked to other blocks not only in a chain but also have branches and be cross-linked to other blocks. This is important to link the blocks of institutes and employers to every end user, which will prove useful further in the work. Proof of Work, Proof of Stake, Delegated Proof of Stake, Proof of Elapsed Time, Proof of Authority, Proof of Space, etc. are some of the prominent consensus algorithms used by blockchain use cases.

For our work, we are not using consensus protocols because each of them has their own drawbacks like usage of too much electricity for mining, making network prone to 51% attacks especially in the initial phases, and requirement of large amount of computing resources. Also, we do not want to keep our blockchain completely transparent to public, as this can lead to privacy issues since data of all the students is stored on those blocks. Furthermore, we want to assign administrating and governing responsibilities to the central authority/government. We propose a distribution of power and trust where some permissions are present only with trust authorities and not every participating entity has the trusting credibility. For instance, the institutions who will be providing the documents are considered as trusting authorities whereas students/learners are considered as end users with limited permissions.

4.4 Using Blockchain as a Solution in Education

Instead of keeping a record of transactions, the blocks in a chain could also be used to store information about other things like the degrees of students or certificates that could be easily verified just like the transactions are verified in Bitcoin. This work aims to provide a model for the same. This could, in theory, reduce the strain on students to keep an up-to-date resume, as a simple pull from the blockchain could act as a record of all the courses and classes they have completed along with their marks.

This would also put pressure on the institutions to produce smart and field ready graduates as the performance of the institutes could be easily measured by how many of the graduates from one college are performing well in the field. Decentralized and available-to-all nature of the blockchain would mean that anyone could check the performance of an institute, any specific student, or any company based on the data and would not have to rely on the information provided by the institute as everyone cherry picks the best results to show their institute or themselves in the best light. In this chapter, the aim is to expand on the Blockcerts of the MIT and Oxford, where the degree and certificates of some courses are not paper based but digital, which makes them more secure as they are tamper-proof and easily shareable.

4.5 Role of Timestamps and Hashing Verification

Timestamp is an important part of a block as it acknowledges the creation time of transactions in a block and their sequence in a blockchain. Furthermore, in our work we need timestamps for the process of result document generation for timestamping the achievements and academic credentials of a student. While generating results, if there are multiple events of same levels (tokens), then the priority for the sorting algorithm will be calculated as per chronological ordering of timestamps for those events. We will be using hashing techniques like SHA-512 for verification of the documents. When a student generates their results, it gets generated in a form of document and also a hash of that document is generated. So, if an employer wants to verify the document provided by the student, they can match the hash of the document that the student provided with the hash given by the blockchain for that student's record.

5 Proposed Model

In this section, we propose our layered model which is based on a careful distribution of trust and power between participating entities. The assumptions are provided first followed by the concept of tokens, proposed model, workflow of model, token generation process, and the role of smart contracts. The notations used in this work are presented in Table 1.

We aim to expand on the idea of tokens which were introduced by [15] and turn them into a currency which helps to immediately evaluate the caliber of the candidate and inform the interviewer if they are the right fit for the job. This will also help find the diamonds in the rough who were earlier rejected because they were not able to create a resume which reflected their skills well enough.

The idea for using a token system in a decentralized manner was inspired by Bitcoin which started the revolution of using decentralized solutions. Bitcoin, at its core was aimed at replacing physical currency by creating a ledger that records all transactions and equates them to each other using bitcoins. A similar idea has been used here where we have tried to measure the intelligence and education level of

Table 1 Notations used in the proposed work

TA	Trust authority
insID	Institute's Unique Identification Number
userID	User's Unique Identification Number
prevHASH	Previous Block's Hash
T	Timestamp
DS_{inst}	Institution's Digital Signature
pvtK	Private Key
DT	Diamond Token
GT	Gold Token
ST	Silver Token
BT	Bronze Token
BlockID	Block Identification Number
userN	User Name
insN	Institution Name
K	Constant based on level of Difficulty of Exam
TR	Time Required to complete course
UI	User Interface
subDoc	Result Document submitted by user

people and add them to the ledger by equating their knowledge to a certain number of tokens and then providing this information to employers. While working on this model, some assumptions were made which are pointed out in the trailing subsection.

5.1 Assumptions

In the section that follows, the writers have made a few assumptions, and this section is dedicated to the naming of and explanation about the same.

- The first assumption is that each certificate given out by an institute has information pertaining to what the course involved, what it qualifies the student for in the future and details about the student's marks including other prominent and noteworthy achievements during that course.
- Secondly, it is assumed that every authorized institute is permitted to add to the chain and their access to the ledger can be trusted for the foreseeable future. We assume that this list of institutes is provided by the Ministry of Education, Government of India.
- Thirdly, it is assumed that the education system of any place that aims to implement the chain has a proper division between different educational levels which are reputed and accepted across a wide group of people. For this chapter, the levels of education are categorized as: Merit (tenth Grade), High school (12th Grade), Undergraduate, Postgraduate, Doctorate, and Post-Doctorate.
- It is also assumed that online courses will be assessed and given a related value with the government to set up the defining rules and regulations for the same.
- Cooperation from all the institutes involved to give token value to each of their courses is assumed, from colleges and schools to small courses and online webinars. TA will verify every entity when incorporating each institute or course.
- Each exam which will grant the student golden tokens will be allotted the token values similar to the institutes by the TA when adding them to the chain.
- The writers have used the words certificate and tokens interchangeably and it is assumed that the reader will relate to both as per the context after reading the section on tokens.

5.2 Tokens

Tokens can be considered as a form of credit which is stored in the user's account. We propose the use of tokens as a *wisdom currency* which a student earns throughout his or her lifetime. The tokens are categorized as: diamond, gold, silver, and bronze which will each be given when a student completes an educational task equivalent to that token. The associated value for tokens is further segregated and quantified in the subsequent sections to accommodate more clarity and integrity into the model. This will allow the model to segregate top 100 institutes or a student's rank or admission to an accomplished institute, so on, so forth. The education milestones that the tokens mark are elaborated as follows:

1. *Diamond Tokens*: Proves that the student has completed a degree from an institute belonging to the system. The degree provided by central, state, and private institutions affiliated by the Government of India are recognized using these tokens. The number of tokens a student received would depend on the number of years of education they take and also on the ranking of the college or school according to the government ranking. Refer to the Table 2 and Fig. 2 for calculating the number of tokens that will be given. Institute ranking in the chart is according to the Trust Authority, i.e., TA.
2. *Gold Tokens*: Proves that the student has qualified an entrance exam which is being held by certified authorities as part of the system. Some of the prestigious and commonly accepted reliable examinations are CAT, UPSC, IELTS, GRE, GMAT, NEET, and IIT-JEE. Here we will have a constant value K (from 1 to 5) for each exam which is according to the level of relative difficulty or required efforts among all those exams. Since the marking scale is very different for each exam, we can take percentiles into account. The number of tokens awarded to a student will be equivalent to $(K*1000)/(100\text{-percentile})^2$.
3. *Silver Tokens*: Proves that the student has completed an online course or a related apprenticeship by qualifying the associated tests. A user will earn silver token for online courses and skills that they learned by passing relevant examinations for proving the authenticity of the work and efforts being put into them. Some of the leading institutions providing such courses and certifications include AWS, Oracle, Cisco, Google, Microsoft Azure, and Hadoop. We can take into account the official number of hours suggested by them for the exam preparation or the average time. Every 2 h add on to provide one silver token to the user. The values are adjusted as per the rubrics assigned for the evaluation of the course.
4. *Bronze Token*: Proves that the student has completed a webinar or short-term course from a credible and registered medium of informal learning. It is useful to account for online courses and webinars which do not have any reliable validating examination to showcase the skills acquired through them. Every 1 h add on to provide one bronze token to the user.

Table 2 Distribution of diamond tokens for institutes

Institute ranking	Diamond tokens per academic year
1–25	140
26–50	135
51–75	130
76–100	125
101–125	120
126–150	115

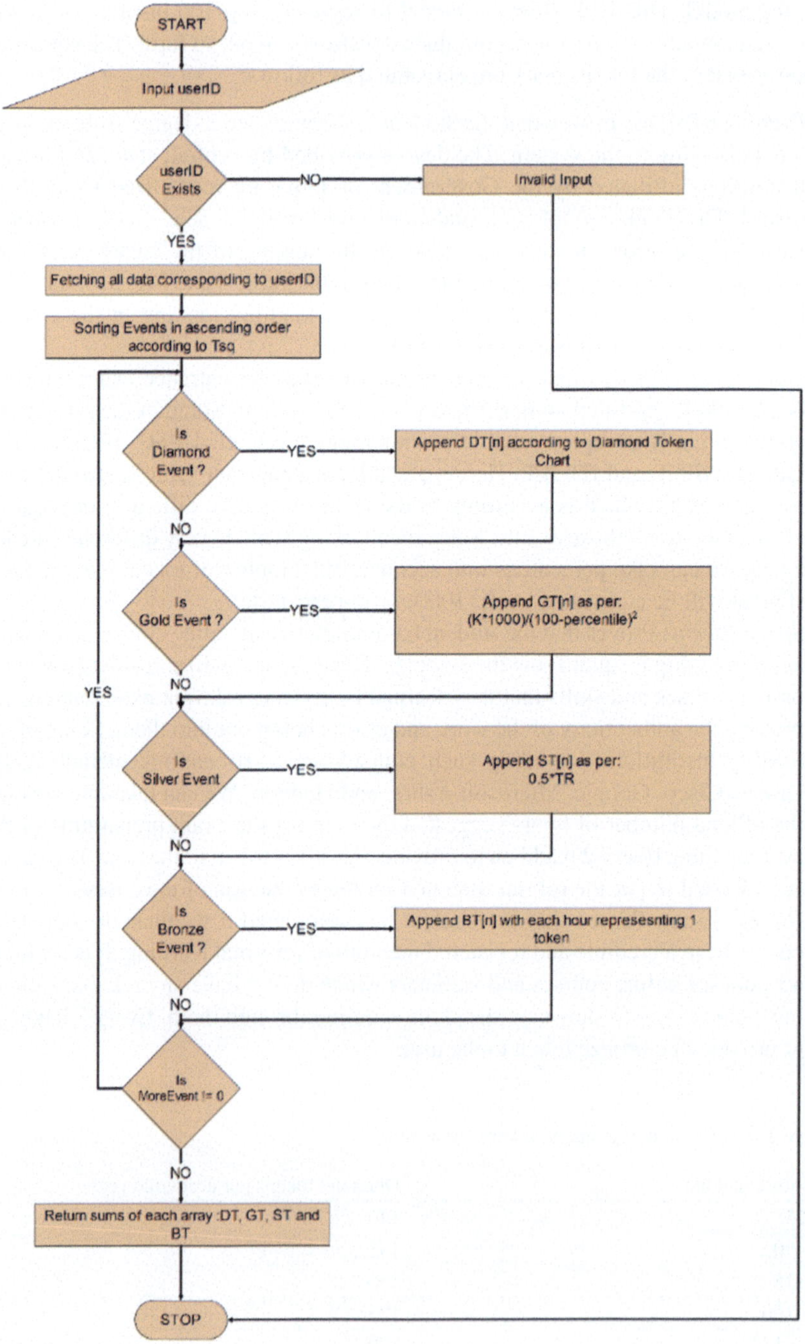

Fig. 2 Token calculation flowchart

For example, if a student were to complete an online course followed by a test which they pass, by a reputable institute part of the system, they would be given a silver token by the parent institute. What would make this system stand out from the old system of certificates is the allotment of different number of tokens from the related level depending on the performance of the student and reputability of the institute. This idea has been properly explained and expanded on further in the upcoming sections.

Employers could easily see the chain, know which candidate has how many tokens, and get an abstract overview of the tokens. This way, the employer can easily sort from the applications to find specific students who fit their requirement and only interview a few people who would be best suited for the position. The token system could be expanded to include all universities, educational institutes, schools and online courses, so that any educational endeavor that the student sets out to can be easily verified and logged.

5.3 Actors

In this section, we briefly overview the participating entities or the actors involved in this model. The actors are further subdivided on the basis of roles. First, we discuss the educational actors involved in this model.

1. *Trust authority*: Trust Authority or TA is the most powerful actor in the chain with the deciding power to add or remove any other actor at any time from the chain. It will create and maintain the chain and decide the formal rules and suggestions for the other actors to follow. TA will be governed by Ministry of Human Resource Development, Government of India, which is responsible for granting power and accreditation to other institutes all over the country.
2. *Centre- and State-level university*: This comprises of all the central- and state-level institutes of the country that are granted accreditation by the trust authority. Since these are already supervised by the TA, they will be easily added to the chain and may even be given the power to add other private universities to the chain which they deem worthy to hand out official degrees.
3. *Private and other institutions*: Private colleges can hand out degrees to the students which will be recognized by the TA, and their associated credibility can be cross checked with the TA.
4. *Schools*: Once a school is recognized by the state or central government, they can add to the chain. Using schools as an entity will help in streamlining the qualification process for teenagers by providing a trustworthy, hassle-free means to view the outcomes of their work. Schools can store other information about their students and register new students to the chain.
5. *MOOCs and Online learning platforms*: Recognized MOOCs can add courses and verify students for the same. The courses will be recognized by the chain and credit tokens to the end user's account. Online platforms can hand out bronze or

silver tokens based on the recognition they get from the TA after proper testing and validation.

6. *Employing authority*: They will have the permission to view and verify the degrees and courses of any interviewee who gives them *userID*. This will be used as an interface to allow validation of claimed documents. It will help in eliminating the hassle of carrying around certificates and proving their validity in person.
7. *Examinations*: Respected Exams such as the UPSC, IIT-JEE, NEET, AIEEE, AFSB, CAT, GMAT, and GRE will be allotted golden tokens as seemed appropriate by the TA.

Apart from the above-mentioned educational participants, we also classify actors on the basis of their role.

1. *Accreditation Authority*: The central- and state-level institutes that have total control over the granting of permissions for the blockchain. In case of error or any complaint, they have the final say.
2. *Certification Authority*: The schools, colleges, MOOCs, and online learning portals giving out tokens and certificates based on the students' performance. This entity is involved in handling of tokens, keys and hashes for making the model work.
3. *Learner*: The students applying for courses and degrees who will get the certificates and will be added to the chain. This is the end user for the model and everyone is allowed to claim this role in order to enhance their wisdom currency.
4. *Employer*: People looking to hire new employees will inspect the blockchain for any interviewees with respect to the keywords which most closely identify with the job at hand.

5.4 Smart Contract

Smart contract as explained earlier can help increase the trust in any system as its actions are predefined and cannot be tampered with by anyone involved in the chain. Thus, smart contracts can be added to this model to increase the trust among the thousands of actors involved. TA can design these smart contracts at the beginning when implementing the basic model and then expand to involve more contracts as their use cases increase. Some of the contracts that would be required in the beginning would be:

- *Addition of new actors*: When a new institute applies to become a part of the model and give out tokens, TA will have a list of requirements for each type of institute. For example, if any institute is giving short courses online or offline and can prove that they take the required exams afterwards and that the students that graduate are proficient in whatever course they have taken, the smart contract can automatically add the institute to the model to hand out silver tokens. Similarly, other institutes and courses can have themselves added faster and TA can make the whole process smoother using legally binding smart contracts.

- *Approval of Tokens*: Each institute can hand out a certain number of tokens to a student based on the length of the course they are taking, ranking of the institute or level of the course itself. This can be implemented using a smart contract, which will transfer the tokens to the student as soon as they complete the course and calculates the number of tokens to be given, making things easier for the institute.
- *Verifying transactions*: This smart contract is responsible for governing the verification of transactions by validating and authenticating the parties involved and then pushing it onto blockchain.
- *Addition and modification of transactions*: Apart from verifying, there are strict smart contracts for governing the addition as well as updation or modification of data. Addition-based smart contract will be frequently used, whereas for modification, only rare cases will use this in case of addressing grievances for end users.

Based on the respective TA's requirements, the model can be tinkered with to make it fit best for them and new smart contracts could be easily designed and added to the model even after its implementation.

5.5 Proposed Layered Model

Any blockchain that is being set up requires a clear definition of the actors involved in the chain with the power and roles of everyone involved clearly laid out. In the proposed model, there will be a layered priority system where any entity on a higher layer will have more power and the power to override the instructions of anyone below them at any point of time. For the implementation of the same, the entities will be divided into layers as follows:

1. Layer 1: The government body which handles education and human resources in any country, such as the Ministry of HRD with the Department of Education in India would be on the top layer and maintain everything in the chain with the power to access and change in blockchain. AICTE, CBSE and ICSE would all be part of the top layer and would thus be able to clearly see what any college or school in the chain was doing. This layer is centralizing the trust towards these parties in the model.
2. Layer 2: This would be the layer having all the schools and colleges along with the websites giving verified courses with tangible value in the real world. To be a part of this layer, any institute would need to be verified from an actor of the top layer and could be subject to checks and regular assessment by the top layer institutes. These institutes would only have the permission to add to the layer for a certain time out of the year when the exam results are declared and this would be monitored by the top layer to prevent the addition of fraudulent certificates by any institute.

3. Layer 3: Independent webinars and courses which have not been verified by the implementers of the chain would be part of this layer and would have the authority to give out a certain number of certificates a year to prevent excess distribution. These would have the ability to move to the second layer if their courses are deemed valuable and vouched for by any actor in the second layer to be considered by a top layer authority.
4. Layer 4: This layer is comprised of the students who would be receiving the token and certificates from the institutes with information about each student in their own block, effectively creating a resume for each student involved from the first certificate they get which is updated with every course they take, every degree they get, and every exam they crack, making it easy to make an industry standard resume system.
5. Layer 5: This is the *preview-only* layer that everyone else is a part of. Anyone with the userID can see a list of the students' tokens and certificates with records of how they got them. Interviewers and institutes would be the main users of this layer admitting new students and employees and wishing to verify their certificates.

Migration in between layers allows this model to flexibly integrate top performers and give them appropriate rights and say. This will help in shifting the institutes up or down the hierarchy to maintain a standard level of teaching (Fig. 3).

The given model would ensure maximum security and ease of use. Firstly, the TA will have absolute control over the system and only it will add new institutes to

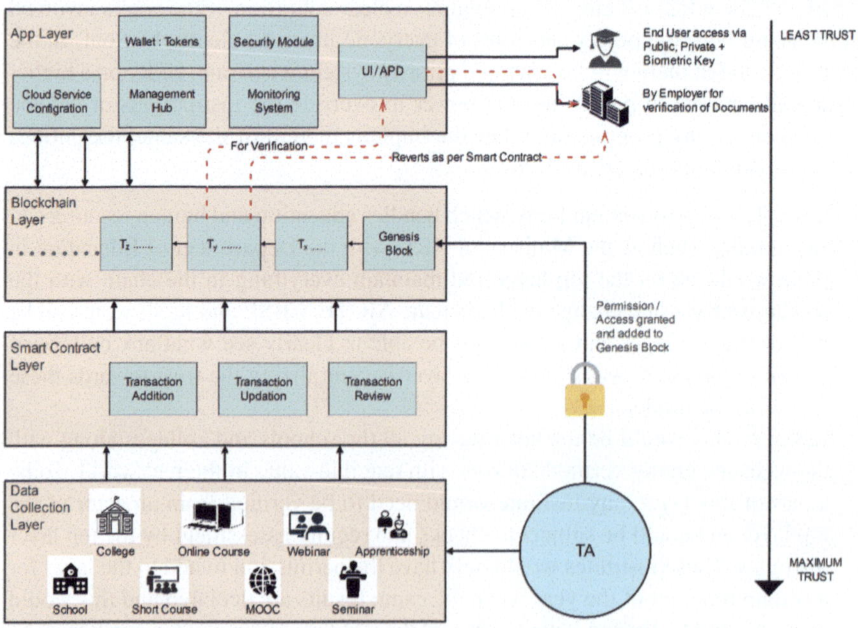

Fig. 3 Layers of the proposed model

the data collection layer who will in turn will have the authority to add new certificates and tokens to the chain using smart contracts. After the smart contracts will approve the addition of any new token or certificate, the userID will be prompted for and the new data will be added to the chain. The chain will also be under the control of the TA who will generate the genesis block and set the precedence for how the chain will be used. The app layer will be the one visible to the user and employers who will use the userID to access the certificates of any user and will ensure a level of security and act as the buffer presenting a user-friendly interface. The result generation when an employer inputs a userID will also be managed by this layer ensuring a smooth experience for all.

5.6 *Working of the Proposed Model*

TA will be on the highest permission level with the ability to add or remove actors at any time. When an actor, approved by TA, wishes to add a new transaction (Certificate or Tokens), the smart contract will help with the same and add it to a block and provide the hash for the same to the user. The token calculation algorithm will help the smart contract to decide how many tokens are to be added for that specific transaction. After their addition, the tokens and certificates can be seen by anyone with the userID as they can request for a Result Generation which will show the whole certificate if the pvtK is provided too but only the hash of the certificate if only the userID is given (Fig. 4). The sequence diagram explaining the working of the proposed model is presented in Fig. 5 followed by the pseudo code for smart contracts.

Algorithm 1: Adding Transactions to a Specific Block
Input: *BlockID, InstituteDigitalSign, userID, document, T, hash*
Call: Via add block by an institute

1. if (BlockID exists):
2. if (BlockID != full):
3. **newTransaction***(DSinst, T, userID, document, hash)*
4. return *transactionID.BlockID*
5. else
6. return "Block overflow"

Algorithm 2: Modifying Transactions in Blockchain
Input: *BlockID, DSinst, userID, document, T, hash*

1. if (BlockID exists):
2. if (BlockID.userID.document != zero):
3. **replace document via Algorithm 1**
4. return $transactionID.BlockID.hash_{modified}$
5. else
6. return "Document not found"

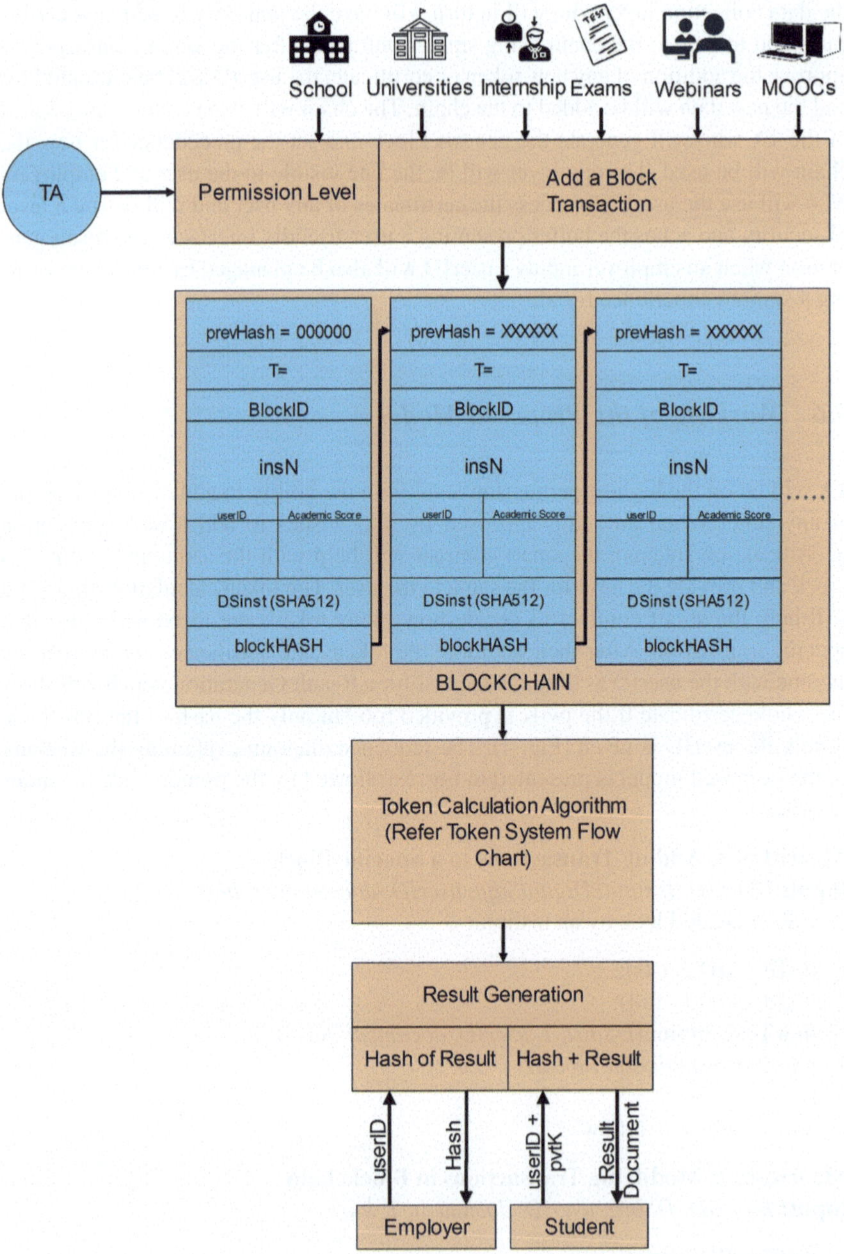

Fig. 4 Working of the proposed model

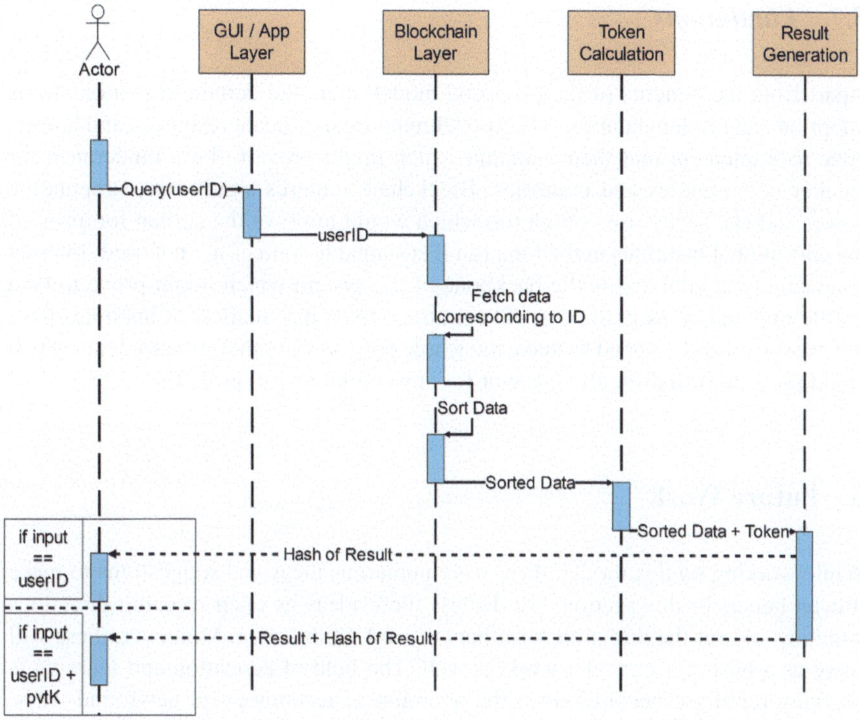

Fig. 5 Sequence Diagram of the proposed model

Algorithm 3: Verifying Transactions
Input: *userID, subDoc*
Call: Via Employer API for verification of document

1. if (userID exists):
2. **return hash of the document;**
3. else
4. return *false*

Use of smart contracts in the working of this model makes the workflow transparent with respect to the institutions and end users. The approach is also capable of leveling the field when it comes to the students of more rural areas trying to compete with learners from modern educational institutes who have better access to the internet or people who can help in resume building. When everything is added to same chain, everyone will be competing with the same starting deck. The use of blockchain and SHA-512 cryptography would automatically qualify for the technology to be virtually tamper proof and it would enjoy all the security benefits seen in Bitcoin and other cryptocurrencies.

5.7 Limitations

Apart from the benefits of the proposed model, there are certain limitations to its adoption and implementation. This model requires significant resources and knowledge to implement and then maintain which might prove to be a hinderance for smaller communities and countries. Blockchain requires significant computation power and electricity use is high too which would improve the carbon footprint of the educational institutes in the long run if sustainable sources are not used. Internet connectivity is vital and is the backbone of the system which might prove to be a hurdle for rural areas in third world countries. Even the smallest of institutes have to be assessed and mapped to make the whole process smooth for the students which could prove to be a difficult endeavor for any community to pull off.

6 Future Work

While working on this model, there were numerous ideas and suggestions to make it even better. In this section, we discuss these ideas as open research directions which would set the direction for others to work in this area. Moreover, these will serve as a basis for our own work as well. The field of education and learning is evolving rapidly especially since the adoption of technology in newfound ways. Some of the open research directions to work on the same are outlined as follows.

- *Patent and intellectual property*: SHA-512 encryption is the highest grade of encryption available today and protection of intellectual property could be easily accomplished by combining features of blockchain and encryption. An individual can add their intellectual property, research, and patents to this blockchain in an encrypted form. These could also be added to the token system to easily distinguish between different types of research and help people easily monetize their life's work and prevent theft.
- *Funding and endowment*: Blockchain could be used to keep a record of all the funding and grants any group or institute receives; this could guarantee transparency and clarity about the spending of the same. How it can be integrated and accounted for using blockchain is a huge topic in itself, which could have aspects of token calculation techniques, economics, and blockchain integrated together to make it work efficiently.
- *Soft skill assessment*: The tokens and certificates in this system are trustworthy only if the said student has actually passed the exam or course they are claiming with information about their marks. The proposed model can be extended to comprehensively incorporate the information related to soft skills of a student which can include comments and remarks by the instructors. This would make the assessment process a more qualitative one as compared to the current process which is based on quantitative measures such as percentile.

- *Vocational training*: The given system is skewed towards the sciences and education with exams that can be easily assessed and boiled down to a number. However, the same system could be expanded to also hold records of students in the field of fine arts, sports, music, dance, etc. The instructor's remarks could be used to create a report of the students' performance over the years and could be used to bring out the potential in these fields as well.
- *Accommodating newly budding education alternatives*: To make the transition as smooth as possible, the writers propose a 3–5-year plan for the transition wherein the students in higher classes would graduate in the old manner but student in fifth grade and lower would be transitioned to the new system. The students would be given their own private keys and blocks when they apply for tenth boards and would be handled by their respective boards in collaboration with the students' schools. The schools could also hold seminars and short courses on how blockchain works and how the students can build a portfolio on blockchain to secure a better future for themselves. The transition period would be difficult for any country and education system that adopts it but would prove to be useful in the long run.

The proposed model serves as the tip of the iceberg of how the education landscape could be revolutionized with the introduction of blockchain. This curious integration has immense potential for research and implementation. Third world countries could finally see their people compete with the leaders in their jobs and subjects. Education is the key to unlocking the true potential of any person, state, or civilization, and blockchain can assist humanity to achieve this goal in future.

7 Conclusion

Blockchain has opened the door for a decentralized governance and distribution of power. In this work, we have used blockchain and applied it to the Indian educational landscape. A layered model based on priority is presented along with the incorporation of tokens which are presented as a form of knowledge currency for end users. The model simplifies the process of calculation of token values for different courses, degrees, certificates, etc., and decentralizes the certification management process to make it easy for students to have their resumes verified by any company they are applying to. The approach was aimed at using blockchain to introduce a standard method of validating the completion of any educational endeavor undertaken by any student throughout his or her life. The work is summarized by pointing out the future research directions for the same.

References

1. R.S. Tomar, Maintaining trust in VANETs using Blockchain. ACM SIGAda Ada Lett. **40**(1), 91–96 (2020)
2. J. Abdella, Peer to peer distributed energy trading in smart grids: A survey. Energies **11**(6), 1560 (2018). https://doi.org/10.3390/en11061560
3. J. Xie et al., A survey of Blockchain technology applied to smart cities: Research issues and challenges. IEEE Commun. Surv. Tutor. **21**(3), 2794–2830 (2019). https://doi.org/10.1109/COMST.2019.2899617
4. A. Sharma, R. Tomar, N.C. Sarishma, B.-G. Kim, Blockchain based smart contracts for internet of medical things in e-healthcare. Electronics **9**(10), 1609 (2020)
5. T.M. Fernández-Caramés, P. Fraga-Lamas, A Review on the Use of Blockchain for the Internet of Things. IEEE Access **6**, 32979–33001 (2018). https://doi.org/10.1109/ACCESS.2018.2842685
6. S. M. Idrees, P. Agarwal, M. A. Alam (eds.), *Blockchain for Healthcare Systems: Challenges, Privacy, and Securing of Data* (CRC, Boca Raton, FL, 2021)
7. Rural population (% of total population)—India|Data. https://data.worldbank.org/indicator/SP.RUR.TOTL.ZS?locations=IN. Accessed 13 Jun 2021
8. M.L. MIT, Blockcert. https://www.blockcerts.org/about.html. Accessed 15 May 2021
9. G. Blossey, J. Eisenhardt, G. Hahn, Blockchain technology in supply chain management: An application perspective, in *Proceedings of the 52nd Hawaii International Conference on System Sciences*, vol. 6, (2019), pp. 6885–6893. https://doi.org/10.24251/hicss.2019.824
10. D. Di Francesco Maesa, P. Mori, Blockchain 3.0 applications survey. J. Parallel Distrib. Comput. **138**, 99–114 (2020). https://doi.org/10.1016/j.jpdc.2019.12.019
11. K. Christidis, M. Devetsikiotis, Blockchains and smart contracts for the internet of things. IEEE Access **4**, 2292–2303 (2016). https://doi.org/10.1109/ACCESS.2016.2566339
12. D. Lizcano, J.A. Lara, B. White, S. Aljawarneh, Blockchain-based approach to create a model of trust in open and ubiquitous higher education. J. Comput. High. Educ. **32**(1), 109–134 (2020). https://doi.org/10.1007/s12528-019-09209-y
13. B. Hameed et al., A review of blockchain based educational projects. Framework **28**, 30 (2019)
14. F. Vidal, F. Gouveia, C. Soares, Analysis of blockchain technology for higher education, in *Proceedings of the 2019 International Conference on Cyber-Enabled Distributed Computing and Knowledge Discovery, CyberC*, (2019), pp. 28–33. https://doi.org/10.1109/CyberC.2019.00015
15. M. Turkanović, M. Hölbl, K. Košič, M. Heričko, A. Kamišalić, EduCTX: A blockchain-based higher education credit platform. IEEE Access **6**, 5112–5127 (2018). https://doi.org/10.1109/ACCESS.2018.2789929
16. S. Kolvenbach, R. Ruland, W. Gräther, W. Prinz, Blockchain 4 Education, in *ECSCW 2018: Proceedings of the 16th European Conference on Computer Supported Cooperative Work*, (2018), pp. 17–20. https://doi.org/10.18420/ecscw2018
17. W. Appelt, With the BSCW System. System, 66–78 (1999)
18. X. Sun, J. Zou, L. Li, M. Luo, A blockchain-based online language learning system. Telecommun. Syst. **76**(2), 155–166 (2021). https://doi.org/10.1007/s11235-020-00699-1
19. J.S. He, *A Novel Blockchain-Based Education* (2018), pp. 178–183
20. H. Sun, X. Wang, X. Wang, Application of blockchain technology in online education. Int. J. Emerg. Technol. Learn. **13**(10), 252–259 (2018). https://doi.org/10.3991/ijet.v13i10.9455
21. H. Haugsbakken, I. Langseth, The Blockchain challenge for higher education institutions. Eur. J. Educ. **2**(3), 41 (2019). https://doi.org/10.26417/ejed.v2i3.p41-46
22. N. Satoshi, S. Nakamoto, Bitcoin: A Peer-to-Peer Electronic cash system. Bitcoin, 9 (2008). https://doi.org/10.1007/s10838-008-9062-0

23. S. Haber, W.S. Stornetta, How to time-stamp a digital document, in *Conference on the Theory and Application of Cryptography*, (1990), pp. 437–455
24. S.M. Idrees, I. Aijaz, R. Jameel, M. Nowostawski, Exploring the Blockchain technology: Issues, applications and research potential. Int. J. Online Biomed. Eng. **17**(7) (2021)
25. S.M. Idrees, M. Nowostawski, R. Jameel, Blockchain-based digital contact tracing apps for COVID-19 pandemic management: Issues, challenges, solutions, and future directions. JMIR Med. Inform. **9**(2), e25245 (2021)

Towards a Distributed Record of Measurement Adapters Powered by Blockchain Technology

Mario Diván and María Laura Sánchez-Reynoso

1 Introduction

Blockchain technology has allowed the implementation of different distributed databases, fostering the participation and interoperability of contents among different stakeholders and nodes [1, 2]. This constitutes an interesting advantage as no one has exclusivity around the content, the content is distributed among nodes, and it is regenerable from the transactions kept in the chain. Each block contains a relationship about the previous node and each transaction needs to be approved by the involved nodes. In this way, all participants provide consent about a modification or incorporation before approving any change, otherwise, the proposal is rejected. In this way, each participant can verify the chain in terms of consistency, integrity, authenticity, and reliability [3–5].

The measurement constitutes a comparison in which a concept's attribute is quantified through a comparison between it and a reference pattern [6]. The underlying suppositions about the measures are associated with the comparability essentially. For example, when an outpatient's heartbeat is monitored over time, the comparison of values is performed to analyze its evolution. In other words, each value is supposed that was obtained based on compatible methods and instruments, with given accuracy and precision. This implies that the measurement procedure

M. Diván (✉)
National University of La Pampa, C. Gil 353, Santa Rosa, Argentina
e-mail: mjdivan@eco.unlpam.edu.ar

M. L. Sánchez-Reynoso
National University of La Pampa, C. Gil 353, Santa Rosa, Argentina

INCITAP, National Scientific and Technical Research Council (CONICET), Santa Rosa, Argentina
e-mail: mlsanchezreynoso@eco.unlpam.edu.ar

S. M. Idrees, M. Nowostawski (eds.), *Transformations Through Blockchain Technology*, https://doi.org/10.1007/978-3-030-93344-9_5

should be consistent and repeatable over time, but also it should be extensible when new requirements are necessary [7, 8].

The measurement frameworks propose different alternatives to agree on the terms, concepts jointly with their relationships for describing a quality model able to be quantified. Such quantification is conceptually consistent in terms of a monitoring objective and it is based on an experimental design [9, 10].

The Internet of Things (IoT) has emerged as a low-cost, cheap, and certain alternative to increase the coverage area and resolution in a lot of monitoring applications. The monitoring needs have been addressed through the articulation of IoT-based devices jointly with well-defined measurement processes deriving in the implementation of different real-time data collecting systems [11, 12].

The Data Stream Processing Strategy (DSPS) emerged as an architecture in which a measurement process is implemented based on a measurement ontology for online monitoring. The implementation considers a set of IoT devices connected to sensors around an entity under monitoring. The heterogeneity of data sources is addressed through a Measurement Adapter (MA), which is responsible for articulating the data heterogeneity from sensors to a common and interoperable measurement interchange schema. The gathering component responsible for centralized data collecting receives a homogenized stream of measures understandable through the measurement project definition. The measurement project definition is supported by a measurement ontology to agree on concepts, terms, and relationships among them [13].

Because a measurement project could have a set of devices distributed in the monitoring field, a set of measurement adapters associated with each one allows becoming raw data into a homogenized stream of measures, using a measurement ontology. However, a measurement adapter could require indirect transmission through another MA close to it, for that reason the Node Unified Record (NUR) was introduced. In this record, the MA are identified and characterized by the following roles: (a) Cooperative: it can collaborate with another MA, (b) Exclusive: it informs its own measures only, (c) Gateway: it does not collect measures, it is only an intermediary in the data providing chain, and (d) Blocked: the node is not allowed to transmit if some suspicious action or identity was detected. Currently, the NUR is managed by DSPS centrally, while the measurement adapters have passive participation [14]. In case of corruption on the unified record, each MA lacks information for verifying it. It is a concern because the risk could limit the possibility to transmit directly or indirectly from the measurement adapter without the knowledge of its neighbors.

As contributions, a blockchain-based distributed record of measurement adapters is proposed for articulating the distributed database with nodes in the data collecting system. Thus, the neighborhood related to the different measurement adapters is known for each one dynamically, taking active participation in any change proposal. Also, the functionality related to the main processes around the distributed database is detailed through BPMN (Business Process Model and Notation) diagrams. Finally, an Open-source library implementing a prototype associated with this proposal is described.

This chapter is organized into six sections. Section 2 describes the main related works. Section 3 introduces the measurement adapter and its role in the data stream processing architecture. Section 4 describes the blockchain-based distributed record through different diagrams. Section 5 describes the prototypical reference implementation related to the proposal. Finally, some conclusions and future addresses are outlined.

2 Related Works

In [15], the authors describe the use of blockchain for implementing a distributed database based on microcontrollers, using a Merkle tree for integrity verification of measures coming from sensors. The microcontrollers gather data from sensors, informing measures to the cloud, and keeping local participation in terms of the distributed data collection schema. As a similarity, the distributed database for keeping content integrity is articulated among the different microcontrollers. In this work, the measurement adapters manage the relationship between sensors and the gathering component responsible for centralized data collecting in a homogenized way. As a difference, this proposal focuses on the metadata describing the participation of each measurement adapter (i.e., the microcontroller itself) to regulate the role of each one in the wireless sensor network. Even, the measurement record of each measurement adapter is managed through a Merkle Tree based on measurement metadata [16].

In [17], it is described as a distributed and digital disease information storage. This allows fostering decentralized access to the information, contemplating the traceability and integrity verification. The strategy is outlined around the reporting systems for avoiding a central dependency and fostering the participation of any kind of device. It constitutes a very interesting proposal to improve a distributed information collecting system. As a similarity, this work proposes a distributed ledger around the nodes' unified record, fostering integrity verification and transaction traceability. As a difference, the distributed ledger is applied on measurement adapters performing automatic collecting of measures based on a measurement ontology. Thus, the current state of each measurement adapter is known or rebuilt through the blockchain.

In [18], a schema is proposed where the transaction involves a set of parties for reaching an agreement instead of two only. Thus, when a given transaction needs to be carried out, the involved stakeholders should agree on it for confirming the operation. As a similarity, each measurement adapter proposes an operation exclusively on its own record (i.e., its own metadata) but the operation needs to be confirmed for each measurement adapter incorporated and active in the blockchain, otherwise, the operation would be rejected. As a difference, the interaction schema here is focused on the current state of the measurement adapters and their ability to transmit measures directly or indirectly.

In [19], the authors describe an interesting proposal for a distributed organization of the Electronic Health Record (EHR). The concern in this area is not only about integrity verification but also of information confidentiality. A blockchain-based proposal is described for describing a distributed ledger, involving hospitals, government agencies, and insurance providers. As a similarity, each node in the blockchain could provide new data and it is responsible for ensuring its authenticity and integrity (e.g., the generated data in a given HER comes from a given hospital). As a difference, this proposal is limited to data collecting systems where the measurement adapters need to be informed about neighbors for coordinating data transmissions from sensors. The involved measurement adapters and measures are organized and guided by a measurement project definition supported by a measurement ontology.

In [20], a blockchain-based technique is introduced for record verification in the context of heterogeneous wireless networks. Records describe the delivery events on a decentralized network. Each agent responsible for data delivery is who generates the data block when a direct transmission is performed. As a similarity, the focus is put on how the transactions and associated data are managed. In this proposal, the role of the measurement adapters and any change from their own data in the unified record is essential for data tracing to sensors jointly with data integrity and authenticity.

In [21], a proposal for applying blockchain-based technology aimed to build a distributed database for a logs system is introduced. It is applied to ensure that network works properly even in the presence of malicious nodes. Logs are stored in a distributed way to increase security, while a voting-based consensus algorithm is employed to ensure the participation of the whole nodes. As a similarity, no changes are approved without the consensus of all the participants (independently of the chosen algorithm) to ensure the participation of each member over the blockchain. As a difference, in this work, each node could start transactions about incorporation, modification, or deletion of its own data. Even there, such transactions need to be approved by member nodes according to a scored profile (e.g., a node A cannot pretend to start a transaction modifying/updating/deleting any another node record). The transaction to be approved needs to gather a score upper or equal to a known threshold. Each node according to the Record Permanence Time, Number of Started Transactions, and Number of Signed Transactions will receive a weighting based on its work. It is computed using the same blockchain. Thus, when the sum of weighting is equal to or exceeds the threshold, the transaction becomes approved, even when some nodes did not approve it yet. The ideal is that all the nodes impress the signature about all transactions; however, when some node died, this policy becomes dangerous, stopping all the transactions on the network.

Table 1 describes a synthesis of aims and approaches associated with the related works introduced in this section.

Table 1 A comparative perspective of the related works

Article	Year	Aim	Approach
S.-J. Hsiao and W.-T. Sung [11]	2021	Integrity verification of measures	A Merkle tree-based implementation for keeping trace of measures transmitted coming from sensors connected to microcontrollers
P. Zhu et al. [13]	2021	Traceability and integrity verification for disease information reporting	A distributed and digital disease information storage is described, fostering the distributed information reporting about diseases
H. Hong and Z. Sun [14]	2021	An agreement schema on transaction is proposed	For conforming to each operation, the agreement schema should be satisfied, involving all of the stakeholders
Y. Xiao et al. [15]	2021	Protection of the integrity and confidentiality of electronic health records (EHR)	A blockchain-based proposal for implementing a distributed EHR database is described, involving hospitals, government agencies, and insurance providers like stakeholders
Y. Watanabe et al. [16]	2020	Event monitoring in wireless sensor networks	A blockchain-based technique for record verification. Each data transmission is incorporated through a block into the blockchain for keeping a data transmission trace in the network
J. Huang et al. [17]	2019	Protection of the log system integrity	A blockchain-based database for logs is implemented. Logs are stored in a distributed way to increase security, while a voting-based consensus algorithm ensures the participation of the whole nodes

3 Measurement Adapters in the Data Processing Architecture

The Data Stream Processing Strategy (DSPS) is focused on the automatization of measurement projects. Measurement projects are defined following a set of concepts, terms, and relationships established in an extended measurement ontology [22]. Thus, the project aim is described jointly with the entity category and their context. They are described through their attributes and context properties respectively. A set of entity's attributes could describe different entity states, while a set of context properties could describe different scenarios. Both attributes and context properties are quantified through metrics that indicate the associated scale, unit, instrument (i.e., sensor), method, among other aspects. Metrics are interpreted through a set of indicators that incorporate decision criteria. Decision criteria are based on previous experience and knowledge from domain experts. Thus, each decision criteria could have an associated classifier that can provide recommendations accordingly. All of this information are contained in a single, hierarchical, and self-contained file known as Measurement Project Definition (MPD).

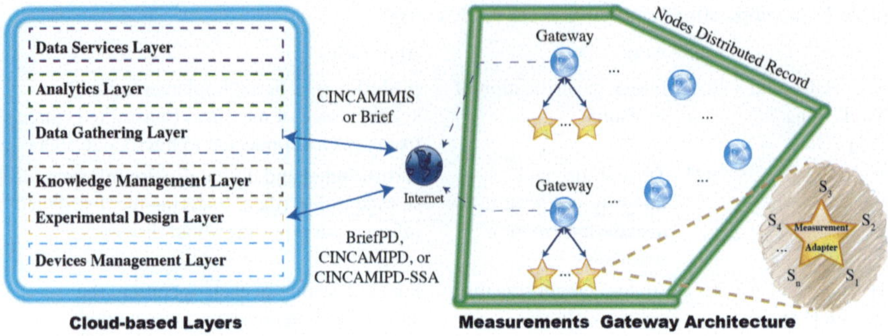

Fig. 1 A global perspective of the data stream processing strategy

Each component participating in a measurement project requires the measurement project definition to be initialized [23]. Briefly, in addition to reserve the local resources, the initialization associates each sensor with the corresponding metric definition in the MPD. Thus, when measures need to be informed, traceability from the sensor to the cloud could be kept. Figure 1 describes the whole data processing architecture to put it in perspective.

The DSPS can be described, on the one hand, in terms of the Cloud-based layers, on the other hand, about the data collecting organization of the field. The first ones refer to how the data are centrally processed and analyzed integrating all the distributed sensors. The seconds describe how the integration from sensors to the cloud is performed.

The Cloud-based perspective is integrated by six layers (1) *Device Management*: It is responsible for keeping track of the distributed measurement adapters and gateways, managing alarms bidirectionally; (2) *Experimental Design*: it stores and makes the measurement project definitions accessible, considering the scenarios and entity states established for different projects; (3) *Data Gathering*: It receives, processes, and derivates the measurement streams from measurement adapters and gateways to the rest of the layers; (4) *Knowledge Management*: It analyzes measures coming from sensors using indicators to orient the search for incremental classifiers that can provide online recommendations accordingly. Also, it establishes how the expert's knowledge could be incorporated into the architecture; (5) *Analytics*: Through the measures coming from the data gathering layer (i.e., by replication), it is responsible for online updating of descriptive statistics, association, and behavioral data analysis for all of the projects; (6) *Data Services*: It establishes how the historical and current data could be accessed. On the one hand, historical data are available through different filters and services. On the other hand, raw data are provided through subscriptions. Thus, data coming from sensors for chosen projects are derived as they arrive.

The Measurements Gateway Architecture (MGA) relates from data sensors to gateways to provide a homogenized version of the data streams. Gateways and measurement adapters use the measurement project definition to articulate data sensors with the underlying concepts in the project. For example, the project definition indicates that a given sensor will provide the environmental temperature related to a metric. Thus, when the measurement adapter finds such a sensor, it matches the sensor with the metric ID contained in the project for informing the measures. The project definition could be informed through two versions. The first version describes the project definition and scenarios/entity states as separated files (CINCAMIPD and CINCAMIPD/SSA respectively). The second version provides the whole definition (i.e., project plus scenarios and entity states) as a unique file (i.e., BriefPD). Once the measures are collected, the MA is responsible for incorporating tags properly for describing each concept surrounding a value. Such information will be provided to the Data Gathering Layer through two possible alternatives (1) *CINCAMIMIS*: A Schema-based file incorporating measures jointly with tags describing their semantics, and (2) *Brief*: An integrated and hierarchical file based on the measurement ontology, that allows optimizing the data transmission. Merkle trees are employed to keep track of them [16].

On the one hand, gateways gather a set of measurement adapters, concentrating common actions to all of them (e.g., the project definitions' updates). On the other hand, measurement adapters deal with data sensors directly guided by the project definition. They are responsible for data synthesis, load shedding techniques, data change detection, and cooperation with other measurement adapters (e.g., for indirect data transmissions).

The proposal of the blockchain-based Distributed Record of Measurement Adapters allows integrating different stakeholders in the measurements gateway architecture, making transparent, reliable, and available the node information to be used for data transmission (e.g., when an MA wants to transmit through a neighbor). Each node can modify its data only (e.g., data updating, deletion, or registration), no transactions about other MA or gateway are allowed. This is useful because MGA is distributed on the monitoring field, out of any central control. For that reason, a blockchain-based alternative is useful especially.

4 A Blockchain-Based Distributed Record

As it was introduced in the architecture previously, the context where the blockchain-based Distributed Record is oriented is associated with low-cost IoT-based devices. In such a context, measurement adapters and gateways should share a distributed database where the nodes' information is available for all the stakeholders. Thus, when a node needs to perform an indirect transmission, it could choose between the available nodes, taking into account whether each one supports the cooperative behavior or not.

This section is organized into two. The first section refers to the architectural proposal associated with the implementation of the distributed database suitable for low-cost devices. The second section describes the behavioral perspective of the operations from nodes. It is described through BPMN diagrams, pointing to each kind of interchanged message between nodes for consolidating or not an operation.

4.1 An Architectural Approach

From the Measurements Gateway Architecture oriented to low-cost hardware, there are two key roles. On the one hand, the Gateways (GW) acting as a concentrator of data for minimizing the dependence of each Measurement Adapter (MA) from the cloud. On the other hand, the measurement adapter dealing with sensors directly, collecting measures, and generating the measurement interchange stream (i.e., data plus tags describing its meaning based on the project definition). Data transmissions could be performed directly or indirectly from the measurement adapters according to the connectivity availability. Typically, as a reference, gateways are associated with hardware as powerful as a Raspberry Pi, while the measurement adapters are related to Arduino boards or similar.

Such cooperative behavior between measurement adapters and gateways requires a whole knowledge of all of the components involved in data collection in the monitoring field. Figure 2 describes a deployment perspective, where both gateways and measurement adapters require a *LocalResources* component for getting access and interaction with the blockchain. This component allows access to the current blockchain, the transactions record jointly with a Merkle Tree describing the transactions sent from the local node. As a difference to the Merkle Tree, the transactions record (i.e., local) contains all the transactions sent but also those received for endorsement from the sensor network. Using such information, the component provides details

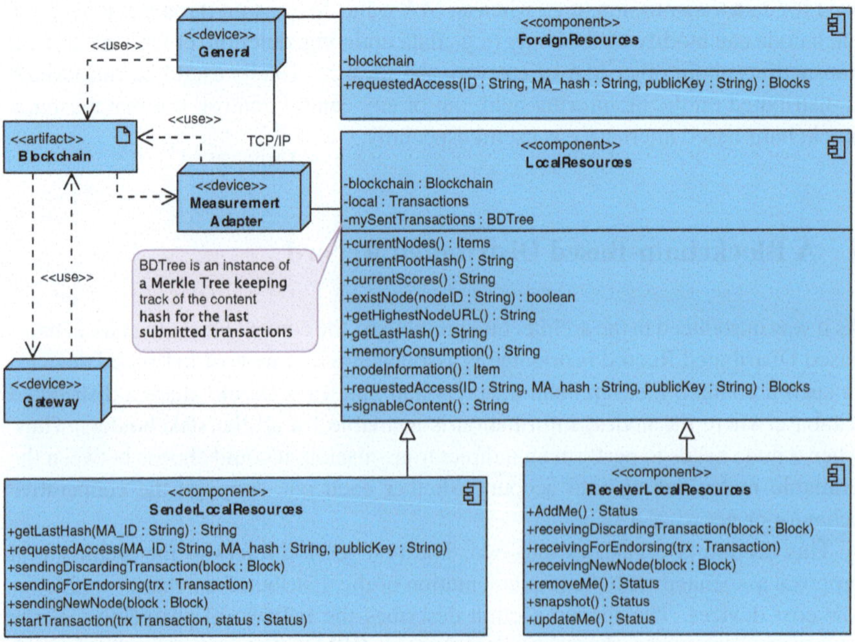

Fig. 2 A deployment perspective of the Blockchain-based Distributed Record

about the current nodes, the current root hash related to the Merkle tree, scores for each node, assessment about the node existence, the Uniform Resource Locator (URL) related to the node with the highest score, last content hash of the last block, current memory consumption per node, local node information, signable content, and the way to read the complete blockchain (i.e., requestedAccess).

On the one hand, the *SenderLocalResources* component is a specialization of the *LocalResources* component that provides additional services such as the transaction sending for endorsement, the possibility to discard not approved transactions, the incorporation of new blocks verifying the corresponding endorsement, and the start of a transaction (i.e., Add, Remove, Snapshot, or Update operation). On the other hand, the *ReceiverLocalResources* component provides the logic for processing each kind of request (i.e., discarding and endorsing a transaction, or incorporating a new node into the blockchain).

Measurement adapters and gateway only can modify, incorporate, or delete their own data. This implies that nobody could pretend to start a transaction in the name of another node. Thus, all the participants have access to the last information provided by each node to know, for example, whether or not it wants to collaborate with indirect data transmissions. Also, it is worthy to mention that the blockchain-based database is independent of the cloud architecture, providing a unified and self-controlled record available in the monitoring field. From the Cloud-based architecture, the important thing is that measures arrive, while from the monitoring field, the important aspect is to transmit measures as clever and soon as possible. Thus, this distributed record provides a certain level of autonomy to each MA or GW to decide about how to send their measures to the cloud layers.

Also, as it is possible to appreciate in Fig. 2, it is contemplated that a general device could access the blockchain database for reading it only, taking a passive behavior. That is to say, it could get the database for knowing the whole nodes participating in the data collection strategy, but it is not allowed to introduce any kind of operation. This is useful when some audit behavior needs to be implemented on the distributed database and the measures transmission.

Figure 3 describes the main concepts for implementing the blockchain-based and Distributed Nodes Record. The blockchain is composed of a set of blocks, where each one has the previous node and hash, the content, timestamp in which it was incorporated, and the root hash related to the requester's Merkle tree at the moment in which the request was performed (See the *Block* and *Blockchain* classes). The blockchain class provides the functionality associated with the information obtained from it, for example, a materialized view with the last known state of the whole nodes (Main functionalities are synthesized in Fig. 2). Each block has one transaction associated (i.e., See the *Transaction* class) where the requester is identified jointly with the timestamp, requester signature, and previous hash related to the block content. For each transaction, an operation is associated with the corresponding item's information (See the *Operation* class). The operation class describes the base class from where the *Add*, *Remove*, *UpdateRecordItem*, and *Snapshot* classes inherit, describing the four basic operations on the blockchain. That is to say, an item can add, remove, or update its information, but only the highest score can perform a snapshot.

The score per node is calculated considering the number of consolidated transactions, the time from which the node is active, and the number of questioned transactions as it is indicated in Eq. (1). The max priority is close to 1, when the node has active participation and permanence in the network, while it is near zero when its participation and permanence are low.

$$\text{score}(\text{node}) = \left(\frac{\#\text{Trx}_{\text{approved}}}{\#\text{Trx\_total}} + \frac{\text{Elapsed\_Time}}{\#\text{Total\_Time}} \right) / 2 \quad (1)$$

where:

- **#Trx$_{\text{approved}}$**: it represents the total number of approved transactions for the node.
- **#Trx_total**: it indicates the total number of transactions performed on the blockchain.
- **Elapsed_time**: it contemplates the whole time in which a node has belonged to the blockchain.
- **#Total_time**: it describes the total time in which the blockchain has been active.

The highest score node is dynamically established over time based on its activity and permanence. It has a meaningful role during the consolidating operation. Because the involved nodes correspond to low-cost hardware, they do not have enough resources to keep all the history associated with the blockchain. For that reason, each a certain operation volume, the current highest score node can request a consolidation operation which implies synthesize all the blockchain's history in a single node with the integrated information of whole nodes (i.e., the last state of the record derived from its own history contained in a single transaction). It seems to be something like a restart of the blockchain, however, this operation contains the lasts node information without any modification (i.e., a snapshot). This is important in this context because helping to release resources in the nodes due to the limited capacity. It is worthy to mention that the snapshot operation can be started by the highest score node, however, it needs to be approved by all nodes before synthesizing the blockchain (In another way, the snapshot operation is rejected and the blockchain is kept).

Thus, each requested transaction needs to be endorsed by the rest of the nodes (See the *EnsorsementDetails* class in Fig. 3) and gather a consensus. When the number of connected nodes is lower than 11, the required consensus is 100%. However, when the number of nodes exceeds 10, 75% of the global score (using Eq. 1) associated with all the nodes is required for approving a transaction.

The *LocalRecord* class (See Fig. 3) contemplates the basic operations described by the *LocalResources, SenderLocalResources,* and *ReceiverLocalResources* components (See Fig. 2). Thus, it keeps track of the requested endorsements and the reached decision for each node. Also, it keeps a record of the started transactions up to they are approved or not (See *Transactions* and *Transaction* classes in Fig. 3). As it was said before, the transaction can contemplate one of four operations (1) *Add*: It requests the incorporation of a record item for an MA or GW; (2) *Remove*: It

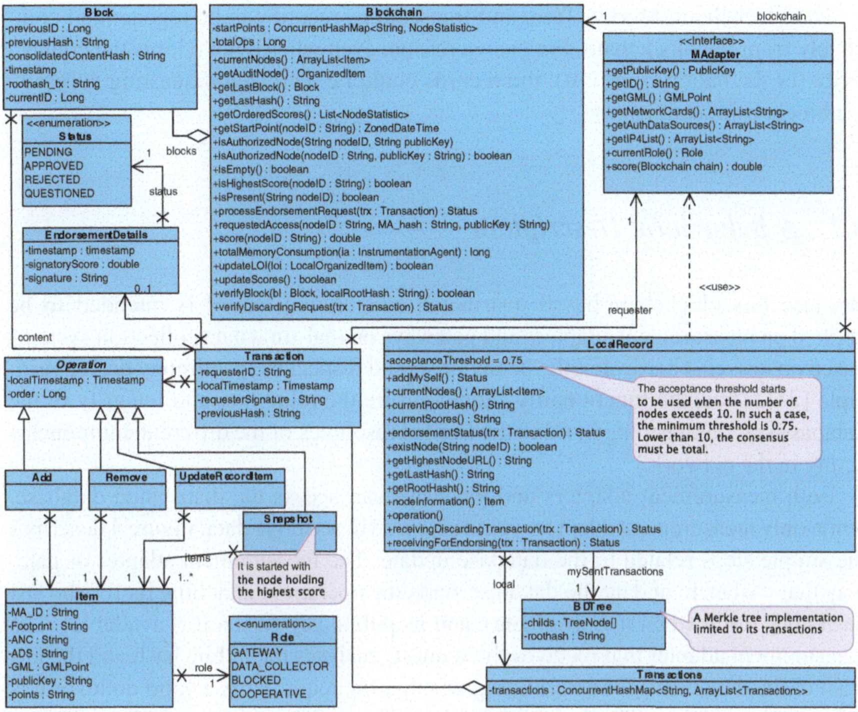

Fig. 3 Main concepts related to the Blockchain-based and Distributed Nodes Record

requests the deletion of the own record from the record (e.g., it could be possible when the MA leaves the sensor network); (3) *UpdateRecordItem*: it is performed when some data from the record item has been updated (e.g., the positioning); (4) *Snapshot:* it is started by the highest score node only. It requests a consolidating operation over the complete blockchain (See *Operation*, *Add*, *Remove*, *UpdateRecordItem*, and *Snapshot* classes in Fig. 3).

Each GW or MA stores a record item (See *Item* class in Fig. 3) in the distributed database, detailing its information. This item contains the node ID (i.e., MA_ID), a Comma-separated list of authorized networks card (i.e., ANC) through which measures could be transmitted, a Comma-separated list of authorized data sources or sensors to inform measures (i.e., ADS), a synthetic geographic markup language describing its current positioning (i.e., GML), and its public key [14]. This information is useful especially because it allows knowing the proximity between nodes, but also the similarity in terms of the used sensors, characterizing the monitored regions.

Each transaction authored per node is stored in a local Merkle tree for integrity verification (See *BDTree* class in Fig. 3). The tree will contain the hash related to each past transaction up to a consolidating operation be performed. Also, the root's hash is indicative of the integrity of the past transactions for a node and will be incorporated when the node wishes to start a new transaction. The Merkle tree is updated each time that a node's transaction is sent.

The blockchain, Merkle Tree, and transactions record can be regenerated completely from the blockchain. For such a reason, even when a node loses the connectivity (or the battery runs out), the records could be recreated requesting access to the blockchain again.

4.2 A Behavioral Description

Because this blockchain-based distributed and unified record is intended to be applied on measurement adapters and gateways on real-time data collection systems based on low-cost hardware, the behavior should respect the parsimony design principle [24]. Thus, the functionality should ensure the reliability and integrity of the database content, keeping in mind the limited resources of the different components acting in the network.

Both measurement adapters and gateways can access the distributed database, while only measurement adapters could act on its descriptive data. Figure 4 describes the simple steps related to the database update. The measurement adapter or gateway that wishes to update the database starts the operation, attaching its ID, the last hash node (only when it is available), and its public key (when it is available). The measurement adapter that receives the request, analyses the public hash and the last hash of the data block for determining whether the requester is a valid node, or it is a foreign component. This is useful for determining the attention priority, prioritizing the registered nodes (MA or GW), and putting the rest of the requests at the end of the queue. According to the informed node hash, the pending data blocks are provided. When the requester receives the set of pending data blocks, it integrates and consolidates them into the blockchain. However, when the requester does not

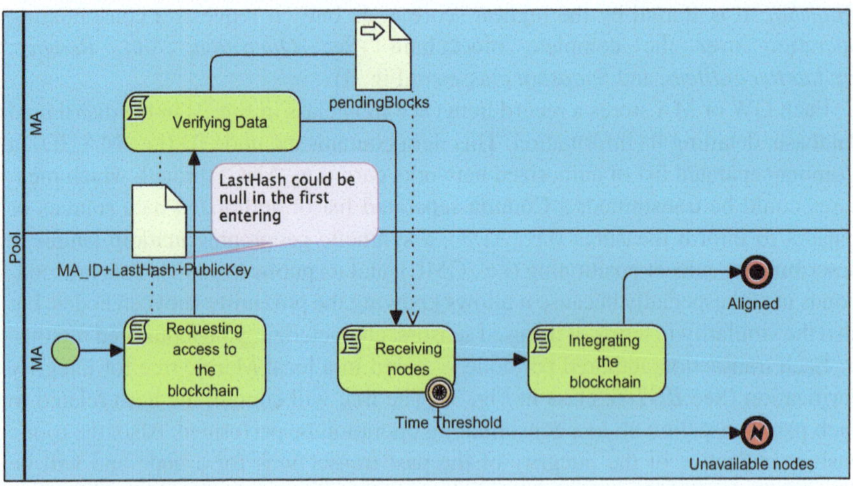

Fig. 4 A BPMN diagram describing the update of the Blockchain-based distributed database

receive an answer after a given time (i.e., a threshold), the request is cancelled understanding that the node is not available. This situation could occur when there is no answer or when the node responsible for answering has other requests with higher priorities.

However, not all the operations can be started from any node. The exception is the snapshot operation (or consolidating), which is performed to release resources from the different nodes and generating the last integrated view based on the database content. The snapshot operation can be started only by the highest score node according to Eq. (1) (See the top-left area in Fig. 5). It is based on the traditional Bitcoin approach related to the Proof-of-Work (PoW) [25]. Because the score is dynamic (i.e., it could change over time), each time a node requests a transaction, the scores are updated to avoid unnecessary calculus.

The rest of the operations can be started by any registered node (See Fig. 3), taking actions limited to the own data exclusively. When a measurement adapter or gateway is not registered, it can start a transaction with an *add* operation to incorporate its data in the distributed database. After that, it is responsible for keeping updated its data record.

When a transaction containing one or more operations is started, it must be signed by the requester using its private key. After that, the transaction is sent to other nodes in the database to be endorsed. Each node starts the endorsement

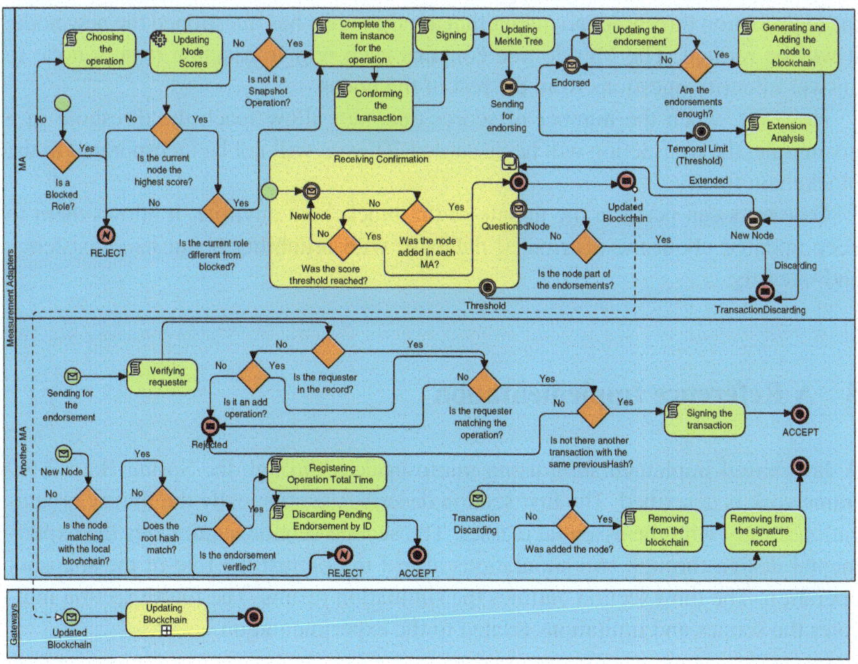

Fig. 5 A BPMN diagram describing the Main Operations related to the Blockchain-based distributed database

process when they receive the message. The requester is verified to determine whether or not it is present in the database. When the operation is different to add and the requester is not present, the transaction is rejected. Otherwise, when the requester is present or the transaction is associated with an add operation, the transaction is verified for determining the consistency between the transaction and requester. When another previous transaction has the same hash (for the last data block), the current transaction is rejected. When the received transaction matches the last hash of the local blockchain, then the transaction is endorsed.

The requester waits for the arrival of endorsements. When the sum of scores related to those approved is upper or equal to an authorization threshold, the transaction becomes approved finally. Then, the requester will create the new node to be distributed and integrated into the blockchain. However, when the authorization threshold is not reached, the requester node waits for a pre-established while. If the waiting time is exceeded, then the transaction is marked to be discarded. When a node receives a transaction discarding message, it removes the transaction from blockchain and local record.

On the other hand, when the new node was generated, it is communicated through an *AddedNode* message. The requester acts as a coordinator, waiting for the message from the rest of the nodes. When the requester node reaches the necessary score threshold, the transaction becomes consolidated independently from other pending nodes. This is for avoiding situations in which a registered node keeps out of a given process because of energy concerns. Even more, if the transaction reaches the approbation threshold, no matter that another node has questioned the new node, it will be ratified. The Merkle tree contains each sent transaction from a node to answer integrity questions from the rest of the nodes.

However, when the number of scores does not allow reach the threshold in a given time, the transaction will be discarded. Thus, it will not be incorporated in the blockchain finally.

Gateways can perform the update of the blockchain anytime. It allows them to keep updated about the distributed database with reliability about its consistency and integrity.

5 A Reference Implementation

A Java-based implementation using microservices through the Spring Boot 2.5.0 framework is described. The first section describes synthetically the library organization jointly with the essential classes. The second section synthesizes the experimental design through two simulations aimed to contrast the role of the snapshot operation. The third section outlines the simulation results. The fourth section indicates the scopes and limitations related to the experimentation.

5.1 *Library Organization*

The library implementing the concepts, simulations, and techniques described here is freely available at https://github.com/mjdivan/nrchain under the terms of the Apache License version 2.0.

Figure 6 describes the role of each main introduced concept of previous sections about the Spring application. The *LocalRecordController* class implements the *RestController* in the SpringApplication, while the NRChainApplication implements the *SpringBootApplication* profile. Both are available under the package org. ciedayap.nrchain. The *Blockchain* class uses instances of the *NodeStatistic* class for keeping information about the first access of a node jointly with the last score and number of approved transactions. The *LocalOrganizedItem* class is a specialization of *OrganizedItem* class that can manage the private key of the local node (e.g., for signing transactions). The *GMLPoint* class allows expressing the position such as a GML point or string (transforming between them according to the necessity).

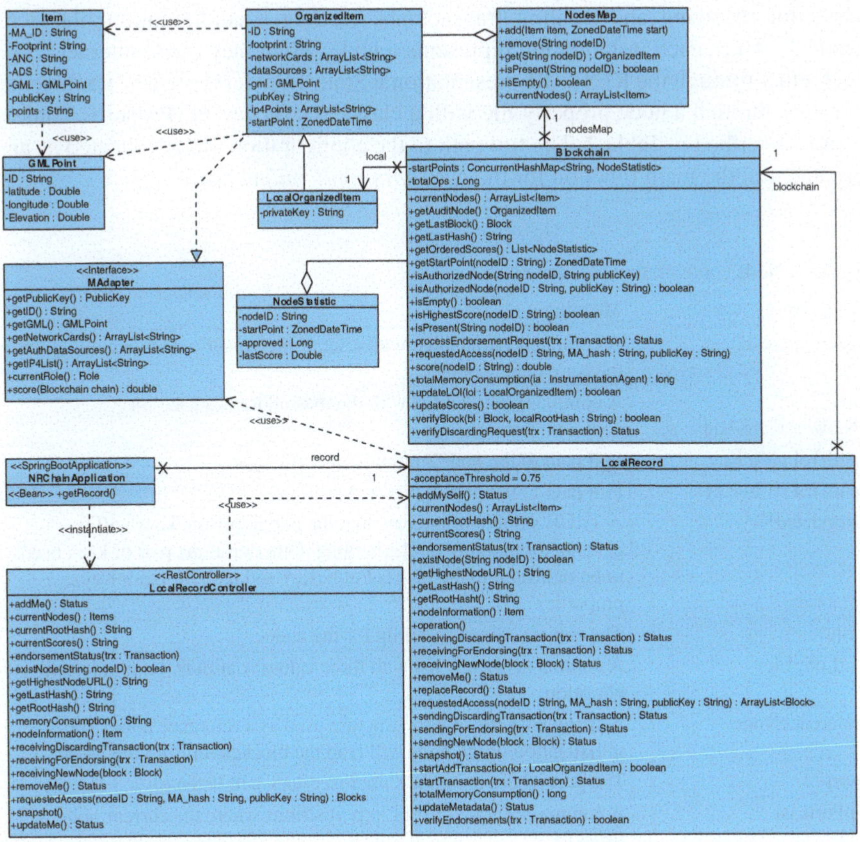

Fig. 6 Internal Organization Articulated with the Spring Boot framework

The library is organized by six packages as follows:

- *org.ciedayap.nrchain*: It contains the SpringBootApplication and RestController classes jointly with the class responsible for implementing the simulations.
- *org.ciedayap.nrchain.exceptions*: It contains classes specializing from the Exception class from Java. It creates new derived classes for specific exceptions, for example, BlockException and EndorsementException.
- *org.ciedayap.nrchain.interfaces*: It describes the interfaces to be implemented by a class about the nodes record context, for example, the *MAdpater* interface.
- *org.ciedayap.nrchain*.merkle: It details the classes involved in the Merkle tree implementation based on a unidimensional array mapping.
- *org.ciedayap.nrchain*.model: It outlines the different classes involved in the blockchain implementation for the distributed nodes record.
- *org.ciedayap.nrchain*.utils: It contains different utility classes used along with the blockchain implementation, for example, list manipulating, public and private key management, among others.

Each node starts as an independent Spring Boot Application running its microservices for receiving and sending transactions, reading a configuration file (see Table 2). Thus, each instantiation represents a node, while they could interact with each other through the REST (REpresentational State Transfer) services. Each node is set up through a node.property file with a classical structure of (Property, Value) as it is described in Table 2. The full path to the configuration file is indicated as an argument of the main function for the *NRChainApplication* class.

Table 2 Node configuration parameters

Property	Meaning
dataSourcesID	A comma-separated list with a set of the sensor's ID mounted on the measurement adapter
mac_ NetworkCardAddresses	A comma-separated list with the mac addresses used for communications
pathToPublicKey	Full path to the node's public key
pathToPrivateKey	Full path to the node's private key
simulatedKey	A TRUE value will indicate that the public/private keys will be generated randomly on the booting. Otherwise, the pair of keys needs to be indicated in the pathToPublicKey and pathToPrivateKey properties
role	The current role to be set up for the node
latLongElev	A comma-separated list with three values: Latitude, longitude, and elevation
referenceNode	A string with the ip:port structure used as a reference node. The initialization of the node will read the blockchain from this node
port	The port where the node's microservices will listen to
ipPortList	A comma-separated list of ip:port strings where the current node will listen to
nodeID	The node ID

5.2 Experimental Design

Two simulations were designed for analyzing the node's required memory size and startup times. The first one considered the snapshot operation, while the second one did not.

Figure 7 describes the simulation where the number of nodes is increased from 1 to 200 without snapshot operation. The first node is used as a reference for the blockchain initialization of the successive nodes. Once the new node is created, the blockchain is updated through the reference node, and after that, the node is self-added into the blockchain. Meanwhile, the final required memory is measured jointly with the total startup time. The operation is repeated until reaches 200 nodes (it is a parameter, and it could be changed). After the max number of nodes is reached, each node is stopped, and the measures are stored in a file for its analysis.

Simulation 2 incorporates the "each" parameter that aims to perform a snapshot operation on the blockchain each time the cycle is reached. Thus, the underlying idea is to contrast the effect on the memory consumption by such an operation in each node. Thus, simulation 2 defines the max number of nodes to simulate (i.e., 200 in this case) jointly with the "each" parameter (i.e., 20 in this case). After that, the rest client is created for invoking the snapshot operation from the highest score node according to a given simulation instant (i.e., the highest score node could change throughout the simulation). Thus, the nodes initialize progressively. The first node acts as the reference node for invoking services and blockchain initialization. Once a cycle is reached (i.e., the rest for the division between the sequence and the "each" parameter is zero), a snapshot operation is invoked from the highest score node. After the snapshot operation was run, the new node is incorporated into the blockchain. In the last case, the snapshot operation time is incorporated into the node startup time for analyzing the effect on it.

In simulation 2, obtained measures incorporate into results for being stored in a file later (See Fig. 8). Before finishing the simulation, a parameter allows keeping alive the reference node for the case in which the blockchain wants to be verified or queried (among other microservices introduced in Fig. 2).

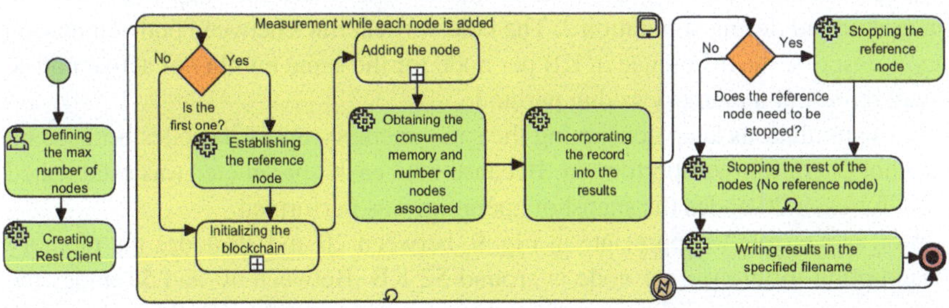

Fig. 7 A BPMN diagram describing the simulation 1. Increasing the number of nodes without snapshot operation

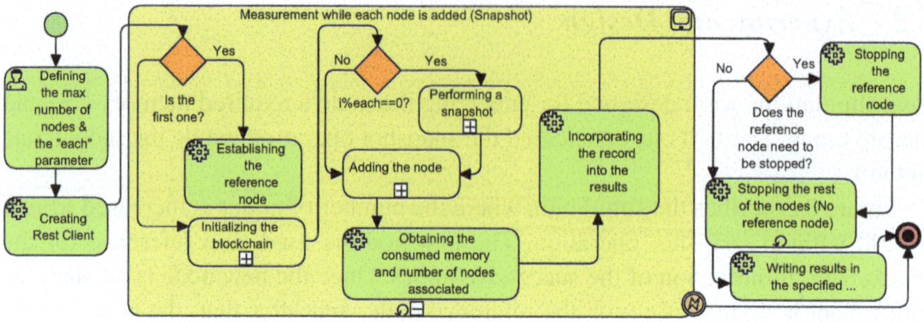

Fig. 8 A BPMN diagram describing the simulation 2. Increasing the number of nodes, activating the snapshot operation with the "each" parameter equal to 20

Both simulations use the com.github.jbellis.jamm package (version 0.3.3) that provides a java agent for measuring the object memory consumption. It is suitable for many java versions, and it is available at https://github.com/jbellis/jamm. Thus, the minimal arguments for initializing the simulation should incorporate:

-Dexec.args=-javaagent:<Path-To-JAMM0.3.3>/jamm-0.3.3.jar -classpath %classpath org.ciedayap.nrchain.Sim /Users/mjdivan/node.properties

The first one refers to the location of the jamm package; while the second one indicates the main class for implementing the simulation (i.e., org.ciedayap.nrchain.Sim). The last element corresponds to the node properties file path. The simulation was performed on a MacBook Pro with 16 GB RAM and Mac OS Big Sur 11.4 with a 2.9 GHz Intel Core i7.

5.3 *Results and Discussions*

Figure 9 describes the memory consumption per node expressed by Kilobytes (KB) for keeping in memory all the required instances by the distributed record. The abscise axis indicates the number of active nodes throughout the respective simulations. The dotted vertical lines describe the instant where the snapshot operation was performed during simulation 2. The bold vertical lines between both simulation curves describe the difference in KB per node for the same number of active nodes. It may represent a memory saving per node.

Both simulations keep consuming the same memory up to 20 nodes because no snapshot operation was performed. Because the "each" parameter was established in 20, for each 20 nodes the snapshot operation was performed.

As it is possible to appreciate in Fig. 9, between 20 and 39 nodes the memory consumption difference per node is around 52 KB. Between 40 and 59 nodes the memory consumption difference per node reached around 99 KB. After that, with each snapshot operation, the difference progressively was increased through around 146 KB, 195 KB, 240 KB up to 424 KB with 200 nodes. This represents an

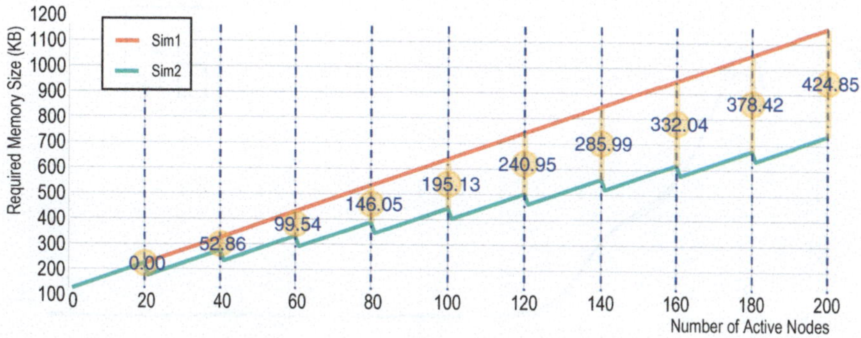

Fig. 9 Evolution of the Memory Consumption, incorporating the Snapshot Operation (i.e., Simulation 2) and avoiding it (i.e., Simulation 1)

important saving that varies with the number of nodes in the record. For example, simulation 1 indicated 1161 KB per node for a record containing 200 active nodes, versus 737 KB from simulation 2 (snapshot included). This represents a saving in simulation 2 of 36.52% (i.e., 424/1161) in memory consumption per node.

It is worthy to mention that even when the snapshot discarded the previous history of the blockchain to save memory (due to the limited resources of the involved hardware), the consistency was kept in the integrated node keeping a materialized view of the previous history and allowing the reconstruction from zero in case of being necessary. Also, this library introduces the snapshot operation as an optional operation, considering the limited resources allocated to the data collectors. Thus, the record can avoid it when blockchain history is required.

Figure 10 shows the density curve related to startup times per node for both simulations. Thus, times associated with simulation 1 are subtle higher than simulation 2 in the beginning. It is expected due to the overhead produced by the snapshot operation into the startup time. To analyze the magnitude of the differences, the Hellinger distance [26–28] allows analyzing the data distribution and quantifying the differences between probability distributions. Using the *statip* package (version 0.2.3) in *R* software (version 3.6.2), the Hellinger distance was calculated using the simulation results for the startup times among data series. The distance could between 0 and 1, where a value close to zero presents low differences, while a value near to one represents high differences. In this case, the obtained distance was 0.1047 which is consistent with Fig. 10 describing low differences.

Figure 11 describes the violin chart for both simulations [29]. Through them, it is possible to appreciate the values concentration and its distribution. The data distributions are similar, introducing subtle variations in simulation 2 (even some outliers as it is possible to see in the max value of simulation 2) that produce a subtle increment in the arithmetic mean.

Independently of the mentioned differences, the interquartile range and its constitution are quite similar, while the introduced outliers are associated with the additional time required by the snapshot operation. Thus, the worst scenario in simulation

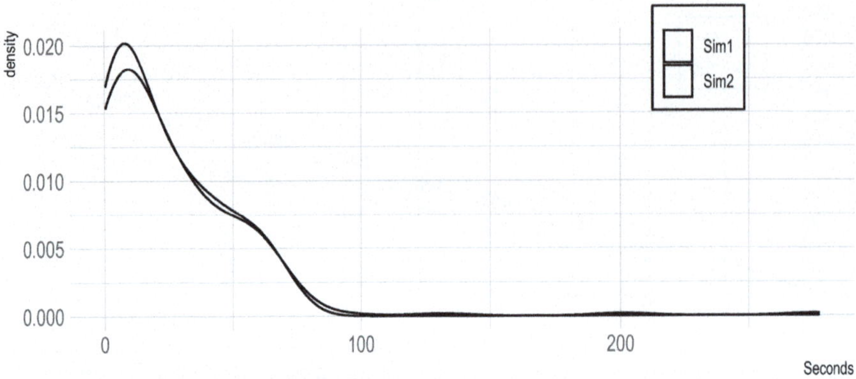

Fig. 10 Density Curves of the Startup Times per Node for each Simulation

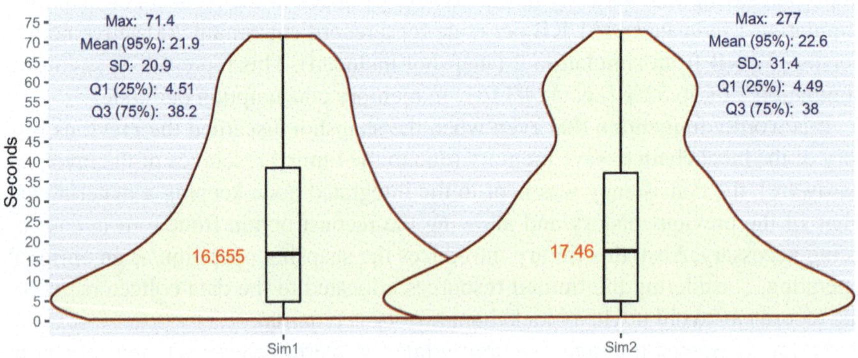

Fig. 11 Violin Diagrams of the Startup Time per Node for each Simulation

2 was that a node requires 277 s to be initialized (See Fig. 10 near to 200 s), while in simulation 1 it was 71.4 s. However, it is important to mention that the blockchain synchronization from the reference node is incorporated in the startup time.

5.4 Scope and Limitations

Results in this section were analyzed considering measures obtained in the discrete simulation. The library introduced is a prototype released under the terms of the Apache License version 2.0. Simulations were performed on an Oracle Java Virtual Machine JDK 1.8.0_271. The obtained results could vary according to the underlying platform.

6 Conclusions

In this work, a blockchain-based distributed record of measurement adapters was proposed. Thus, a decentralized and distributed nodes' record is coordinated among low resources devices independently of the data processing architecture. This provides more independence to the data collector devices, being able to choose nodes by proximity using its local record and minimizing the communication latency (i.e., the communication with a central authority is avoided). To provide reliability, the database is based on blockchain technology. Thus, each node can perform one of four operations: Add, Remove, Update, or Snapshot. The first three operations are limited to their own descriptive information, while the last one is an alternative for optimizing the memory consumption per node associated with the number of blocks retained.

As an advantage, the neighborhood related to the different measurement adapters is known for each one dynamically (synchronizing the blockchain), taking the node active participation in any change proposal. An open-source library named *nrchain* was synthetically described and released under the terms of the Apache License 2.0. This library introduces a prototypical implementation of this blockchain-based distributed database for data collector nodes.

Two discrete simulations were performed to analyze the involved times in the node start-up and the effect of snapshot on the node initialization. The memory saving per node obtained through snapshot operation reached 36.52%, while the max required memory (without snapshot operation) reached 1161 KB per node for keeping a full description of 200 nodes interacting through a sensor network. The maximum initialization time (incorporating the snapshot operation and the blockchain synchronization) was 277 s per node.

As future work, strategies oriented to in-memory data compression on a node will be analyzed as a new alternative for the snapshot operation, avoiding discard historical blocks.

References

1. Y. Inedjaren, M. Maachaoui, B. Zeddini, J.-P. Barbot, Blockchain-based distributed management system for trust in VANET. Veh. Commun. **30**, 100350 (2021). https://doi.org/10.1016/j.vehcom.2021.100350
2. S.M. Idrees, I. Aijaz, R. Jameel, M. Nowostawski, Exploring the Blockchain technology: Issues, applications and research potential. Int. J. Online Biomed. Eng. **17**(7), 48–69 (2021). https://doi.org/10.3991/ijoe.v17i07.22803
3. S.F. Aghili, H. Mala, C. Schindelhauer, M. Shojafar, R. Tafazolli, Closed-loop and open-loop authentication protocols for Blockchain-based IoT systems. Inf. Process. Manag. **58**(4), 1–19 (2021). https://doi.org/10.1016/j.ipm.2021.102568
4. H.S.A. Fang, T.H. Tan, Y.F.C. Tan, C.J.M. Tan, Blockchain personal health records: Systematic review. J. Med. Internet Res. **23**(4), e25094 (2021). https://doi.org/10.2196/25094

5. S.M. Idrees, M. Nowostawski, R. Jameel, A.K. Mourya, Security aspects of Blockchain technology intended for industrial applications. Electronics **10**(8), 951 (2021). https://doi.org/10.3390/electronics10080951
6. K. Krechmer, Relational measurements and uncertainty. Measurement **93**, 36–40 (2016). https://doi.org/10.1016/j.measurement.2016.06.058
7. R. Omnès, Measurement theory, in *The Interpretation of Quantum Mechanics*, (Princeton University Press, Princeton, NJ, 2018), pp. 324–377
8. S.M. Idrees, M. Nowostawski, R. Jameel, Blockchain-based digital contact tracing apps for COVID-19 pandemic management: Issues, challenges, solutions, and future directions. JMIR Med. Inform. **9**(2), e25245 (2021). https://doi.org/10.2196/25245
9. A. Shams, H. Sharif, M. Helfert, A novel model for cloud computing analytics and measurement. J. Adv. Inf. Technol. **12**(2), 93–106 (2021). https://doi.org/10.12720/jait.12.2.93-106
10. J. Mancebo, C. Calero, F. Garcia, M.A. Moraga, I. de Guzman, FEETINGS: Framework for energy efficiency testing to improve environmental goal of the software. Sustain. Comput. Inform. Syst. **30**, 100558 (2021). https://doi.org/10.1016/j.suscom.2021.100558
11. D. Singh, M. Dahiya, R. Kumar, C. Nanda, Sensors and systems for air quality assessment monitoring and management: A review. J. Environ. Manage. **289**, 112510 (2021). https://doi.org/10.1016/j.jenvman.2021.112510
12. S.M. Idrees, M. Nowostawski, R. Jameel, A.K. Mourya, Privacy-preserving infrastructure for health information systems, in *Data Protection and Privacy in Healthcare*, (CRC, Boca Raton, FL, 2021), pp. 109–129
13. M.J. Diván, M.L. Sánchez Reynoso, An architecture for the real-time data stream monitoring in IoT, in *Multimedia Big Data Computing for IoT Applications*, vol. 163, (2020), pp. 59–100
14. M. Diván, M.L. Sánchez Reynoso, M.H.A. Wahab, Dynamic switching in the measurements' collecting from heterogeneous data sources. J. Phys. Conf. Ser. **1529**, 022058 (2020). https://doi.org/10.1088/1742-6596/1529/2/022058
15. S.-J. Hsiao, W.-T. Sung, Utilizing blockchain technology to improve WSN security for sensor data transmission. Comput. Mater. Contin. **68**(2), 1899–1918 (2021). https://doi.org/10.32604/cmc.2021.015762
16. M.J. Diván, M.L. Sánchez-Reynoso, Metadata-based measurements transmission verified by a Merkle tree. Knowl. Based Syst. **219**, 106871 (2021). https://doi.org/10.1016/j.knosys.2021.106871
17. P. Zhu, J. Hu, Y. Zhang, X. Li, Enhancing traceability of infectious diseases: A Blockchain-based approach. Inf. Process. Manag. **58**(4), 102570 (2021). https://doi.org/10.1016/j.ipm.2021.102570
18. H. Hong, Z. Sun, A secure peer to peer multiparty transaction scheme based on blockchain. Peer-to-Peer Netw. Appl. **14**(3), 1106–1117 (2021). https://doi.org/10.1007/s12083-021-01088-4
19. Y. Xiao, B. Xu, W. Jiang, Y. Wu, The healthchain blockchain for electronic health records: Development study. J. Med. Internet Res. **23**(1), e13556 (2021). https://doi.org/10.2196/13556
20. Y. Watanabe, W. Liu, A. Abbas, Y. Andreopoulos, M. Hasegawa, Y. Shoji, Verifiable event record management for a store-carry-forward-based data delivery platform by Blockchain, in *2020 IEEE 31st Annual International Symposium on Personal, Indoor and Mobile Radio Communications*, (2020), pp. 1–6. https://doi.org/10.1109/PIMRC48278.2020.9217149
21. J. Huang, H. Li, J. Zhang, Blockchain based log system, in *Proceedings—2018 IEEE International Conference on Big Data, Big Data 2018*, (2019), pp. 3033–3038. https://doi.org/10.1109/BigData.2018.8622204
22. M. Diván, M.L. Sánchez-Reynoso, Monitoreo de entidades en tiempo real basado en estados y escenarios. CLEI Electron. J. **23**(1), 2:1–2:25 (2020). https://doi.org/10.19153/cleiej.23.1.2
23. M.J. Diván, M.L. Sánchez Reynoso, An architecture for the real-time data stream monitoring in IoT, in *Multimedia Big Data Computing for IoT Applications*, ed. by S. Tanwar, S. Tyagi, N. Kumar, (Springer Nature, Singapore, 2020), pp. 59–100

24. T. Gavrilova, A. Alsufyev, M. Gladkova, Perceptual factors in knowledge map visual design, in *ACM International Conference Proceeding Series*, vol. 21–22-Octo, (2015). https://doi.org/10.1145/2809563.2809599
25. L. Li, Y. Jiang, G. Liu, Consensus with voting theory in blockchain environments, in *Proceedings—10th IEEE International Conference on Big Knowledge, ICBK 2019*, (2019), pp. 152–159. https://doi.org/10.1109/ICBK.2019.00028
26. M.K. Garba, T.M.W. Nye, R.J. Boys, Probabilistic distances between trees. Syst. Biol. **67**(2), 320–327 (2018). https://doi.org/10.1093/sysbio/syx080
27. W. de Lima, A. do Nascimento, G. do Amaral, Distance-based tests for planar shape. J. Multivar. Anal. **184**, 104743 (2021). https://doi.org/10.1016/j.jmva.2021.104743
28. I. Goldenberg, G.I. Webb, PCA-based drift and shift quantification framework for multidimensional data. Knowl. Inf. Syst. **62**(7), 2835–2854 (2020). https://doi.org/10.1007/s10115-020-01438-3
29. P. Mallick, M.B. Rao, Graphics miscellanea. Handb. Stat. **32**, 93–142 (2014). https://doi.org/10.1016/B978-0-444-63431-3.00003-6

Blockchain Technology: A Breakthrough in the Healthcare Sector

Lakshmi Narasimha Gunturu, Girirajasekhar Dornadula, and Raghavendra Naveen Nimbagal

1 Background

In the contemporary world, numerous scientific and technical innovations have changed the lives of the public in various aspects like money transactions and decision making. Among those technical innovations, Blockchain had become most popular since the age of internet availability [1]. There has been increased attention among the public on Blockchain technology considering the rise of digital currencies like bitcoin, ripple, and litecoin [2]. It is described as a digital and decentralized technology that can be made public or private depending upon the type of user utilized for the money transaction without the involvement of third parties. The information is present as nodes without any central node. It has applications in the educational sector, cybersecurity areas, exit polls, health care sector, and industrial side [3]. Of late it has gained popularity in the health care domain due to its unique nature and property termed as a Decentralized database that can carry out operations independently. The principal benefit of blockchain technology is its decentralized nature, due to which the dealings are captured and recorded in nodes as real-time updates in a network model [4] as shown in Fig. 1. As a result, the information present in each node is similar without any variations. Therefore, it is translucent, self-governing, data stored is permanent and cannot be falsified. Further, this improves the information standards connecting various stakeholders [5]. The authenticity of transactions made by using this technology was done by the use of

L. N. Gunturu (✉) · G. Dornadula
Department of Pharmacy Practice, Annamacharya College of Pharmacy, Rajampeta, Andhra Pradesh, India
e-mail: gunturunarasimha007@gmail.com

R. N. Nimbagal
Department of Pharmaceutics, Sri Adichunchanagiri College of Pharmacy, Adichunchanagiri University, Mandya, Karnataka, India

S. M. Idrees, M. Nowostawski (eds.), *Transformations Through Blockchain Technology*, https://doi.org/10.1007/978-3-030-93344-9_6

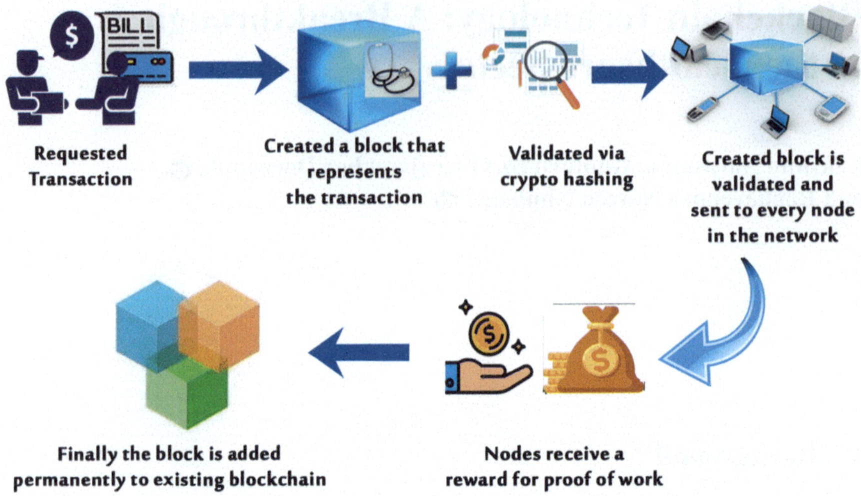

Fig. 1 Flow of information through blockchain technology

cryptogenic algorithms. It also overcomes the limitation of single-point failure which can be routinely encountered by centralized systems [6]. Blockchain technology provides special features that lack in centralized health systems such as data accuracy, transparency, cost-effectiveness, and information privacy with security [7].

According to IBM prediction, blockchain had used majorly as 70% in the healthcare domain to improve the areas such as clinical trials, regulatory management, and electronic health records maintenance (EHR). Furthermore, the value of blockchain technology in the healthcare sector is going to cross 500 million dollars by 2022. Despite potential improvements in the health care sector by blockchain technology, the modern promotion adjoining this technology likewise necessitates the ideas of current evidence to provide less information. Blockchain is considered as a distributed ledger in the health care system because of the various tasks or applications it performs. Many studies explained the use of blockchain in the storage of healthcare data for genomics and precision medicine [8, 9], Patient outcome data [10], clinical trial data [11–13], pharmaceutical information [14], and biomarker data [15]. The complex nature of blockchain technology in healthcare system allows money transactions by stakeholders. It is a useful approach to deal with different parties that can permit entry to the same type of institution. Subsequent importance to blockchain technology helps to gain a prioritized value to medical treatments. A project named Gem Health Network provides therapy-related data to various health care providers. This possibility can overcome the problems of information availability and accessibility, thus reducing the medical negligence caused by outdated information or source. Apart from this the operational expenses are also reduced by the health care settings that use the blockchain application. Health care workers can track the patients frequently, review their health condition, and manage the treatment. All these further improve the quality of health care and improve the services provided by the healthcare staff. For all these mentioned reasons blockchain technology is considered to be a distributed ledger in the healthcare system [16].

Blockchain technology of healthcare system allows patients and the public by directing medical information into disruptive matters. A suitable example to deal with a statement is MedRec which can provide all the information to the patient's hand. Information present is associated with smart contracts which use the patient-provider relation to share all the information [17]. Instead of directly acting as a repository for information these smart contracts also possess the references to separate all the information accumulated in a single record [18]. This facility allows comprehensive information on medical records which can be accessed by physicians and patients so that tracking and data audit can be easily handled. This source also provides an option for the patients to develop a relationship with the physicians so that physicians can spend extra time with their patient to educate and provide information regarding the therapy, which can finally improve therapy outcomes [17]. By incorporating information into a steady database without any direct storage, MedRec supports the existing infrastructure. Specific characteristics associated with blockchain technology include the following:

1. Distributed ledger.
2. Decentralization.
3. Cryptographic sealing.
4. Digitalization.

1.1 Distributed Ledger

Different network nodes of blockchain allow sharing the information among peers and individuals on a daily basis. All the money payments are placed together in a block and all the blocks are interconnected with each other. Every node stores a copy of the transaction so that all nodes had similar information. This facilitates the running of the technology independently without any authorization parties. Therefore, even if one node fails to perform its function other nodes continue to function [18].

1.2 Decentralization

In centralized management, each and every transaction is mediated by a third party or centralized author. However, blockchain ignores the presence of third-party institutions, which serves as a boon for the health care system. All the transactions are validated and checked by participants in each node. Therefore, data is present same all over the nodes which can help the healthcare systems to maintain accurate health information of patients and their electronic records [19].

1.3 Cryptographic Sealing

Blockchain technology is built by the utilization of an algorithm called Cryptographic hashing. This mathematical algorithm can convert the input into fixed output irrespective of the input length. A chain is formed that links every node. This chain involves components like a root hash connected to every block and a hash pointer. Whenever there is a change in the transaction a new root hash is produced and invalidates the consecutive blocks. This hashing technology act as a secure system for blockchain technology because every node in a link chain can detect even a small change [20].

1.4 Digitalization

Digitalized blocks help to store the information so that manual entry is completely avoided or restricted. Apart from this, each participant is authorized with a digitalized signature during every transaction to maintain transparency. Asymmetric encryption is used to authorize the digital signature with a pair of keys such as private and public keys. A private key is used for the data encryption whereas a public key is given to participants to check the messages of the recipient and verify them [20]. Transactions are proceeded forward by the digitalized signature of the participants that grants all the rights. This feature of blockchain helps to improve and reinforce privacy and cybersecurity features. Figure 2 represents the key characteristic features of blockchain technology.

This chapter focus on blockchain technology features and their importance in the current medical scenario to overcome the conventional healthcare issues. All the related articles were extracted from search engines like Google Scholar and PubMed.

2 Structural Components Associated with Blockchain

The Genesis block is a term given to the first block of blockchain technology. Once started at any block and follow the chain lineage as per chronological order finally genesis block arrives. The client software is associated with the genesis block which cannot be changed as it is fixed. The algorithm is fixed in such a way that each node can identify the genesis block information and structure and also the transaction time. Due to this possibility, blockchain is the most trustworthy technology in use in the present times [21].

This technology usually consists of four elements; information or data block, hash block, hash information concerned with the previous block, and a timestamp. Every block is connected to the previous one. This blockchain is usually classified

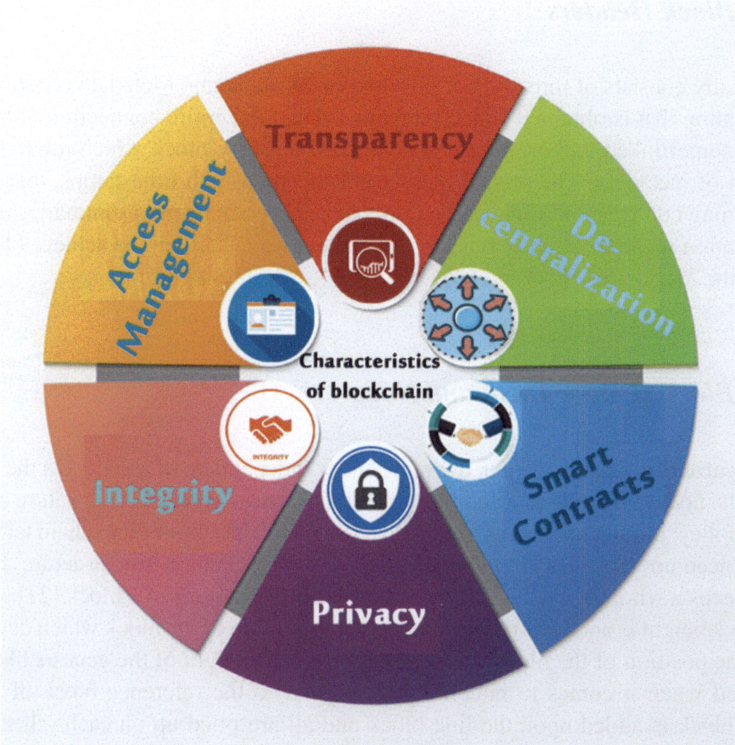

Fig. 2 Special features of Blockchain technology

as follows: public, which is not concerned with any permissions, private, which needs permissions to access such features, and a hybrid model technology blockchain. The public feature of blockchain helps to access this feature to the public domain. The best example for the public domain is Proof-of-work. This is a widely accepted algorithm in the public blockchain models [21–23]. Apart from these transactions done in the public blockchain are transparent to all hence every participant had the permission to view the message without any involvement of a third party or single entity user. In contrast to the public domain, to utilize the private feature of the blockchain network, it is mandatory to take respective permissions from the authorized person by the individuals. In the hybrid blockchain model, some amount of information is made available to use as public, whereas a certain amount of data is made private hence named as hybrid blockchain [21–23]. The key structures of blockchain includes:

1. Block headers.
2. Block header hash and node.

2.1 *Block Headers*

It typically consists of three sets of information as metadata. Metadata is defined as information that is obtained concerning other data. The entire connection of blockchain is interlinked and each hash is connected to every block which refers to the first set of metadata. The second dataset is concerned with timestamps, difficulty, and mining competition. Finally, the third dataset is associated to summarize all the information regarding transactions in an efficient way. This can be achieved by the use of the Merkle tree [21].

2.2 *Block Header Hash and Node*

It recognizes the hash of a block uniquely and is independent of a node. On the other hand, the node is defined as a complete client. Due to the characteristic nature of full client policy, it transfers the information across all the blocks of blockchain technology. It need not be considered to be a miner of the blockchain infrastructure. Every node keeps a whole replica of information about transactions in a block [21].

Block height is another criterion for the identification of the block which depends upon the position of the block in the blockchain. The height of the genesis block is zero and when it comes to bitcoin it was similar to the reference block of hash. Every block is added upon the first block and all are piled up on each other. One thing that must be kept in mind while considering block height was it does not represent a single block. Sometimes two or more blocks had a similar block height which indicates that both blocks are competing for the same position in a block [21].

3 Need and Various Blockchain Types

At present healthcare system adopts the conventional strategies in maintenance, therefore a lot of problems have to be addressed. Figure 3 represents the problems faced by the conventional healthcare system in various domains.

Generally, blockchain is highly secured and utilized to perform transactions or payments. But, different people use this technology in different ways. For example, consider Bitcoin where the blockchain had mainly been implemented in the mainstream. Bitcoin is considered a digital currency that uses the blockchain to perform the transaction [24]. It is a public network where any person can access the feature of blockchain to perform a transaction. Therefore, a person can become a node, confirm other nodes, and carry out the business payments. In another case for instance bank uses a private network model so that only an authorized person has permission to view information. So, no other person can have the chance to get the data. This type of private network consists of limited nodes maintained by a network administrator. Such a type of data shared by the private network remains within the

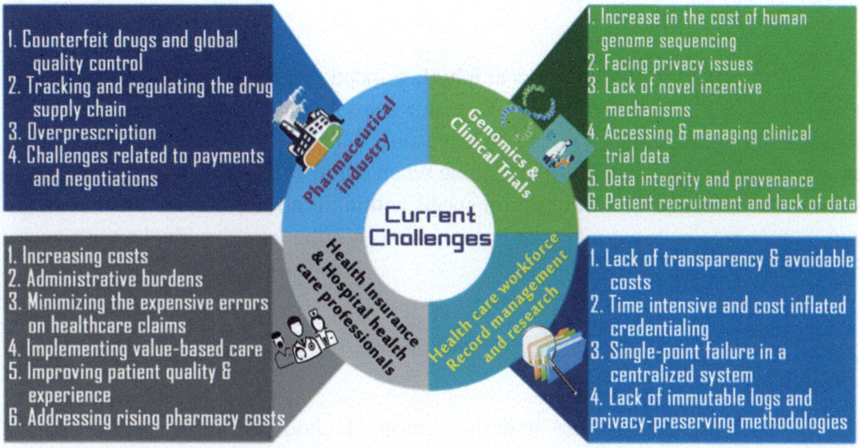

Fig. 3 Current issues need to be resolved in healthcare domains

Table 1 Types of blockchain in use

	Public	Private	Consortium	Hybrid
Type	Non-restrictive	Restrictive	Semi-decentralized	Private and public
Application	Mining and information exchange	Organizational sites like voting and supply chain management	Banks, other government organizations	Private and public institutions
Examples	Bitcoin, Litecoin, and Ethereum	Corda, multichain project	Energy web foundation	Dragon chain best suits the hybrid network
Type	Open type	Closed type	Not specific	Not specific

specific network. In case any node or participant wishes to join the network he needs permission from the authorized person. Therefore, depending upon the requirement blockchain is used in different ways [18]. This led to the development of various blockchains, namely as shown in Table 1.

1. Private.
2. Public.
3. Consortium blockchain.
4. Hybrid blockchain.

3.1 *Merits of Private Blockchain*

1. Speed.
2. Scalability.

3.1.1 Speed

This private network model works at a higher speed in contrast to the public blockchain. This is a direct indication of greater transactions per second (TPS) in the private blockchain model. The possible reason for this is the availability of a limited number of network nodes in private mode when compared to a public network. This feature also enhances the transaction speed within all nodes of a network. In addition, sometimes new payments are also added into the existing blocks with accuracy [24].

3.1.2 Scalability

The private blockchain users had the option of choosing their network which depends upon their needs. Hence it provides pretty much comfort to its users around the world. For example, consider an organization that needs 30 nodes for the transaction and further processing of information. Then in such a case one or two nodes are enough to complete the process. Later if they need some more nodes, additional nodes are added by taking permission from the authorized person. Therefore, private blockchain provides the participants with an option of scalability which can increase or decrease the network size depending upon the need of the user [24].

3.2 Demerits of Private Blockchain

1. Trust building.
2. Centralization.
3. Lesser security.

3.2.1 Trust Building

In a public blockchain network, it is considered an open type without any restrictions. This feature can increase the network security of participants. However, in private networks participants in a network are limited and access is restricted. Particularly within an institution where participants know each other, trust is necessary to proceed with transactions and maintain the information in a private mode [24].

3.2.2 Centralization

As private networks are the restricted type, they use the Identity and Access Management system (IAM) to work efficiently. This system has a complete monitoring capacity. This can allow the new nodes in the block and decide whether to permit the new user or not to do so. This also means that information access is also restricted by IAM [24].

3.2.3 Lesser Security

There are increased chances for data leakage or security breach because this model uses a less number of nodes or users. If any unknown user gains access to one node, he can extract the information present in all the nodes. Hence, these are easily prone to hacking or security breach.

3.3 Merits of Public Blockchain

1. Trustworthy.
2. Transparency.
3. Security.

3.3.1 Trustworthy

In contrast to a private network where limited nodes are present, public types have many nodes. So participants need not worry about transparency or accessibility. Because of this feature, it is not possible for any fraud or misleading of information because each node carries the same data. Hence, public blockchain is trusted more without any fear [24].

3.3.2 Transparency

A public blockchain is an open type without any restriction. At every node, a copy of the transaction is available digitally. This allows the public blockchain an open system of network. There is no chance of fraud, because at any given point in time information regarding transactions remains constant in every node [24].

3.3.3 Security

The presence of more participants in the public networks enhances the security feature. Work distribution is also more, and it becomes difficult for hackers to hack the security system. Apart from this, each node can perform the transaction verification along with work proof this makes a legitimate block. All these features make the public blockchain safer and secure when compared to a private network.

3.4 Disadvantages of Public Blockchain

1. Reduced transaction rates.
2. Higher consumption of energy.
3. Scalability problems.

3.4.1 Reduced Transaction Rates

The number of transaction rates per second is very low in the public networks. This is due to the reason that the public type has many nodes, and at each node, there can be a verification process that takes place for the work proof of transaction. Hence it is time-consuming with less transaction rates. Therefore, bitcoin technology can allow only seven transactions per second (TPS) and Ethereum allows 15 transactions per second. But, when considering an example of a visa as a private network it can process nearly 24,000 TPS, which is a huge difference [24].

3.4.2 Higher Energy Consumption

As proof of work takes place at each node it needs more time and energy consumption. For this process to take place some, specialized algorithms and hardware should be installed to work efficiently. Therefore, it is highly economic.

3.4.3 Scalability Problems

As mentioned previously, public networks have less TPS, which disrupts security and increases the chances of information leakage. In case if we try to enhance the network or block size, the rate of transactions still falls. Therefore, to overcome this problem Bitcoin Lightning Network was introduced to improve the transaction rate [24].

4 Health Care Domain Applications

Blockchain technology is used in the health sector at varied organization levels at different places as shown in Fig. 4. It solves the problems in the areas such as the following:

1. Patient related.
2. Organization related.
3. Medical data management.
4. Clinical trials.

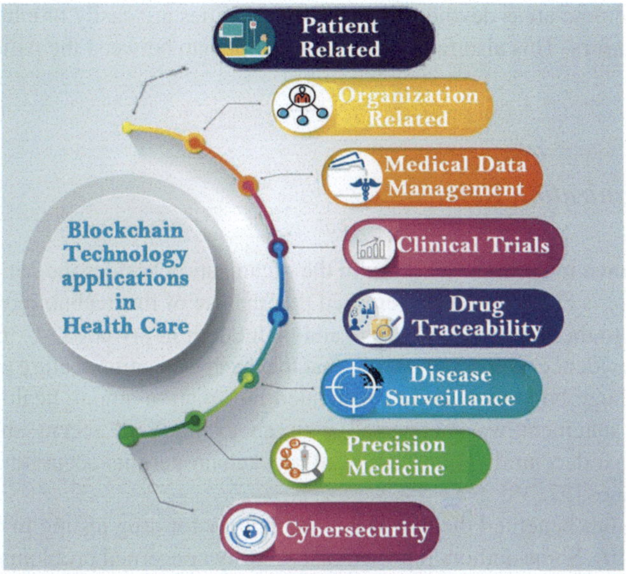

Fig. 4 Healthcare domain applications of Blockchain technology

5. Drug traceability.
6. Disease surveillance at a community level.
7. Precision medicine.
8. Cybersecurity.

4.1 Patient Related

Access and security are the prime benefits to the patients using this blockchain technology. Many studies have concluded that the implementation of blockchain can increase the security and privacy of patient information by adopting a decentralization process on a peer-to-peer basis [21, 23, 25]. There are other studies that concluded that blockchain can reduce the hacking and data breach as it avoids the single point of failure [26, 27]. Apart from these functions, it was also considered as a confirmation provider to check participant's entry for health-related queries information [28]. Other than the security concerns, there is evidence reported that blockchain can enhance personalized medicine in healthcare by gaining access to its participants to share the medical health data and their personalized plans [29–32]. Timestamps used in blockchain can help to record the information for every transaction. Now, this feature helps the physicians to detect and track the patient's data more easily [29, 33]. This can also provide a big sigh of relief for severely ill patients because doctors can monitor their condition very closely in any emergency situation. The best example for this is blockchain-dependent eHealth applications where patients in remote areas devoid of health care facilities are easily monitored by the use of biosensors. This also facilitates bridging the gap between the patient and the physician.

4.2 Organization Related

It is well known for its renowned use in the organization besides the patient-related benefit. There is evidence which reported that the use of this technology can transmit the data to the organization's users and maintain it in privacy [34–36]. There is some other evidence which had proven that the decentralization feature of this technology is a huge boon to share health-related information among health organizations. This statement was supported with an example of recent study which concluded that decentralization feature of blockchain permits accurate analysis of medical images [37, 38].

Other studies benefited the use of this technology for drug testing in clinical trials [34, 39, 40]. Some authors reported that security concerned problems in clinical trials are resolved by the implementation of blockchain technology [34, 37]. The traceability application of blockchain is found to be used by some other authors in the pharmaceutical management of organizations. This technology also prevents the use of fake drugs in the pharmaceutical market [24]. Apart from having the clinical benefits it also provides non-clinical benefits to its users in providing medical insurances. This feature of blockchain is considered a prominent one which had become a boon for medical insurance industries. It also permits the organizations to store, backup the information, and papers of claim across their computers [41, 42]. It connects the stakeholders like doctors, patients, organizations, hospitals, pharmacists, and testing centers to maintain secure, reliable, and authentic health records. The summary of blockchain uses in medical insurance policies is represented in Table 2.

Table 2 Organization benefits of blockchain in medical insurance

Feature	Use
Security	There are fewer chances for hackers to steal information which allows the increased security and privacy
Availability	This ledger allows the use of health-related information among the different organizations of people
Immutable audit trail	The immutable feature of blockchain helps users and authorities to detect frauds
Decentralization	This helps real-time claim processing by removing the third parties involved in the transaction
Data origin	The main problem involved in insurance is the verification process at different sites. This technology address all such problems through data provenance

4.3 Medical Data Management

At present many countries are industrialized and adopting advanced features in health organizations. Implementation of blockchain technology can help to maintain the Electronic Medical Records (EMR) in the health care systems. Advanced features like robustness, security, and decentralization help to manage, share, and protect the information present in electronic health record databases. This is well explained by the application of hyper ledger fabric in blockchain-based electronic health record management. This optimizes the use of EMR and improves the features such as adjustability and security [43, 44]. Possibly it also improves the transaction data in billing sections, insurance-related claims, and surveillance data for nosocomial infections. The merit of this technology is that huge amounts of information can be stored, shared, and processed among various stakeholders and patients at high speed without any need for links. A study by Peterson et al. concluded that blockchain can transform the health care system interaction with extraordinary features like authentication controls, which ultimately lowers data hacking [43, 44]. Table 3 depicts the summary of medical data management with blockchain technology.

4.4 Clinical Trials

Blockchain has a huge impact on clinical trials because it can store, share, and access the patient remotely. Technologies such as Health banks, data lakes, and other sharing networks can enhance the process of clinical trials. These tools are used in clinical and biomedical research to facilitate the smooth functioning of

Table 3 The blockchain advantages in Medical management

A key feature of blockchain	Advantage
Robustness	The decentralization feature of blockchain allows the storage of all the data in its database. So, there is less chance for hackers to steal the information It also reduces the manual health records maintenance of patients
Security	Information is encrypted in blockchain and it can be accessed only by the user with the use of a private key. This allows increased security for health records
Immutable audit trail	As data stored cannot be interchanged by users such as physicians and patients which paves the way for unaltered patient health records
Data origin	There is no source for the existence of false documents as each and every stage is verified. Therefore, medical records are source-verified
Decentralization	Here patients can access the data on their own, thus there is no need for health care records copies of patients. It ultimately provides a source for patient-managed health care records

Table 4 Key advantages of blockchain in clinical trials

Features of blockchain	Use
Robustness	It increases access to real-time health care information. This provides improved therapeutic care and coordination among the clinical staff and patients. By assessing the real-time health care data scientists can predict the pattern of epidemics emergence
Immutable audit trail	This facilitates tracking of the patients in emergencies and the research process as it is time-stamped generated
Security	This technology can connect different stakeholders, patients, organizations, and researchers to share their valuable information with the public regarding health, environmental, genetic, and lifestyle modifications with enhanced security and privacy
Data provenance	Platforms like MedRec have improved features in terms of provenance and data origin. This origin of the information is used in medical research areas
Decentralization	As it is devoid of third parties, organizations involved in clinical trials can share the information with their participants and analyze the information for research purposes

projects [16, 45]. A study by Zhuang et al. utilized features like smart contract systems and trial-based contract mechanisms to improve the patient recruitment process of clinical trials. These features in turn save time, increase scrutiny, and improve patient empowerment [46].

Phase I of clinical trials is usually carried out in a small number of healthy patients. This stage access the acute toxicity and side effects associated with experimental moiety. Nearly 70% of drugs are passed through this phase. The common problems that are resolved by the implementation of blockchain in this phase are the recruitment of patients and retention, clinical trial management, trial analysis, and transactions. Phase II of clinical trials goes with an estimation of drug efficacy. This is done in a randomized fashion to avoid bias. In this phase blockchain technology use can improve the patient selection process, reduce manual errors, and maintain data integrity effectively. Phase III is the most expensive and time-consuming with years to end up. Several thousands of patients are involved and this phase analyzes drug effectiveness. In this phase, blockchain databases can minimize the issues such as patient tracking, laboratory management, security of payment portal, and information integrity [Blockchain and clinical trials: driving efficiency through data management clinicaltrialsarena.com].

Apart from this it can track every transaction and share the patient information among various platforms in a secure manner as shown in Table 4. Further, it provides immutable and intrinsic proof for regulatory organizations regarding every activity and transaction. It also notices a change in users, information, and modification time which are essential for the effective running of clinical trials [Blockchain and clinical trials: driving efficiency through data management clinicaltrialsarena.com].

4.5 Drug Traceability

Drug traceability is an important concern in the pharmaceutical market when it comes to over-the-counter drugs and counterfeit medicines. An authorized process is present from the start of medicine manufacturing to the distribution of the drug to end users like patients. The onus lies on the healthcare stakeholders to distribute safe and effective drugs to the public to improve health care and quality of life. The contemporary drug distribution system is large and more complex [47]. It has limitations like limited data, lack of proper ownership responsibilities, and variations in transactions of stakeholders etc. All these limitations provided a need for third-party surveillance to monitor the drug supply chain. As a part of this different technological approaches such as the Internet of things, bar codes, and e-pedigree had been in use in the market to improve the scrutiny and security of pharmaceutical supply chain management. Despite these services, there are serious demerits when it comes to privacy, security, and scalability towards avoiding the fake drugs in the market [47].

In this concern, blockchain had gained enormous use in drug traceability and supply chain management. Its features like immutability a transparent nature helps to record the payments between non-trusting participants. This technology can trace and track the transaction details with its distributed ledger system and time stamp recordings thus enabling direct access to participants without any third-party involvement. This can be integrated with existing technologies like bar codes and e-pedigree to get better solutions to avoid fake medicines in the market. In the study of Uddin et al., they have described the two blockchain architectures namely Hyper ledger fabric and Hyper ledger Besu architecture [47].

A pharmaceutical supply chain mainly consists of manufacturers, distributors, market owners, and patients. During the manufacturer stage, all the items produced contains QR code with time stamp, item name, manufacture date, and expiry date, etc., stored in the blockchain. This produces a specific hash ID for future tracking of these goods. Once distributers receive the goods, they check for identity, origin, and presence of counterfeit drugs with the help of hash ID of blockchain, and proceed for signing. At the third stage market owners check the items with unique hash ID for their purity. These hash IDs cannot be accessed by any other third person. A private key is present to access these items through hash ID. This feature allows no addition of counterfeit drugs to the market. Finally, the end-user patients can purchase these products. While purchasing there is an option of scanning the QR code which not only allows purchasing of products addition and it also helps to store the feedback given by the patients regarding their safety and efficacy. In this manner blockchain technology helps to trace the drugs at each level of Pharmaceutical supply chain management. [https://www.interwork.biz/blog/blockchain-based-software-for-drug-traceability].

4.6 Drugs Surveillance at the Community Level

Surveillance is structured, continuing, and timely collection of information and dissemination of gathered information to the public. It is essential to monitor the patient infected with either communicable diseases or non-communicable diseases. For example, past pandemics like Nipah had reached and spread among the public at a very fast rate probably due to increased globalization. The present pandemic Severe Acute Respiratory Syndrome −2 (SARS CoV-2) had also reached the wider public since its emergence. Therefore, it is a continuing, difficult, and inefficient method because most of the reporting system follows centralized flow [48]. Hence, it becomes a challenging task to provide the information available throughout the process. To overcome this situation blockchain technology has been in use in tracking patients and the common public during these pandemic times. It helps the non-governing bodies to collect and analyze the information in a precise manner, particularly during these pandemic times. It acts as a tracking source for persons met with road accidents, people addicted to narcotics, and so on [49]. Therefore, implementation of this technology in health care domains helps to track the public and patients during these crisis times. It also helps to track the information about bioterrorism. Therefore, disease preventive measures are adopted properly using this technology. In recent days there has been a widespread of false information related to medicines and other communicable diseases which makes the public panic. A report by Huckle et al. used blockchain as a basic technology measure to find the origin of such fake information and the population at risk. This application highlights the importance of blockchain technology in the digital era [50].Figure 5 describes the healthcare surveillance applications of blockchain technology.

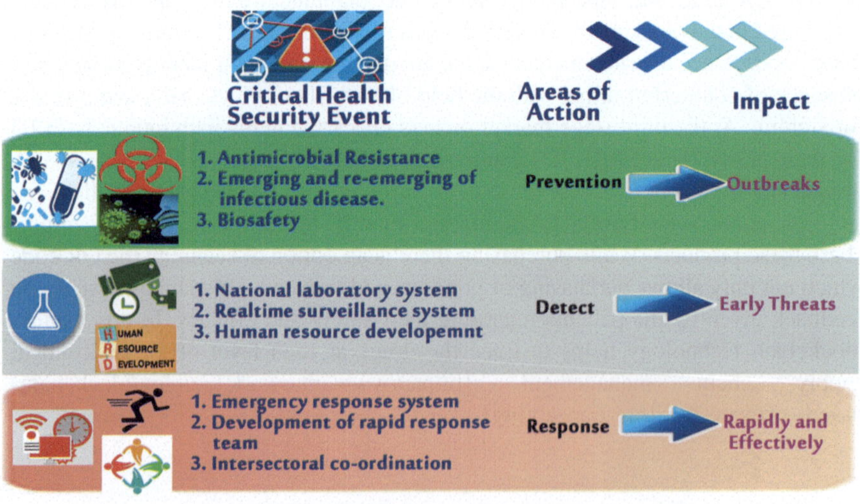

Fig. 5 Areas of blockchain technology applications in health surveillance

In present days, many countries use Artificial Intelligence techniques in surveillance. Techniques like blockchain are added advantage over others such as machine learning. They, especially prevent data replication, hacking, and curbs fake news. Furthermore, Blockchain technology integrated with Artificial Intelligence is an added advantage to healthcare and research areas [51, 52]. There are enormous opportunities for the incorporation of Blockchain with Geographical Informational System (GIS) to accelerate the pandemic investigations and drug discovery process. Its transparent nature allows recording the deaths that occurred during pandemics and prevents data duplication in health care databases.

4.7 *Precision Medicine*

Deloitee in his study concluded precision medicine integrated with blockchain technology helps to improve the clinical outcome of a patient. Information such as genomic data, previous health records are transferred between the patient, physician, and health care management team. This improves personalized medicine and patient-physician relation. Apart from this ownership of data is controlled by a patient on aspects such as who can view their information. The users can protect their health information privately by allowing entry to selected physicians or Pharmaceutical companies. This feature makes the participants to involve in the clinical trial aspects and responding to follow-up during the trials [53].

In the contemporary world, getting access to the genetic data of patients is an essential need for medical authorities to discover and develop new therapies and diagnostic aids. Support is needed from our side if we require to hasten the advancements in the medical field. Implementing all these blockchains is a potential technology that can store the genetic data of patients. It connects the healthcare staff, patients, and Pharmaceutical companies for mutual benefit. For researchers, this is a huge benefit because they can get an unlimited amount of stored personalized data that can be used for ongoing research studies. This also helps to access the degree of risk between the genetic traits and existing co-morbidities [53].

4.8 *Cybersecurity*

The features of hashing, maintaining multiple copies of data, and decentralization allow to increase the security features of blockchain and protect the data from hackers and thieves. However, protecting the healthcare data of a patient is still elucidated clearly. Authorities of healthcare changed to electronic management records to reduce the costs and have multiple copies of information [54]. Other problem in present times is there is an increased need for implantable devices such as cardiac implants and insulin pumps the fear of dangerous hackers being capable of theft and

Table 5 Key problems solved by blockchain in the health sector

Issues	Conventional healthcare system	Blockchain integrated health domain
Data segregation	Decentralized information is produced in a fragmented manner. Information received from patients, doctors, and other heal staff is produced separately	Computer sensors and decentralization features create a huge network among different stakeholders which helps to receive the information as a whole
Sensor-based information handling.	A huge number of internet of things (IoT) devices are available to gather information, which becomes a challenging problem in tracing and tracking	Blockchain has a feature to protect the IoT and protect the information. This is made possible by forming a private blockchain network among themselves
Users privacy	Presently, the information of all the participants is open which is easily traced by a third party	It provides a secure way of information handling where information processing is made open to its users. It provides an option for a smart contract
Processing costs	Presently, due to the involvement of third parties, there is an increased cost for processing and analysis of information	Blockchain architecture is capable of reducing the costs associated with information processing
Stakeholder communications	Currently, by the involvement of third parties change in data is possible which results in false transactions to make profits	Blockchain technology does not provide a chance for third-party organizations. So, there is no chance for the fake transactions

analyze the wireless signals and control their function. Hence, risk management is minimized by installing enhanced features in implantables such as

(a) To have an inspecting process of complete details of implantable and its activity to aid in the diagnosis or the detection of dangerous activities of hackers.
(b) Quick report of software malfunctions to manufacturers to avoid the exploitation of security concerns and their drawbacks.
(c) Providing multi-factor access keys to implantable devices.
(d) Providing patient education about cybersecurity risks and attacks [54].

5 Issues Resolved by Blockchain Technology in Healthcare

See Table 5.

6 Threats Associated with Blockchain Technology

According to Abu-elezz et al. threats concerned with blockchain are broadly classified into three categories as technical-, social-, and organization-related dangers as presented in Fig. 6. Under technical category, issues like scalability, authorization

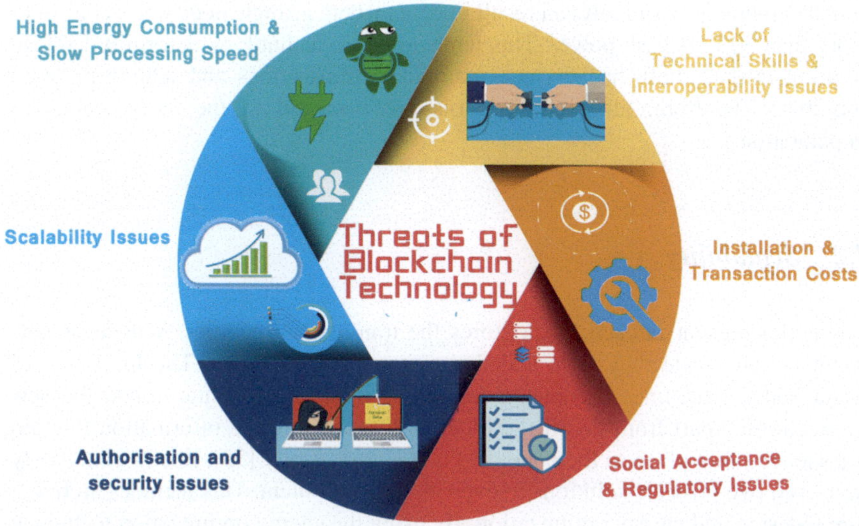

Fig. 6 Threats associated with Blockchain technology

issues, and high energy consumption are included. Under social threats, regulatory issues are included. Finally, in the organization-related category, issues like costs associated with payments, lack of technical quality, and operational issues are involved [55].

7 Future Perspectives and Challenges

Challenges that need to be overcome for the implementation of blockchain in healthcare are

1. Security.
2. Scalability.
3. Legality.
4. Storage problems.

7.1 Security

Security concerns are a huge challenge, which pull back the implantation of this technology in specific applied domains. Mendling et al. in their study reported issues concerned with security are privacy, integrity, and network availability. Information present in the nodes is same and every user can possess the same set of transactions this creates no space for confidentiality [56]. A study from Efanov et al.

concluded that 51% of theft can result in cases where a single node takes control of more than 50% of hash power. This tends to entire technology deformity [57]. In addition to data theft, hackers can manipulate transactions and other validation activities. Therefore, all such concerns are managed at the early stages of implantation.

7.2 Scalability

The nodes present in each block stores the transaction activities. Whenever new payments are made they have added automatically to the list. The limitation of fewer nodes sometimes becomes a demerit as it takes more time to add the new transactions. Apart from this if the block is filled with more information then no transaction takes place as the node rejects it automatically [58]. Zheng et al. study described two possible solutions to overcome this problem. This includes architecture redesign and energy optimization. By using the energy optimization technique new schemes are generated which erase all the old transactions and stores the new information obtained from the users. By redesign technology, traditional blockchain is separated into the micro block and key block that holds different sets of information [19]. To fix these scalability issues lightning networks, Plasma cash, and Forks are used.

7.3 Legality

There is a limitation to legal conditions for blockchain since this is a newly installed technology. A study by Yoeh et al. concluded the implementation of blockchain is restricted by the non-appearance of governing instructions that follow rules and restrictions to improve trust. Different Bitcoin scams concerned with money had impacted the real nature of blockchain [59].

7.4 Storage Problems

Health systems and medical records generate huge amounts of electronic data from patients and wearable devices. In contrast, blockchain structure holds very limited data depending on the number of nodes in architecture. The hash code along with decentralization architecture is too expensive. If the data storage is high, it finally leads to increase in block size which is directly proportional to cost. In other ways, management and operations done with blockchain are also expensive if the data size is bigger. Therefore, all these factors are kept in mind while designing it [59].

8 Conclusion

The applications of blockchain technology in the healthcare domain are still in the budding stage and hence a lot of transitions need to take place. Blockchain is being used in many organizations, hence to strength its effectiveness its problems and issues are yet to be addressed. Data storage of patients is considered to be a critical aspect concerned with blockchain because of security reasons. Resolving all these challenges improves the security concerns and durability of blockchain technology.

In this chapter, we have addressed an overview of blockchain technology, its applications along with issues resolved in the healthcare system and pharmaceutical supply chain management. Apart from this information, we also discussed the potential threats associated with blockchain and their future prospective. However, there are certain limitations in our work such as the following: Based on the search strategy we included only the data collected from search engines such as PubMed and Google scholar and other databases were not screened. No statistical analysis and tools were used to analyze the data of study results. However, by searching only peer-reviewed articles and excluding gray literature in our search, we analyzed the articles with healthcare areas of interest integrated with blockchain technology.

It would be interesting if further research is carried out to evaluate the utility and address the challenges of scalability, latency, interoperability, security, and privacy concerning the use of blockchain technology in healthcare. Researchers should be encouraged to supplement the ongoing efforts with new solutions for existing problems related to blockchain technology. A thorough analysis of our data related to blockchain technology makes it easy for the target audience and readers to understand the basic concepts, healthcare domain applications, threats, and future challenges.

References

1. G. Chen, B. Xu, M. Lu, N.-S. Chen, Exploring blockchain technology and its potential applications for education. Smart Learn. Environ. **5**, 1 (2018). https://doi.org/10.1186/s40561-017-0050-x
2. E. Meinert, A. Alturkistani, K.A. Foley, et al., Blockchain implementation in health care: Protocol for a systematic review. JMIR Res. Protoc. **8**, e10994 (2019). https://doi.org/10.2196/10994
3. M. Campbell-Verduyn, *Bitcoin and Beyond: Cryptocurrencies, Blockchains, and Global Governance* (Routledge, London, 2017)
4. G. Rathee, A. Sharma, R. Kumar, R. Iqbal, A secure communicating things network framework for industrial IoT using Blockchain technology. Ad Hoc Netw. **94**, 101933 (2019). https://doi.org/10.1016/j.adhoc.2019.101933
5. M. Crosby, Nachiappan, P. Pattanayak, et al., *Blockchain Technology—Beyond Bitcoin*. Berkley Engineering (2016)
6. I.C. Lin, T.C. Liao, A survey of blockchain security issues and challenges. Int. J. Netw. Secur. **19**, 653–659 (2017). https://doi.org/10.6633/IJNS.201709.19(5).01

7. A.A. Vazirani, O. O'Donoghue, D. Brindley, E. Meinert, Implementing blockchains for efficient health care: Systematic review. J. Med. Internet Res. **21**, e12439 (2019)
8. K.J. McKernan, The chloroplast genome hidden in plain sight, open access publishing and anti-fragile distributed data sources. Mitochondrial DNA A DNA Mapp. Seq. Anal. **27**, 4518–4519 (2016). https://doi.org/10.3109/19401736.2015.1101541
9. A.A. Shrier, A. Chang, N. Diakun-Thibault, et al., Blockchain and health IT: Algorithms, privacy, and data, in *ONC/NIST Use of Blockchain for Healthcare and Research Workshop*, (2016)
10. J. Goldwater, *The Use of a Blockchain to Foster the Development of Patient-Reported Outcome Measures* (National Quality Forum, Washington, DC, 2016)
11. IBM, *Blockchain: The Chain of Trust and Its Potential to Transform Healthcare—Our Point of View* (IBM, Armonk, NY, 2016)
12. T. Nugent, D. Upton, M. Cimpoesu, Improving data transparency in clinical trials using blockchain smart contracts. F1000Research **5**, 2541 (2016). https://doi.org/10.12688/f1000research.9756.1
13. E.J. Topol, Money back guarantees for non-reproducible results? Br. Med. J. **353**, i2770 (2016)
14. P. Taylor, *Hyperledger—Applying Blockchain Technology to Medicine Traceability*. Webpage (2016)
15. J. Jenkins, J. Kopf, B.Q. Tran, et al., Bio-mining for biomarkers with a multi-resolution block chain, in *Independent Component Analyses, Compressive Sampling, Large Data Analyses (LDA), Neural Networks, Biosystems, and Nanoengineering XIII*, (2015)
16. M. Mettler, Blockchain technology in healthcare: The revolution starts here, in *2016 IEEE 18th International Conference on e-Health eNetworking, Applications and Services*, (Healthcom, 2016)
17. L. Bell, W.J. Buchanan, J. Cameron, O. Lo, Applications of Blockchain within healthcare. Blockchain Healthc. Today **1**, 1–7 (2018). https://doi.org/10.30953/bhty.v1.8
18. A. Azaria, A. Ekblaw, T. Vieira, A. Lippman, MedRec: Using blockchain for medical data access and permission management, in *Proceedings—2016 2nd International Conference on Open and Big Data, OBD*, (2016)
19. Z. Zheng, S. Xie, H. Dai, et al., An overview of Blockchain technology: Architecture, consensus, and future trends, in *Proceedings—2017 IEEE 6th International Congress on Big Data, BigData Congress*, (2017)
20. T.T.A. Dinh, R. Liu, M. Zhang, et al., Untangling Blockchain: A data processing view of Blockchain systems. IEEE Trans. Knowl. Data Eng. **30**, 1366–1385 (2018). https://doi.org/10.1109/TKDE.2017.2781227
21. B. Shah, N. Shah, S. Shakhla, V. Sawant, Remodeling the healthcare industry by employing Blockchain technology, in *2018 International Conference on Circuits and Systems in Digital Enterprise Technology, ICCSDET*, (2018)
22. A.A. Monrat, O. Schelén, K. Andersson, A survey of blockchain from the perspectives of applications, challenges, and opportunities. IEEE Access **7**, 117134–117151 (2019)
23. A. Shahnaz, U. Qamar, A. Khalid, Using Blockchain for electronic health records. IEEE Access **7**, 147782–147795 (2019). https://doi.org/10.1109/ACCESS.2019.2946373
24. S. M. Idrees, P. Agarwal, M. A. Alam (eds.), *Blockchain for Healthcare Systems: Challenges, Privacy, and Securing of Data* (CRC, Boca Raton, FL, 2021)
25. T. McGhin, K.K.R. Choo, C.Z. Liu, D. He, Blockchain in healthcare applications: Research challenges and opportunities. J. Netw. Comput. Appl. **135**, 62–75 (2019)
26. H. Chen, X. Huang, Will Blockchain technology transform healthcare and biomedical sciences? EC Pharmacol. Toxicol. **6**, 910–911 (2018)
27. P. Esmaeilzadeh, T. Mirzaei, The potential of blockchain technology for health information exchange: Experimental study from patients' perspectives. J. Med. Internet Res. **21**, e14184 (2019)
28. J.H. Beinke, C. Fitte, F. Teuteberg, Towards a stakeholder-oriented blockchain-based architecture for electronic health records: Design science research study. J. Med. Internet Res. **21**, e14184 (2019). https://doi.org/10.2196/13585

29. K. Shuaib, H. Saleous, K. Shuaib, N. Zaki, Blockchains for secure digitized medicine. J. Pers. Med. **9**, 35 (2019)
30. G. Leeming, J. Cunningham, J. Ainsworth, A ledger of me: personalizing healthcare using blockchain technology. Front. Med. **6**, 1–10 (2019)
31. T. Zhou, X. Li, H. Zhao, Med-PPPHIS: Blockchain-based personal healthcare information system for National Physique Monitoring and scientific exercise guiding. J. Med. Syst. **43**, 305 (2019). https://doi.org/10.1007/s10916-019-1430-2
32. M. Greenberger, Block what? The unrealized potential of blockchain in healthcare. Nurs. Manage. **50**, 9–12 (2019). https://doi.org/10.1097/01.NUMA.0000557782.47673.80
33. P. Zhang, D.C. Schmidt, J. White, G. Lenz, Blockchain technology use cases in healthcare, in *Advances in Computers*, (Elsevier, Amsterdam, 2018)
34. R.H. Hylock, X. Zeng, A blockchain framework for patient-centered health records and exchange (healthChain): Evaluation and proof-of-concept study. J. Med. Internet Res. **21**, e13592 (2019). https://doi.org/10.2196/13592
35. S.M. Idrees, M. Nowostawski, R. Jameel, Blockchain-based digital contact tracing apps for COVID-19 pandemic management: Issues, challenges, solutions, and future directions. JMIR Med. Inform. **9**(2), e25245 (2021)
36. P. Pandey, R. Litoriya, Implementing healthcare services on a large scale: Challenges and remedies based on blockchain technology. Health Policy Technol. **9**, 69–78 (2020). https://doi.org/10.1016/j.hlpt.2020.01.004
37. T. Kumar, V. Ramani, I. Ahmad, et al., Blockchain utilization in healthcare: Key requirements and challenges, in *2018 IEEE 20th International Conference on e-Health Networking, Applications and Services*, (Healthcom, 2018)
38. V. Patel, A framework for secure and decentralized sharing of medical imaging data via blockchain consensus. Health Informatics J. **25**, 1398–1411 (2019). https://doi.org/10.1177/1460458218769699
39. T.K. Mackey, T.T. Kuo, B. Gummadi, et al., "Fit-for-purpose?"—Challenges and opportunities for applications of blockchain technology in the future of healthcare. BMC Med. **17**, 68 (2019). https://doi.org/10.1186/s12916-019-1296-7
40. M.N. Kamel Boulos, J.T. Wilson, K.A. Clauson, Geospatial blockchain: Promises, challenges, and scenarios in health and healthcare. Int. J. Health Geogr. **17**, 25 (2018)
41. E. Gökalp, M.O. Gökalp, S. Çoban, P.E. Eren, Analysing opportunities and challenges of integrated blockchain technologies in healthcare, in *Lecture Notes in Business Information Processing*, (2018)
42. S.G. Alonso, J. Arambarri, M. López-Coronado, D.I. de la Torre, Proposing new blockchain challenges in eHealth. J. Med. Syst. **43**, 64 (2019)
43. J.M. Roman-Belmonte, H. De la Corte-Rodriguez, E.C. Rodriguez-Merchan, How blockchain technology can change medicine. Postgrad. Med. **130**, 420–427 (2018)
44. K. Peterson, R. Deeduvanu, P. Kanjamala, K. Boles, A blockchain-based approach to health information exchange networks, in *NIST Workshop Blockchain Healthcare*, (2016)
45. L.A. Linn, M.B. Koo, Blockchain for health data and its potential use in health IT and health care related research, in *ONC/NIST Use of Blockchain for Healthcare and Research Workshop*, (2016)
46. Y. Zhuang, L.R. Sheets, Z. Shae, et al., Applying blockchain technology to enhance clinical trial recruitment, in *AMIA Annual Symposium Proceedings*, (American Medical Informatics Association, Washington, DC, 2019)
47. M. Uddin, K. Salah, R. Jayaraman, et al., Blockchain for drug traceability: Architectures and open challenges. Health Informatics J. **27**, 14604582211011228 (2018). https://doi.org/10.1177/14604582211011228
48. S.U. Khan, N. Islam, Z. Jan, et al., An e-health care services framework for the detection and classification of breast cancer in breast cytology images as an IoMT application. Futur. Gener. Comput. Syst. **98**, 286–296 (2019). https://doi.org/10.1016/j.future.2019.01.033

49. V.K. Chattu, A. Nanda, S.K. Chattu, et al., The emerging role of blockchain technology applications in routine disease surveillance systems to strengthen global health security. Big Data Cogn. Comput. **3**, 25 (2019)
50. S. Huckle, M. White, Fake news: A technological approach to proving the origins of content, using blockchains. Big Data **5**, 356–371 (2017). https://doi.org/10.1089/big.2017.0071
51. M. Rahouti, K. Xiong, N. Ghani, Bitcoin concepts, threats, and machine-learning security solutions. IEEE Access **6**, 67189–67205 (2018). https://doi.org/10.1109/ACCESS.2018.2874539
52. P. Mamoshina, L. Ojomoko, Y. Yanovich, et al., Converging blockchain and next-generation artificial intelligence technologies to decentralize and accelerate biomedical research and healthcare. Oncotarget **9**, 5665–5690 (2018). https://doi.org/10.18632/oncotarget.22345
53. A.W. Peters, B.M. Till, J.G. Meara, S. Afshar, Blockchain technology in health care: A primer for surgeons. Bull. Am. Coll. Surg. **12**, 1–5 (2017)
54. L. Pycroft, T.Z. Aziz, Security of implantable medical devices with wireless connections: The dangers of cyber-attacks. Expert Rev. Med. Devices **15**, 403–406 (2018)
55. I. Abu-elezz, A. Hassan, A. Nazeemudeen, et al., The benefits and threats of blockchain technology in healthcare: A scoping review. Int. J. Med. Inform. **142**, 104246 (2020)
56. J. Mendling, I. Weber, W. Van Der Aalst, et al., Blockchains for business process management—Challenges and opportunities. ACM Trans. Manag. Inf. Syst. **9**, 1–16 (2018). https://doi.org/10.1145/3183367
57. D. Efanov, P. Roschin, The all-pervasiveness of the blockchain technology. Proc. Comput. Sci. **123**, 116–121 (2018)
58. A. Prashanth Joshi, M. Han, Y. Wang, A survey on security and privacy issues of blockchain technology. Math. Found. Comput. **1**, 121–147 (2018). https://doi.org/10.3934/mfc.2018007
59. P. Yeoh, Regulatory issues in blockchain technology. J. Financ. Regul. Compliance **25**, 196–208 (2017). https://doi.org/10.1108/JFRC-08-2016-0068

Digital Transformation of Healthcare Sector by Blockchain Technology

Dudala Sai Sushma, Varun Jaiswal, and Tarun Pal

1 Introduction to Blockchain Technology

Blockchain is an immutable shared ledger that enables tracking assets and recording of transactions that can be tangible or intangible in any business network. In general, virtually whatever thing of value can be traded and tracked on a blockchain network, risk reduction and cost cutting for all is included. It provides greater trust, security at the same time and provides various efficiencies for doing smart work by reducing time wastage in record reconciliations. The stored data can be automatically executed. The blockchain stores information in the form of blocks and these are linked in chronological order and when new information is added then new blocks can be inserted and the information is secured. This is a digital ledger system sorted across a variety of network systems that can be programmed to record and track information needed. This includes information of any value such as from medical to food supply. It works with transactions with trust, accountability, and transparency [1]. Bitcoin first implemented this blockchain technology that works by

D. S. Sushma
Department of Biotechnology, Vignan's Foundation for Science, Technology and Research, Vadlamudi, Guntur, Andhra Pradesh, India

V. Jaiswal
School of Electrical and Computer Science Engineering, Shoolini University, Solan, Himachal Pradesh, India

Department of Food and Nutrition, College of BioNano Technology, Gachon University, Gyeonggi-do, South Korea

T. Pal (✉)
Department of Biotechnology, Vignan's Foundation for Science, Technology and Research, Vadlamudi, Guntur, Andhra Pradesh, India

Zuckerberg Institute for Water Research, Blaustein Institutes for Desert Research, Ben-Gurion University of the Negev, Midreshet Ben Gurion, Israel

S. M. Idrees, M. Nowostawski (eds.), *Transformations Through Blockchain Technology*, https://doi.org/10.1007/978-3-030-93344-9_7

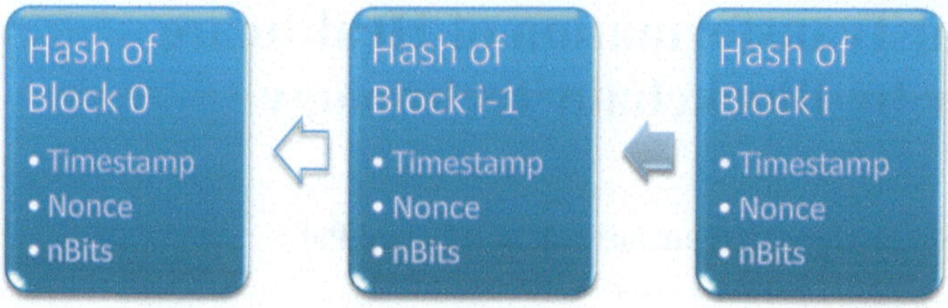

Fig. 1 Example of blockchain hashing

using the secure hash algorithm by SHA-256 hashing. The general flow of blockchain hatching is presented in Fig. 1 representing a sequence of blocks. It uses peer to peer payment processing technique. In this there is no involvement of third party and the payments may be transferred to anyone. Here the transactions of payments once generated cannot be revoked [2]. The chief elements of blockchain include smart contracts, distributed ledger technology, and immutable records. It utilizes functionalities such as data protection, point-to-point transmission, data rights, and transaction processing [3]. The frameworks which are the software solutions used by the blockchain include the ethereum, bitcoin, corda, IOTA, and hyperledger fabric [4, 5]; these are implemented based on the requirements. That is in the private, public, consortium blockchain [6]. The components that work together in blockchain technology are the ledger, cryptography, peer-to-peer network, assets, merkle trees, and consensus algorithms. There are various consensus algorithms used in blockchain and each has its specific role. These consensus algorithms are the proof-of-work (PoW), proof-of-stake (PoS), Proof of authority (PoA), proof-of-elapsed time (PoET), proof-of-capacity (PoC), proof-of-burn (PoB), delegated proof-of-stake (DPoS), and IOTA. Merkle trees are used for quick and efficient validation of data [7].

The blockchain benefits include enhanced security, greater transparency, instant traceability, increased efficiency speed and automation.

2 Key Elements of Blockchain and Role in Healthcare Sector

2.1 Distributed Ledger Technology

It is a database where the data entries are maintained and validated by distributed consensus protocol and executes in a peer-to-peer network by nodes. Here all peers contribute to maintain database integrity, thereby replacing a central administrator and the distributed ledger is accessed by all participants. With this shared ledger, there is only onetime recording of transactions excluding the repetition of effort that is typical of traditional business networks. This technology is divided into two layers that are present above the networking layer. These are fabric layer (distributed database structure, public key infrastructure, communication, and consensus) and above it is the application layer (governance, user controlled services, autonomous services) [8].

In healthcare sector, this distributed ledger can be used to store the data that includes sharing, recording, exchanging, analyzing, and validating purposes between stakeholders [9]. Brogan et al. [8] demonstrated based on the case study, the use of distributed ledger technology in authenticating health activity data for advanced monitoring of health activity. It incorporated the IOTA protocol using masked authenticated messaging (MAM) extension module for securely sharing, storing, and retrieving the encrypted activity data with a distributed ledger (tamper proof). Among the two fundamental layers that are the fabric layer and decentralized application layer, the IOTA protocol make available the fabric layer for an immutable audit trail of data activity broadcasted from wearable devices. The extension module (MAM) is retained in the application layer of the IOTA stack. The use of MAM facilitates three ways of privacy that control access and visibility to private, public, and constrained channels. This overall protocol enables effective conduit for authentic activity data and provides design and improved performances enabling patients to outline granular access controls that may be timely updated.

Liu et al. [10] anticipated Blockchain and Distributed Ledger-based Improved Biomedical Security system (BDL-IBS) focused on security for both user and sharing levels along with improved privacy of Electronic Health Records (EHRs). The EHR comprises of wide health records than Personal Health Record (PHR) as they are produced, maintained, and bring about by the healthcare suppliers whereas PHR is managed by the proprietor of the data, the security relays through linear decision-making. It allows to share trusted data and sharing is based on classification instances.

Hawig et al. [11] developed two distinct deviations of proof-of-concept system for blood glucose data. It includes interoperable, immutable, and general data protection regulation–compliant exchange of this data from patients and providers. The two variations used in the study are the public IOTA distributed ledger and the other one used is the same public IOTA ledger in union with interplanetary file system cluster, privately. These are accessed according to risks associated with data reversal and data linkability, time for processing, size compatibility, and entire complexity.

2.2 *Immutable Records*

The ability of a blockchain ledger to remain indelible and unaltered is referred to as immutability and is achieved by using blockchain technology. The records once entered in it cannot be changed by any participant. In case of any mistake, a new transaction needs to be included to reverse the mistake, and both transactions are then noticeable. In order to achieve this immutability in blockchain, the cryptographic hashing needs to be understood. This assures maintenance of records by authorized trustworthy user using blockchain technology. The records are certain due to public-key cryptography because this allows each transaction to be digitally signed prior to bringing verified and noted onto the ledger. It enables data security because immutable history of transactions is fundamentally recorded by the distributed ledger system. Along with this the encryption of data and authorization with a private identification key data ensure the data security All this can be applied to healthcare sector for meeting the requirements by the healthcare sector that includes sharing of EHR, patient-centered health data management, cost-effectiveness, high performance, data security to personal health records, privacy, availability, and transparency. This also enables effective use of technologies like telemedicine, telehealth for remote accessing of patient's data, supply chain, traceability, automated payments, and trustworthy monitoring during pandemic situations [12, 13]. Med Rec is one among many which utilize blockchain to regulate access to patient's data that provides patients an immutable records and easy access to medical information at the same time managing crucial considerations such as confidentiality, authentication, accountability, and data sharing of sensitive information [14]. A crypto-spatial coordinate system is applied in geospatial blockchain to build an immutable spatial context. Immutability confers security and is a byproduct of decentralization and cryptographic security. Healthcare tracking devices that are based on blockchain technology also can utilize immutability to prevent device theft, loss, and various means of malicious tampering [15].

2.3 Smart Contracts in Healthcare

The smart contracts implement the terms of a contract and are computerized transaction protocol that was introduced back in 1994 by Nick Szabo. The platforms in smart contract are Ethereum, Stellar, Fabric, Corda, EOS, and Rootstock. The application of Corda, Stellar, and Rootstock is a digital currency and the others are general. In healthcare sector, the use of smart contacts had gained importance in enabling secure monitoring of patient's data. Griggs et al. [16] proposed blockchain-based smart contracts system to handle the Protected Health Information (PHI) that is obtained through Internet of Things (IoT) devices and various remote patient monitoring systems. Solidity is a programming language introduced by Ethereum network that first provides appliance to write smart contracts. This is based on this Ethereum protocol based on a separate private chain. This total system enables support to healthcare interventions and real-time patient monitoring at the same time maintaining secure record. The health insurance portability and accountability (HIPAA) act of 1996s privacy rule includes electronic data transmission. Here the data cannot be covered under HIPAA if it fails to identify that it belongs to a specific patient.

The phases and challenges in smart contracts (Fig. 2) are provided by Zheng et al. [17]. Among various applications provided by the smart contracts, the data provenance benefits the capturing of malicious data falsification, improving data reliability and preserving privacy will be helpful in cases that include scientific

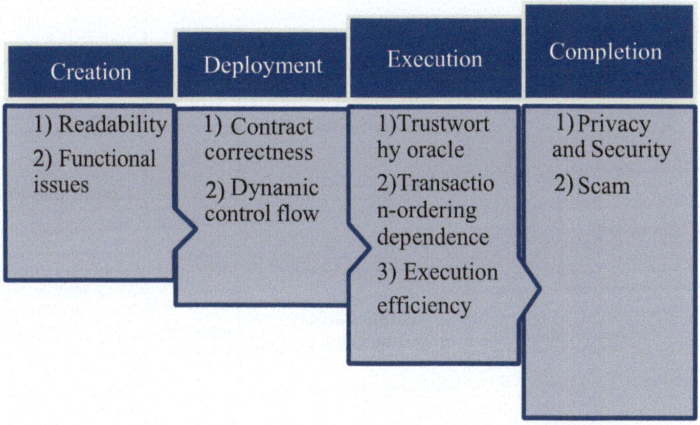

Fig. 2 Phases and challenges in smart contracts

research, public health and cloud data provenance. Hang et al. [18] proposed a project stating it is required to enhance data-accessing at the same time providing privacy and security. The proposal utilizes smart contract to secure the collection of patient's health information digitally in Electronic Medical Record (EMR) management. It enables to host the ledger functions across the network and maintains the data consistency of the individual health data, and also offers controlled access to the ledger. It receives all data transactions and runs various queries and updates the data. In this platform, in order to keep medical data private, they can create own subnetworks where only few can access data. This will be useful in case they don't want to reveal discounts offered to few numbers but not to everyone. The proof of concept has been implemented with a case study on a permissioned network for hospital, along with series of experimentation to determine usability and efficiency of the designed proposal.

These abovementioned key elements in blockchain are broadly used in healthcare to eliminate the need for trust in data transfer, securing patient identities, pharmaceutical and healthcare supply chains managing, fraud detection, surveillance, clinical research and data monetization, authorized control, and nationwide interoperability.

3 The Blockchain Network

The blockchain network can be built in several ways such as the public, private, and consortium blockchains (Fig. 3) [19]. These ways are discussed as follows:

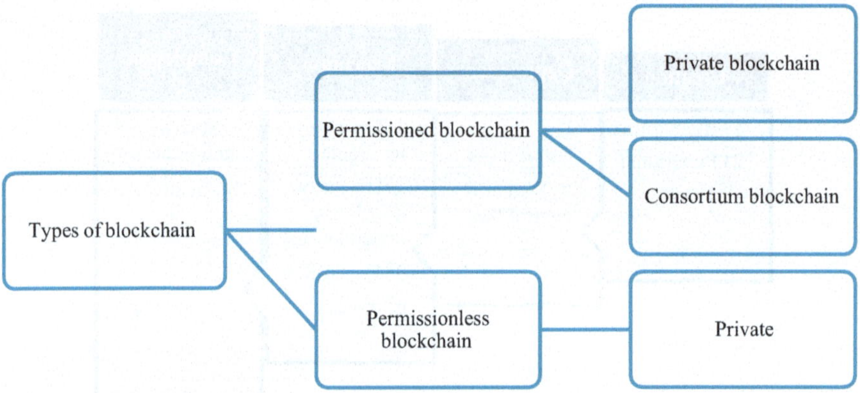

Fig. 3 Types of blockchain networks

3.1 Public Blockchain

In public blockchain, only selected nodes are responsible for verification of block and every node can take part in the consensus process. The read permission is given to public and the data is nearly impossible to tamper. It is not centralized and the efficiency is low in this. One such example that uses public blockchain is bitcoin and it has been proven to be a very safe platform for exchanging cryptocurrency. Transparency is provided by the platform and anyone can access the blockchain to view transactions and balances for any bitcoin address even though some of the identities behind few bitcoin transactions remain unknown.

Due to lack of privacy for data and security concerns, bitcoin public blockchain is unsuitable in healthcare sector as it requires data privacy, auditable and controlled access. There are also limitations in block size and number of transactions per second. Due to this it is unsuitable for large-scale and widely used system such as healthcare.

3.2 Consortium Blockchain

The consensus process is to set of selected nodes and read permission granted could be restricted or public. There is chance that it could be tampered. It is partially centralized, and its efficiency is high. The problem with respect to security, scalability, and privacy concerns in public chain can be addressed by using consortium blockchain in healthcare sector [20]. Du et al. [6] proposed a platform using consortium blockchain for medical information sharing and recording. The algorithm proposed will improve the security and efficiency. Consortium blockchains are implemented in healthcare sectors in Block Insure and MedChain [21].

3.3 *Private/Permissioned Blockchain*

The consensus process is to single organization, permission to access could be restricted or public. There is chance that it could be tampered. It is fully centralized, and its efficiency is high. As privacy and security of data are the primary requisites in maintaining sensitive information related to healthcare, the use of private/permissioned blockchain would be beneficial [20]. Xiao et al. [22] proposed EMR share which is for medical data distribution in a highly cooperative healthcare and it is implemented through permissioned blockchain framework. There exists another one, the Med-PPPHIS model for transforming the health data into on-chain tokens. It is efficient and a safe channel for data privacy and data circulation. It utilizes both permission and permission-less blockchain. The application of this system is to guide and monitor health as it is a personal healthcare information system [23].

4 Life Cycle of Blockchain

The structure of any blockchain includes three core parts that are the blocks (recording transactions into a ledger), chains (linking one block to another called the hash), and network (full of nodes).

After the evolution of blockchain in bitcoin and based upon its potency, the blockchain concept has been widely established that includes the second evolution Ethereum network and the third evolution Factom network. The primary use of the ethereum is to create smart contracts, decentralized autonomous organizations (DAOs) and trade Ether. It has various uses in securing blockchain applications. Here the DAOs are a type of virtual entity within Ethereum used to invite other participants. The factom uses lighter consensus system and helps to build secure data and system. It is a public blockchain and builds immutable asset records and follows peer-to-peer transfer where secure transactions take place between untrusted peers with public validation. The bitcoin, ethereum including many others use proof-of-work consensus that works well in securing blockchain [24, 25]. The general life cycle of blockchain technology in health sector in presented in Fig. 4.

The user requests for a transaction via peer-to-peer network. This was communicated to a network where validation and verification process takes place. After verification, this transaction is confirmed and added to the blocks (lists of transactions) and chained to other transaction in a blockchain ledger and the transaction process is complete [26].

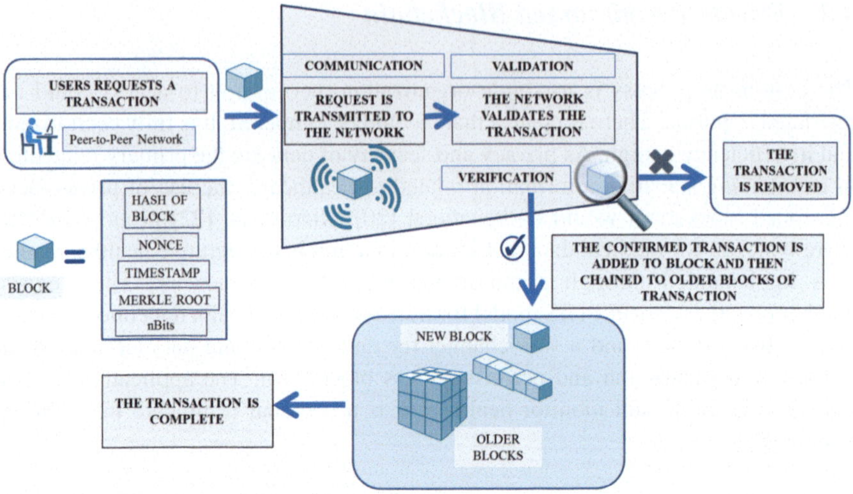

Fig. 4 Life cycle of blockchain technology

5 Evaluation Criteria and Results for Blockchain

The pharmaceutical companies lack data flow of patient's information time to time and also face challenges such as the inconsistency in pharma data, time-consuming data analytics and hefty resource, data inaccuracies, and ambiguity around relevance. These shortcomings can be overcome with the help of blockchain technology where it is utilized to deliver efficient and accurate results; manage large data sets; record, store, and sort data by providing a relevant and unambiguous framework. Along with this it helps to build unalterable, auditable, and distributed databases for accessing and storing drug trial data. The blockchain helps the pharma companies to enhance data provenance, integrity, functionality, and security of a pharma supply chain. It also includes optimizing the efficiency of the Internet of Healthy Things (IoHT), protecting from theft for digitally connected health equipment, and combating counterfeit drugs [27].

With digitalization rapidly advancing, to solve for long-standing problems in different areas and domains, every organization is looking for blockchain platform. Sufficient effort and time are spent on understanding the needs and requirements of the application concerning the blockchain platform features, so that a proper fit is made. Following are the different criteria that are considered to evaluate the blockchain models' impact on healthcare services.

- Patient identity.
- Data security.
- Data monitoring.
- Immutability.
- Consensus.
- Value.

Based upon the fuzzy-ANPTOPSIS method, private blockchain model is considered as the best alternative means for offering robust and effective service in healthcare blockchain technology. The data is secured over a distributed peer-to-peer infrastructure in private blockchain, therefore renovating the way in which the EHRs of patients are maintained and exchanged [28].

6 Blockchain Technology Associated Challenges in Healthcare Sector

Healthcare sector is a wide industry in terms of revenue and employment and there is an abundant scope of digitalization in healthcare sector due to the increase in a number of health organizations. The main challenges faced by the healthcare sector are the data storage and privacy. The protected health information (PHI) transmission and electronic health records (EHRs) data must be easily transferable and manageable at the same time managing patient privacy. This data is solely the personal data in the healthcare sector. According to General Data Protection Regulation (GDPR) EU2016/679, personal data means information related to an identifiable or identified living individual that needs to be well protected using adequate technical solutions from unauthorized accesses [29]. They consist of various valuable sensitive data relating to examination of patients in healthcare sector that is to be shared with doctors, companies which need to be protected from unauthorized access. They contain sensitive information and are not properly protected, and so various proposals are made with blockchain to ensure its security. The healthcare sector includes patients, doctors, pharmaceutical, insurance companies, stakeholders, etc.; the digitalization of healthcare sector with blockchain either public or private enables the users obtain the needed information from any place. The blockchain technology need to be implemented in such a way that it benefits users and at the same time facing challenges and overcoming them to enable safe and secure digitalization. Blockchain enables facing challenges and various concerns in healthcare sector (Fig. 5) like tracking of data supply chain, data sharing between various departments in a healthcare sector and also from telemedicine and traditional care, private sharing of data with permission, maintaining digital identity of patients, registry of medical data and interoperability challenges (immutability, storage, privacy, scalability) [30]. The blockchain enables healthcare sector to overcome tough situations in case of pandemic outbreaks such as COVID-19.

Digital transformation of healthcare sector by blockchain technology. The pharmaceutical companies lack data flow of patient's information time to time and also face challenges such as the inconsistency in pharma data, time-consuming data analytics and hefty resource, data inaccuracies and ambiguity around relevance. These can be overcome with the help of blockchain technology. The block chain technology can be utilized to deliver efficient and accurate results, manage large data sets, record, store, and sort data by providing relevant and unambiguous data. Along with

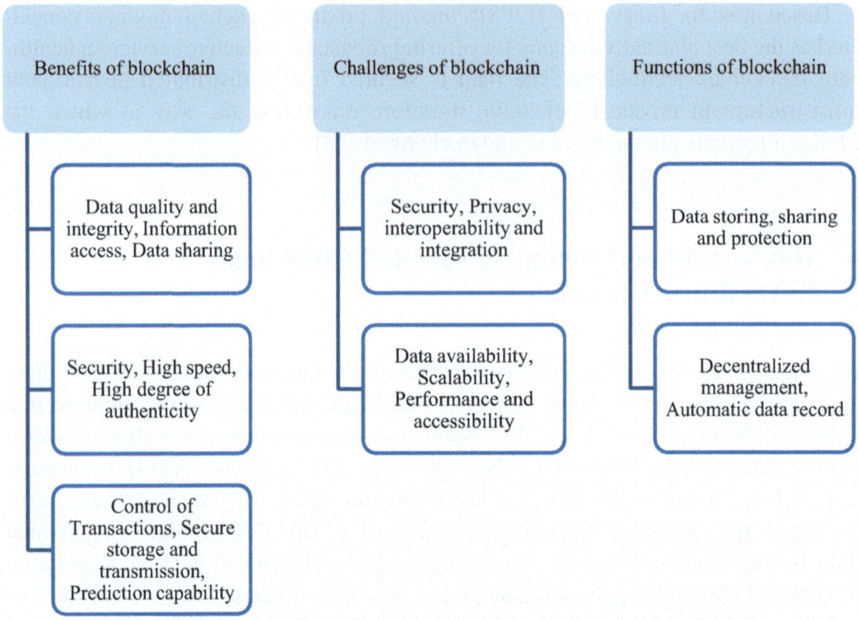

Fig. 5 The illustrative figure depicting holistic view of blockchain in healthcare sector

this it helps to build unalterable, auditable and distributed databases for accessing and storing drug trial data. The blockchain helps the pharma companies to enhance data provenance, integrity, functionality, and security of a pharma supply chain. It also includes optimizing the efficiency of the Internet of Healthy Things (IoHT), protecting from theft for digitally connected health equipment and combating counterfeit drugs.

7 Blockchain Technology Application Example in Healthcare Sector

Blockchain is best suited when there are multiple stakeholders and trust is required between parties that exist. The other projects that can benefit the use of blockchain technology include the elimination capabilities to increase trust or efficiency, data integrity and activity tracking projects [31]. The healthcare system comprises hospitals, patients, pharmaceutical companies, insurance companies, healthcare professionals, policymakers, etc. As there are multiple stakeholders at any point in time, the healthcare sector is one of the projects that can benefit from the use of blockchain technology.

Below example can help to clarify concept of how blockchain technology can be applied. Busting prescription drug fraud:

In this example, Nuco a blockchain company tries to tackle three common ill-uses employed to perform prescription fraud:

1. Changing prescription by changing the quantities.
2. Duplicating the prescription.
3. Swindlers visiting many doctors to collect several original prescriptions as possible also known as doctor shopping.

This fraud is due to an open-ended loop where the prescription is provided by doctor to patient. The patient then delivers that prescription to one or more pharmacists. The pharmacist lacks the knowledge of whether the prescription is accurate, original, or previously filled. To have a closed ended loop, transactions are stored on blockchains. Each participant in the loop can access and add to the blockchains as applicable. As an example, a record can be added to the original prescription by the doctor. The pharmacist then can check if the prescription is unchanged and add the actions on the prescriptions that any other pharmacist or doctor can view.

The solution that was provided by Nuco's is to generate a machine-readable code that can be tagged and can serve as a unique identifier on doctor's prescription to patient. The block of information containing the name and quantity of the dose, timestamp, and the identity of the patient is then tagged to the unique identifier. The symbol then can be scanned by the pharmacist when filling the prescription and it can be compared against the blockchain. This helps pharmacists to be rapidly informed on accuracy and eligibility of the prescription [31].

When multiple stakeholders have involved the copies of blockchain are held by all the stakeholders (Fig. 6). As described above the healthcare sector which has stakeholders will have a decentralized network and the privacy can be maintained by using encryption and the entitled parties can use the data by using correct cryptographic keys.

This example illustrates how a permissioned blockchain can be leveraged for the problem where only specific shareholders can access information and transactions. Among the four common implementations of blockchain technology, permissioned blockchain is one of them.

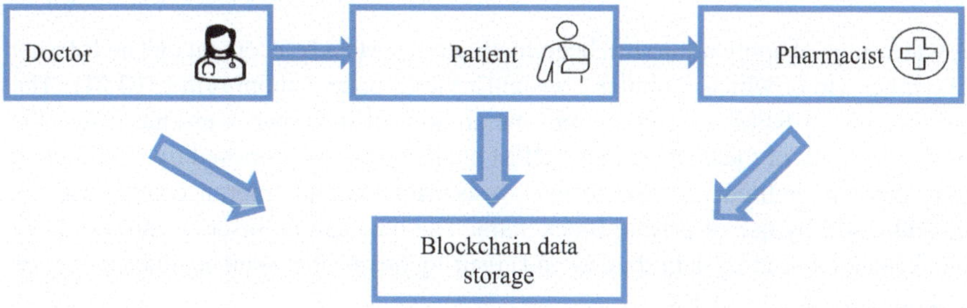

Fig. 6 Multiple stakeholders accessing blockchain data

8 Broad Classification of Healthcare Sector

The healthcare sector can be classified into two broads as telecare medical information systems (TMIS) and E-health system.

8.1 Blockchain Technology in Telecare Medical Information Systems (TMIS)

TMIS is a platform that allows patients and physicians to send and obtain medical information or health amenities from remote sites. The sensors measure the condition of the patient and the data is fed to mobile devices from where it is then transferred to healthcare provider's TMIS. The health records, test results, or the history of medicines prescribed can be viewed by patients by logging into the system. Similarly, the test results and prescribed medicines history can be viewed by the physicians and based on those they can always modify the prescription [32]. The TMIS-based sectors where blockchain can be leveraged are discussed below:

8.1.1 Blockchain Technology in Telemedicine

Medical-on-demand (MoD) service is delivered by telemedicine. The challenges in distant rural communities due to distance barriers can be reduced as the process of accessing medical services is made easy. Cloud Service Provider (CSP) provides the MoD services in the telemedicine. This helps the patient in connecting to the medical staff in different places with fidelity and convenience. However, there are some new security related challenges as the data is outsourced on public cloud platforms. Blockchain and distributed technologies can be used in protecting the data with reliability to avoid the misdiagnosis accident due to tampering of the health records by a malicious authority or user from the inner cloud [33].

8.1.2 Remote Patient Monitoring

A wide range of applications has been facilitated due to advancement of The Internet of Things (IoT) which includes constant remote patient monitoring (RPM). The architecture of RPM is complex and the amount of data that is produced and the power capacity of the devices make RPM challenging. As proposed by Uddin et al. [34], there are bottlenecks with respect to security and fault tolerance and these can be addressed by use of private blockchain. The data can be inserted into personal blockchain facilitating data sharing and ensuring privacy between healthcare experts and integration into EHRs.

8.1.3 Teledermatology

The interchange of distant data about skin conditions helps in promoting the quality and diagnosis of skin for rural and remote areas. Mannaro et al. [35] proposed the usage of private blockchain technology to ensure data maintenance and security. The proposed method allows the patients to have full access to their medical records while maintaining security.

8.1.4 Telesurgery

The healthcare surgical services to the distant/remote locations can be provided with extreme accuracy and quality over the wireless communication. This is beneficial to the society to have diagnosis procedures with improved accuracy and precision. There are privacy, security, and interoperability issues in the existing telesurgery systems. Gupta et al. [36] proposed blockchain technology architecture that uses smart contracts to operate with immutability and interoperability, thus improving the security.

8.2 Blockchain Technology in E-Health System

Prompt expansion of mobile cloud computing technologies and wearable devices led to new opportunities for healthcare systems on large scale. The health information of the patients is remotely sensed by the wearable devices and is sent through wireless devices to a computing system for evaluation and processing. Healthcare agencies will take care of such health information [37].

8.2.1 Electronic Health Records (EHRs)/Electronic Medical Records (EMRs)

The maintenance and sharing of health records are the important tasks in the healthcare sector. The compromise of the confidentiality will lead to security impact and compromise of integrity will lead to significant impact and may also led patient to death. Hence it is of primary importance to secure EHRs. The system is prone to cyber-attacks due to the centralized storage of health data and it is challenging to constantly view patient records. Nagasubramanian et al. [38] proposed searchable and permissioned blockchain technology for ensuring the data integrity. The architecture also has the faster response times compared to traditional systems.

EMRs are extremely significant for treatment of patients, diagnoses of doctors, and development of medical technology. There still exists a series of crucial difficulties with respect to security of patient's information. Fu et al. [39] proposed a blockchain scheme to hide and encrypt sensitive data stored in the mediator storage.

9 Projects/Technologies Developed Using Blockchain Technologies That Benefit Healthcare Sector

They are various blockchain-based technologies proposed to fulfill various applications in healthcare sector. These are implemented in various healthcare sectors (Fig. 7). Al-Karaki et al. [40] proposed a blockchain-based framework that supports healthcare sector by providing decentralized, scalable, accessible and secure access to healthcare records and services. It facilitates the real-time access to updated data by the stakeholders and is named as DASS-CARE. The primary objective of the proposed work is to advance the quality of healthcare and reduce the delivery cost followed by improving medical records management including unification of electronic health records so as to make available users to access their health records irrespective of their history (Fig. 8) It addresses various opportunities such as standardization of medical records, consent management, micropayments, and system interoperability. It also demonstrates the operation using three practical cases that are remote diagnosis of a patient, worldwide and cloud-based collaboration. There are several proposed frameworks in healthcare system that utilizes blockchain, but it is also detected that verification, validation, and testing of these proposed frameworks are not often utilized. The related tests are required to be implemented for evaluating the precise working of the framework.

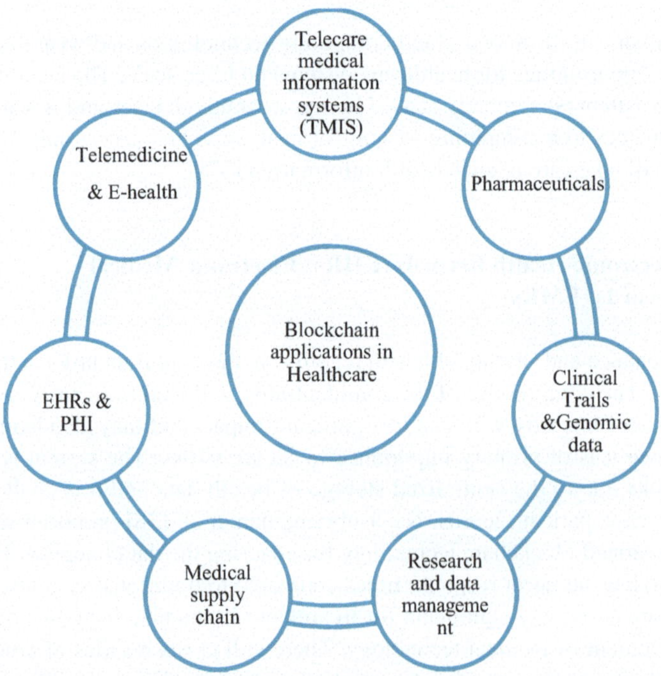

Fig. 7 The applications of blockchain in healthcare sector

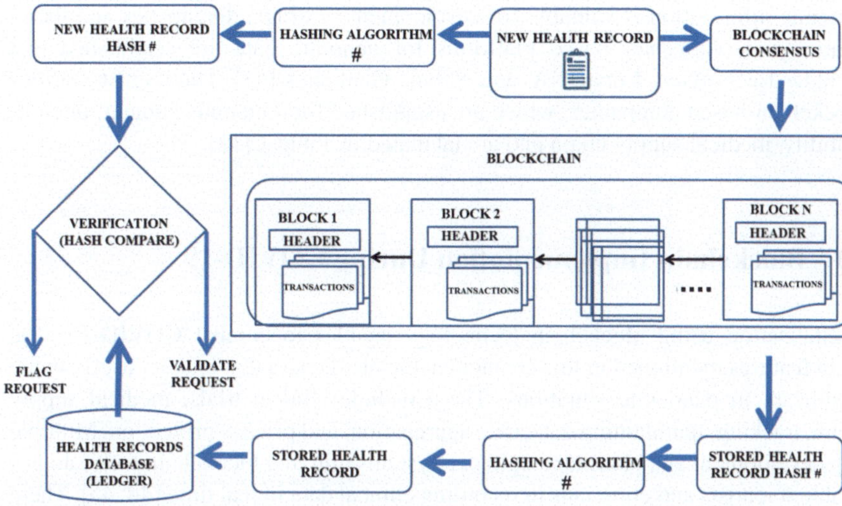

Fig. 8 Health records validation process using blockchain

Dagher et al. [41] proposed another framework named Ancile, it uses six distinctive types of smart contracts for operation in an ethereum-based blockchain and cryptographic techniques. These six contracts are classification, service history, consensus, permission, ownership, and re-encryption contract. It consists of three main software components that are the database manager to navigate present EHR databases; cipher manager for cryptography and ethereum-go client/geth to access private blockchain. The comparative performance analysis was done based on computational cost with MedRec. This proposed Ancile attains a high level of decentralization with high security, offers significant data integrity and privacy preservation. As many are still in development stages, framework rests mainly on their success.

Kumar et al. [42] proposed to design as mart healthcare system through interoperability and integration of Blockchain 3.0 and Healthcare 4.0. The smart healthcare system has explored optimization algorithms related to Healthcare 4.0 and increases the performance of applications based on decentralized blockchain uses. It discusses parameters on comparative analysis of security issues in blockchain-based healthcare systems, i.e., confidentiality, confidentiality, integrity, authentication, non-repudiation, resource availability, users availability, network availability, data storage security, data communication security, and data processing security. The authors presented the role of various security companies in the healthcare domains that utilizes blockchain technology. These companies include BurstIQ, CDC and IBM collaboration blockchain-based solutions, Chronicled, Curisium, MediLedger, and many more.

Another challenge faced is the fragmented data, the projects by the QBRICS and Nuco (Aion) were introduced to develop technologies using blockchain and to translate and consolidate data from multiple sources in order to reconstruct fragmented patient data across platforms [31]. The personal information such as the genomic information is valuable for development of drugs, diagnostics and therapeutics and blockchain-based platforms for genomic data are established that includes EncrypGen, LunaDNA, and Nebula Genomics [43]. There exists various blockchain-based companies, which are established for data management, interoperability, medical supply chain and are tabulated in Table 1 [12].

10 Blockchain Implementation During COVID-19

Digitalization using blockchain technology enables to combat COVID-19. The main features mentioned in this chapter enable blockchain that make it useful in the healthcare in pandemic situations. These include clinical trials, medical supply chain, tracking, maintaining data, data aggregation, and privacy protection. Multiple app development and pharmacovigilance systems utilizing blockchain technologies enable scientists and clinicians in recording clinical data in real time [44, 45]. There exist various examples that use the blockchain technology to combat COVID-19 such as Civit, which is as an app developed for assisting local authorities and managing clinical trials during COVID-19. MiPasa launched by World health organization (WHO) is a blockchain innovation which helps with gathering, processing, and learning information related to the spread and containment of virus. The Coalition is an app designed to track sickness among the users. The work proposed by Marbouh et al. [46] utilizing blockchain-based system will enable tracking reported data regarding new infected individuals, deaths, and recovered individuals obtained from trusted sources. This work proposed known to ensure data integrity, transparency, security, traceability, and feasibility. The hyperchain in blockchain can enable tracking donations of infected victims during pandemic by the various health and government organizations. Some help the public to know about the extent of infection spreading overtime like the HarshLog dashboard of Acoer. It collects information from various sites and brings various models for data visualization related with

Table 1 Blockchain-based solutions in healthcare sector

Blockchain-based solutions in healthcare sector		
Solution	Company	Place of origin
Data management	Blockchain Health	San Francisco, USA
	Bloq	Chicago, USA
	Burst IQ	Denver, USA
	ScalaMed	Sydney, Australia
	Gem Health	Venice, CA, USA
Interoperability	MedicalChain	London, England
	Patientory	Atlanta, USA
Supply chain	Blockpharma	Paris, France
	iSolve	Sandton, South Africa

clinical trial data. Smart contracts help in identifying, recording, and transferring the data. Contract tracing solutions can be implemented with the help of blockchain technology to overcome the pandemic.

11 Conclusion

The problems and challenges faced by the healthcare sector are interoperability, privacy, and supply chain traceability. The blockchain helps to overcome these by providing greater trust, collaboration, security, and easy transactions. Apart from this, the challenges of healthcare sector such as proprietary, electronic health record systems can be made easy and can be benefiting by using blockchain technology. In countries affected by the pandemic, the healthcare sector and professionals need to face many challenges that include supply and delivery of protective equipment and promptly developing treatments, tests and vaccines and also at the same time maintaining and securing healthcare records. It also provides the tracking facilities that will enable tracing of the product till its origin to the consumer there by preventing counterfeiting and enabling recalled products in seconds. Therefore, blockchain is an ideal and upcoming technology to maintain the distributed digital ledger and used in healthcare sector domain and also it has been applied to deal with recent pandemic covid-19 situation as well. It proves the trends of blockchain in health sector at different levels to ensure the integrity, transparency, and security.

References

1. M. Gupta, *Blockchain for Dummies (2nd IBM Limited Edition)* (Wiley, Hoboken, NJ, 2018)
2. F.A. Khan, M. Asif, A. Ahmad, M. Alharbi, H. Aljuaid, Blockchain technology, improvement suggestions, security challenges on smart grid and its application in healthcare for sustainable development. Sustain. Cities Soc. **55**, 102018 (2020)
3. O. Ali, A. Jaradat, A. Kulakli, A. Abuhalimeh, A comparative study: Blockchain technology utilization benefits, challenges and functionalities. IEEE Access **9**, 12730–12749 (2021)
4. C.C. Agbo, Q.H. Mahmoud, Comparison of blockchain frameworks for healthcare applications. Internet Technol. Lett. **2**(5), e122 (2019)
5. S. M. Idrees, P. Agarwal, M. A. Alam (eds.), *Blockchain for Healthcare Systems: Challenges, Privacy, and Securing of Data* (CRC, Boca Raton, FL, 2021)
6. Idrees, S. M., Nowostawski, M., & Jameel, R. (2021). Blockchain-based digital contact tracing apps for COVID-19 pandemic management: Issues, challenges, solutions, and future directions. JMIR medical informatics, 9(2), e25245.
7. L. Ismail, H. Materwala, A review of blockchain architecture and consensus protocols: Use cases, challenges, and solutions. Symmetry **11**(10), 1198 (2019)
8. J. Brogan, I. Baskaran, N. Ramachandran, Authenticating health activity data using distributed ledger technologies. Comput. Struct. Biotechnol. J. **16**, 257–266 (2018)
9. T.T. Kuo, H.E. Kim, L. Ohno-Machado, Blockchain distributed ledger technologies for biomedical and healthcare applications. J. Am. Med. Inform. Assoc. **24**(6), 1211–1220 (2017)

10. H. Liu, R.G. Crespo, O.S. Martinez, Enhancing privacy and data security across health care applications using block chain and distributed ledger concepts. Healthcare **8**(3), 243 (2020)
11. D. Hawig, C. Zhou, S. Fuhrhop, A.S. Fialho, N. Ramachandran, Designing a distributed ledger technology system for interoperable and general data protection regulation–compliant health data exchange: A use case in blood glucose data. J. Med. Internet Res. **21**(6), e13665 (2019)
12. K. Yaeger, M. Martini, J. Rasouli, A. Costa, Emerging blockchain technology solutions for modern healthcare infrastructure. J. Sci. Innov. Med. **2**(1), 1 (2019)
13. R.W. Ahmad, K. Salah, R. Jayaraman, I. Yaqoob, S. Ellahham, M. Omar, The role of blockchain technology in telehealth and telemedicine. Int. J. Med. Inform. **148**, 104399 (2021)
14. A. Azaria, A. Ekblaw, T. Vieira, A. Lippman, Medrec: Using blockchain for medical data access and permission management, in *IEEE.2016 2nd International Conference on Open and Big Data (OBD)*, (2016), pp. 25–30
15. M.N.K. Boulos, J.T. Wilson, K.A. Clauson, Geospatial blockchain: Promises, challenges, and scenarios in health and healthcare. Int. J. Health Geogr. **17**, 25 (2018)
16. K.N. Griggs, O. Ossipova, C.P. Kohlios, A.N. Baccarini, E.A. Howson, T. Hayajneh, Healthcare blockchain system using smart contracts for secure automated remote patient monitoring. J. Med. Syst. **42**(7), 1–7 (2018)
17. Z. Zheng, S. Xie, H.N. Dai, W. Chen, X. Chen, J. Weng, M. Imran, An overview on smart contracts: Challenges, advances and platforms. Fut. Gener. Comput. Syst. **105**, 475–491 (2020)
18. L. Hang, E. Choi, D.H. Kim, A novel EMR integrity management based on a medical blockchain platform in hospital. Electronics **8**(4), 467 (2019)
19. Z. Zheng, S. Xie, H.N. Dai, X. Chen, H. Wang, Blockchain challenges and opportunities: A survey. Int. J. Web Grid Serv. **14**(4), 352–375 (2018)
20. L.A. Linn, M.B. Koo, Blockchain for health data and its potential use in health it and healthcare related research, in *ONC/NIST Use of Blockchain for Healthcare and Research Workshop, Gaithersburg, Maryland*, (2016), pp. 1–10
21. Z. Alhadhrami, S. Alghfeli, M. Alghfeli, J.A. Abedlla, K. Shuaib, Introducing blockchains for healthcare, in *IEEE, 2017 international conference on electrical and computing technologies and applications (ICECTA)*, (2017), pp. 1–4
22. Z. Xiao, Z. Li, Y. Liu, L. Feng, W. Zhang, T. Lertwuthikarn, R.S.M. Goh, EMRShare: A cross-organizational medical data sharing and management framework using permissioned blockchain, in *IEEE 24th International Conference on Parallel and Distributed Systems (ICPADS)*, (2018), pp. 998–1003
23. T. Zhou, X. Li, H. Zhao, Med-PPPHIS: Blockchain-based personal healthcare information system for national physique monitoring and scientific exercise guiding. J. Med. Syst. **43**(9), 1–23 (2019)
24. T. Laurence, *Blockchain for Dummies* (Wiley, Hoboken, NJ, 2019)
25. A. Farouk, A. Alahmadi, S. Ghose, A. Mashatan, Blockchain platform for industrial healthcare: Vision and future opportunities. Comput. Commun. **154**, 223–235 (2020)
26. S. Efendi, B. Siregar, H. Pranoto, Concept designs of patient information security using e-health sensor shield platform on blockchain infrastructure, in *Proceedings Series of MICoMS 2017 (Emerald Reach Proceedings Series)*, vol. 1, (Emerald, Bingley, 2018), pp. 641–646
27. D. Guru, S. Perumal, V. Varadarajan, Approaches towards Blockchain innovation: A survey and future directions. Electronics **10**(10), 1219 (2021)
28. M. Zarour, M.T. Ansari, M. Alenezi, A.K. Sarkar, M. Faizan, A. Agrawal, R. Kumar, R.A.P. Khan, Evaluating the impact of blockchain models for secure and trustworthy electronic healthcare records. IEEE Access **8**, 157959–157973 (2020)
29. General Data Protection Regulation, Regulation EU 2016/679 of the European Parliament and of the council of 27 April 2016. Off. J. Eur. Union (2016)
30. P. Zhang, D.C. Schmidt, J. White, G. Lenz, Blockchain technology use cases in healthcare. Adv. Comput. **111**, 1–41 (2018)
31. M.A. Engelhardt, Hitching healthcare to the chain: An introduction to blockchain technology in the healthcare sector. Technol. Innov. Manag. Rev. **7**(10), 22–34 (2017)

32. F.A. Khan, A. Ali, H. Abbas, N.A.H. Haldar, A cloud-based healthcare frame work for security and patients' data privacy using wireless body area networks. Proc. Comput. Sci. **34**, 511–517 (2014)
33. R. Guo, H. Shi, D. Zheng, C. Jing, C. Zhuang, Z. Wang, Flexible and efficient blockchain-based ABE scheme with multi-authority for medical on demand in telemedicine system. IEEE Access **7**, 88012–88025 (2019)
34. M.A. Uddin, A. Stranieri, I. Gondal, V. Balasubramanian, Continuous patient monitoring with a patient centric agent: A block architecture. IEEE Access **6**, 32700–32726 (2018)
35. K. Mannaro, G. Baralla, A. Pinna, S. Ibba, A blockchain approach applied to a teledermatology platform in the Sardinian region (Italy). Information **9**(2), 44 (2018)
36. R. Gupta, S. Tanwar, S. Tyagi, N. Kumar, M.S. Obaidat, B. Sadoun, Habits: Blockchain-based telesurgery framework for healthcare 4.0, in *IEEE International Conference on Computer, Information and Telecommunication Systems (CITS)*, (2019), pp. 1–5
37. Q. Althebyan, Q. Yaseen, Y. Jararweh, M. Al-Ayyoub, Cloud support for large scalee-healthcare systems. Ann. Telecommun. **71**(9), 503–515 (2016)
38. G. Nagasubramanian, R.K. Sakthivel, R. Patan, A.H. Gandomi, M. Sankayya, B. Balusamy, Securing e-health records using keyless signature infrastructure blockchain technology in the cloud. Neural Comput. Applic. **32**(3), 639–647 (2020)
39. J. Fu, N. Wang, Y. Cai, Privacy-preserving in healthcare blockchain systems based on light weight message sharing. Sensors **20**(7), 1898 (2020)
40. J.N. Al-Karaki, A. Gawanmeh, M. Ayache, A. Mashaleh, DASS-CARE: A decentralized, accessible, scalable, and secure healthcare framework using blockchain, in *IEEE.201915 Wireless Communications & Mobile Computing Conference (IWCMC)*, (2019), pp. 330–335
41. G.G. Dagher, J. Mohler, M. Milojkovic, P.B. Marella, Ancile: Privacy-preserving framework for access control and interoperability of electronic health records using blockchain technology. Sustain. Cities Soc. **39**, 283–297 (2018)
42. A. Kumar, R. Krishnamurthi, A. Nayyar, K. Sharma, V. Grover, E. Hossain, A novel smart healthcare design, simulation, and implementation using healthcare 4.0 processes. IEEE Access **8**, 118433–118471 (2020)
43. E. Ahmed, M. Shabani, DNA data marketplace: An analysis of the ethical concerns regarding the participation of the individuals. Front. Genet. **10**, 1107 (2019)
44. P. Beninger, M.A. Ibara, Pharmacovigilance and biomedical informatics: A model for future development. Clin. Ther. **38**(12), 2514–2525 (2016)
45. M.V. Satwika, D.S. Sushma, V. Jaiswal, S. Asha, T. Pal, The role of advanced technologies supplemented with traditional methods in pharmacovigilance sciences. Recent Pat. Biotechnol. **15**(1), 34–50 (2021)
46. D. Marbouh, T. Abbasi, F. Maasmi, I.A. Omar, M.S. Debe, K. Salah, R. Jayaraman, S. Ellahham, Blockchain for COVID-19: Review, opportunities, and a trusted tracking system. Arab. J. Sci. Eng., 1–17 (2020)

A Systematic Review on Blockchain in Transforming the Healthcare Sector

Medini Gupta, Sarvesh Tanwar, Sumit Badotra, and Ajay Rana

1 Introduction

Blockchain was initially introduced in 2008 by group of people known as Satoshi Nakamoto, for doing financial transactions for Bitcoin cryptocurrency. It is a distributed ledger that stores the data in chronological order. Blockchain can entirely change the way for storage, data security, payments through smart contract. A smart contract by using Ethereum cryptocurrency is a computer program which is stored inside a blockchain. It is similar to contract in real world. Everything is distributed. No one is in control of money. It is immutable. Once smart contract is created, it cannot be edited again. It guarantees reliability and security. **Smart contract** allows patients time-limited access to their electronic medical information. Doctor can store transaction data about prescription and pathology reports. Dispensary provides medicines and records transactions on the blockchain. Insurer can also access treatment and payment details for verification. Doctors remotely give consultation and get rewarded by patients. Patients get rewarded in form of tokens or low-cost insurance premium when they grant access to insurer. Patients also get rewarded when they permit research organization for accessing their records for clinical study. **Data integrity issues** such as corrupted files, manually removing errors, and

M. Gupta · S. Tanwar (✉)
Amity Institute of Information Technology, Amity University, Noida, Uttar Pradesh, India
e-mail: stanwar@amity.edu

S. Badotra (✉)
Department of Computer Science and Engineering, Lovely Professional University, Phagwara, Punjab, India
e-mail: sumit.26152@lpu.co.in

A. Rana
Shobhit University, Meerut, Uttar Pradesh, India
e-mail: ajay_rana@amity.edu

S. M. Idrees, M. Nowostawski (eds.), *Transformations Through Blockchain Technology*, https://doi.org/10.1007/978-3-030-93344-9_8

threat of cyber-attacks hamper fruitful collaboration in healthcare. FireEye reported 68 lakhs healthcare data breaches in 2019. Blockchain provides enhanced **data security** by assuring access control through distributed public or private chains. Public chains are permission less which enables anyone to participate on the network, whereas information on private chains is encrypted and can be accessed by authorized party. Once data is entered in the block, then it is irreversible.

Record Tracking—Health records are the mainstay of every healthcare system. Medical data increases every time when you visit your doctor. Data storage varies from hospital to hospital which makes it quite difficult to obtain when required. Patients don't have complete ownership of their own medical data because of customized treatment. Blockchain requires **consensus** of all the users on the network to add or delete any information, thus increasing privacy. It provides single point access to it for patients and caregivers.

Supply Chain—Pharmaceutical field is known for stability and safety. It can keep track of individuals, cost, and release of waste which is directly reducing human errors and time. Goods can be authenticated by monitoring them from beginning, quality assurance which may get affected due to high or low temperature.

1.1 Overview of Blockchain

Blockchain is a chain of blocks that contains data. It is a distributed ledger which is open to all. It has an immutable nature which makes it highly secure. Once any data is entered, it is irreversible [1]. It is quite impossible to backdate or temper with digital documents due to timestamp. Timestamp is similar to notary. Each block contains data, hash of that particular block and also hash of preceding block. A hash is just like a fingerprint created through cryptographic algorithm. It identifies all the blocks and the information stored inside it. Hash plays a crucial role in detecting changes in the block. If someone has tempered with the data, then hash value also changes. Once it changes, block no longer remains the same. First block is known as Genesis because it doesn't have any previous block. When any change is made with second block, then it leads to change in hash. This causes subsequent blocks invalid. Implementing hashing algorithm is not enough to avoid tempering of data. Hackers can calculate thousands of hash values by using computers to make the blockchain valid again. To overcome this problem blockchain has concept of Proof-of-Work. It is a process which slow down the creation of new blocks. When user enter any record in the block, they have to recalculate the proof-of-work for rest of the subsequent block. There are Confidentiality, Integrity, and Availability which are known as CIA triad for ensuring security mechanisms in the organization.

Confidentiality—It is in reference with privacy of data. Sensitive data should be only accessed by authorized people. Only those for whom information is meant can read or process it. No third party can read private messages. It can be achieved by using username-password, two factor authentication and through encryption techniques. Confidentiality is said to be zero when illegal person has traced the packet and even understood the content inside it.

Integrity—Integrity refers to validity, continuity, and trustfulness of data. Data should not be altered during transmitting it from sender to receiver. Backup should be available to cover any corrupted data. User access control and file permission have to be set so that only authorized user can modify the information. To avoid any unwanted changes by authorized user, version control can be used. Checksum can be applied to ensure integrity such as Message Direct 5 (MD5) or Secure Hashing Algorithms (SHA).

Availability—It ensures that data is available to authorized user when and where required. Operating system and software should be up to date. Maintaining resources, hardware repairs and ensuring backup for availability of data. Proper planning should be done to recover from manmade or natural attacks. Security services such as firewalls and antivirus should be installed for preventing the attack.

Our world is moving towards a new era of decentralization, once quite 20 years of research, significant progress has been made in the cryptography field and localized computer networks, leading to a great technology, which is called blockchain. It has the great potential for shifting the way society functions, but especially banks and currency operations. It allows random people to reach agreement on the prevalence of a particular community activity or occurrence for the primary time, while not the requirement of a dominant authority, in the terms of nomads—it means that no authority governs or supervises the transactions, and that they square measure produced with absolute namelessness.

The blockchain is a distributed system at the most basic level; we can also say, it is a log, a record of a series of transactions, but it contains a unique set of properties such as:

1. **The transactions are chronologically documented—**Each transaction has a timestamp, then those transactions are combined in form of blocks and later they are officially added into the ledger in a sequential way.
2. **Each and every block can be rendered only through importing data from previous block—**In each transaction, the unique data from the last block is used to build the next block, using this the blocks establish a connection. The relation among every block and the block before it creates a concept known as cryptographic hashing.
3. **Blockchain is distributed—**Anyone that makes a transaction and participates has a copy of the ledger.
4. **The encryption is incorporated into each part of the process—**The people who created the blockchain learned much from the initial stage mistakes during the internet was being made. As compared to the previous program, which took several years to directly address the fact that it does not do anything to guarantee privacy or protection, the blockchain uses state of privacy, security, and encryption as its primary functionality.
5. **The system makes consensus automatic—**Blockchain's most notable innovation is that it forces everyone to agree on one meaning of the truth by some very interesting methods. The device will automatically detect and fix the inconsistencies in the data if the nodes begin to disagree about inputs in the database.

6. **The insertion of a transaction into the ledger entails an expense**—The processes are very complex, some participating systems in the network must perform some quite costly extreme tough math equations to insert a new block in the blockchain. They need to use up electricity, human labor, and computer to solve these equations, thus allowing records to be added to the ledger. This criterion for carrying out costly calculations is called "proof-of-work," and is deliberately designed in a way to push miners' costs up.

Blockchain is being distributed. It is a decentralized ledger where no third party is involved for source of trust. It uses peer to peer network where anyone is allowed to join. When user creates a new block, it is shared with everyone on the network. Each node is verified to know that it is not tempered. If all the details are verified, then each node adds that block in their blockchain. All the nodes approve about the blocks which are valid and which are not and creates consensus. Tampered blocks are identified and are rejected by rest of the nodes on the network. If hacker tries to alter with the blockchain, then they have to alter the data of each block and also have to take control of 51% of blocks by calculating proof-of-work for each of them. Then only nodes will agree on tampered data and consider it valid. It is impossible to do so. Blockchain features such as immutable, timestamp, hashing, proof-of-work, distributed, decentralized, and consensus make it much secure [2] (Fig. 1).

1.2 Motivation

Sometimes there are situation of life and death which requires immediate and accurate details of patient's medical history. Even a delay of few minutes can lead to bad consequences. Individual life is very precious. A person can overcome any problem no matter how big it was. But if they have lost the battle of life, then nothing can be done. Poor health not only affects the patients but also affects the quality of life of their family. Healthcare system plays a crucial role for the development of economy. In blockchain network data can be stored, updated, and shared securely in real time

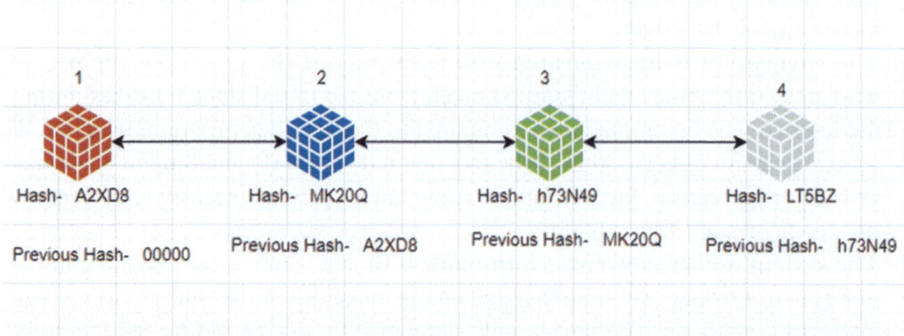

Fig. 1 Working of blockchain through hashing

by authorized users. It enables universal access of medical records. Better and personalized diagnosis of patient by various caregivers can avoid any miscommunication among them. Using smart contract can lead to seamless connectivity. Every user has a pair of public and private key which can be used to unlock for a particular period of time. It minimizes the treatment cost by making the information decentralized. Inaccessibility of data, tempering by third party, inadequate records, and other issues can be resolved by implementing blockchain in healthcare industry.

1.3 Literature Questions

One of the major challenges with healthcare is that organization holds multiple and fragmented records about patients. To know whether the medications prescribed by the doctors are available in the dispensary you want to visit and does it come under your insurance. Long queue at healthcare services is result of inefficient transactions. Securing the data through middleman leads to high cost and risk of tempering with the data. Blockchain puts medical record transactions on the network to address interoperability problem and develops smart healthcare environment.

2 Literature Review

There are various researchers who have proposed blockchain solution in the field of healthcare. Due to its enormous potential and practical relevance, study in the field of Blockchain is still going on [3] with the aim to merge hospitals, medical practitioner and patients on a single platform to provide better degree of service. Below we have discussed about few of them (Table 1).

3 Blockchain in Transforming Healthcare

In present time blockchain is creating wide variety of potential opportunities in enhancing healthcare system. It will protect patient's confidential data and ensure that only legitimate users can access electronic health records [3]. In healthcare system middleman may have connections which result in data breaches and increased management cost [4]. Blockchain eliminates the involvement of middleman for trust which also reduces cost. Blockchain being immutable, decentralized and distributed ledger make it suitable to be implemented in medical field. Security breaches risks can be minimized by encrypting information using asymmetric cryptography where involvement of multiple users on network will be required to decrypt the information [5]. Users have complete ownership of their records and can get benefits from them. Interoperability feature makes data gathering much easier

Table 1 Representation of the work done by other researchers in this field

S. no	Author	Year of publication	Work
1	Tooska Dargahi, Koosha Mohammad Hossein, Mohammad Esmaeil Esmaeili, Ahmad Khonsari	2019	Tooska Dargahi et al. proposed a blockchain-based architecture for securing the privacy of patients through access control. The architecture consists of low energy wearable sensors such as ECG and blood pressure. Smartphones act as gateway to transmit data to healthcare providers and performs cryptographic algorithms. Central server for storing patient's health records and sending the hash values over blockchain network. Patient can finalize who can view their data and up to what extent. They did the security analysis of their architecture
2	Liviu Hirtan, Jordi Mongay Batalla, Ciprian Dobre, Piotr Krawiec	2019	Liviu Hirtan et al. proposed a solution that uses mainchain and side chain for trusted nodes and untrusted nodes, respectively. Digital signatures are also used to protect identity and personal details of patients. Asymmetric key cryptographic is implemented for authentication and message signing. It recognizes that patients should be the sole owners of their confidential data. Ancillary parties such as insurance companies and research institutes can only access health information based on security policies set by patients. All the medical details are validated by trustworthy parties
3	Cosmin Stirbu, Valeriu Manuel Ionescu, Nicu Bizon, Andrei Cirstea, Florentina Magda Enescu	2018	Cosmin Stirbu et al. conducted a research study on advantages of blockchain such as easy identification of inefficient medicines, increasing efficiency of treatment through its cryptographic operations and early diagnosis of diseases. It has various applications in healthcare such as MedBlocks. Patient can access the application from any device and database stores all the medical details in encrypted format. Patient can consult the specialist and learns about the latest drugs or treatment related to their illness. Authorized users can access patients details such as height, weight, blood group, and medical history. Frequency of visit to doctors. They concluded it by mentioning that healthcare system should not depend only on single entity. Technology should be flexible not only for patient but also for doctors and health insurance companies who want to access the data

(continued)

Table 1 (continued)

S. no	Author	Year of publication	Work
4	Ijaz Ahmad, Tanesh Kumar, Vidhya Ramani, Mika Ylianttila, Erkki Harjula, An Braeken	2018	Ijaz Ahmad et al. discussed about the implementation of healthcare applications to provide better quality of service such as maintaining transparent health records, universal data sharing, secure payment, interoperability, confidentiality, and data security. Even patients are reluctant in sharing their medical details due to loss or misuse of their sensitive data. Patients face difficulty when they move to other hospital for the consultation and not able to access their medical history which is in different hospital. Blockchain being immutable, distributed and its proof-of-work makes it suitable to solve these issues. This paper also presents the challenges of blockchain that needs to be overcome before adopting it. These includes high cost, scalability, and lack of standardization
5	Joaqun Ordieres-, Xiaochen Zheng, Ravi Vatrapu, Raghava Rao Mukkamala	2018	Joaqun Ordieres-Mere et al. designed a healthcare data sharing system using blockchain, machine learning, and cloud storage. Healthcare data is divided into static and dynamic data which is collected through wearable sensors and smart phones. The limitation of dynamic health data is solved by merging blockchain with cloud storage. Large amount of medical data in encrypted format is stored on cloud and only the meta data and transactional data is shared on blockchain. Machine learning algorithms are used for validating the hardware and software quality. Security breaches can be controlled by asymmetric cryptography, authentication, and consensus by system users over the network
6	Taylor Hardin, David Kotz	2019	David Kotz et al. in their survey presented the implementation details of healthcare applications, security, and performance analysis. Patients have access to whom they can allow or deny to access different parts of their medical records. Hashes can be used to see any update in data which is prone to change. They concluded by stating that most of the systems are using private blockchain and smart contract consortium. More focus should be on testing, concern related to smart contract

(continued)

Table 1 (continued)

S. no	Author	Year of publication	Work
7	Sudeep Tanwar, Jayneel Vora, Neeraj Kumar, Sudhanshu Tyagi, M. S. Obaidat, Joel J. P. C. Rodrigues	2018	Sudeep Tanwar et al. proposed a framework to store and maintain electronic health records efficiently so that patients, doctors, and other parties can access them without keeping the patient's personal details at stake. Various contracts are being used. Ethereum contract is used for network communication. Consensus contract for validating the information entered on the network. Patients have full ownership of their medical records and transactions. Uncertified access and data alteration is reduced through decentralization and smart contracts. They have shown concern with respect to complete interoperability, privacy, and integrity of data which cannot go hand in hand
8	Sudeep Tanwar, Richard Evans, Karan Parekh	2019	Sudeep Tanwar et al. proposed a permission-based EHR system to enhance the efficiency and security issues in blockchain. Various solutions to overcome the drawbacks of existing healthcare system that involves client server architecture are discussed where adopting blockchain will be more effective. Symmetric key is used for the data security. Effective algorithm to increase access control by patients and caregivers are mentioned. This system also reduces one point failure and involvement of third party for trust which is increasing the security of data. Performance of the system is evaluated based on the metrics such as latency, network security, and throughput

for commercial and research purposes. Blockchain smart contract is a computer program which acts as an agreement and provides trustworthiness Blockchain uses peer to peer network where consensus algorithms are used to validate transactions. Every user on the network should agree upon the transaction which makes it transparent and more secure as compared to other record tracking system. All the blocks are timestamped as transactions are listed in order of occurrence.

3.1 Interoperability

Crucial data is fragmented among different systems and many times it is not accessible when needed the most. Through blockchain digital health records can be easily accessible by caregivers irrespective of their location. It increases transparency by sharing data with hospitals and patients through access control mechanisms. One of the major challenges in existing healthcare system is lack of interoperability. Unavailability of pathology reports delays the treatment, increases cost and conducting the same medical test again and again can have harmful effect on the patient. Absence of global standards is the primary reason for interoperability [3]. Even interoperability may also lead to rise in security vulnerabilities. However, blockchain can mitigate the risks associated with data sharing [6]. It has potential to strengthen electronic health records interoperability in present healthcare system.

3.2 Security and Privacy

Security should be the top priority for blockchain-based healthcare system since many participants are involved in it. A cyber-attack on healthcare system can reveal the sensitive data not only of individuals but about entire group. Duplicate copy of records and transactions are stored on entire network [3]. Unauthorized user has to control 51% records if they want to alter the data on the network which is quite time consuming and impossible. Major reason for high privacy and trust is due to transparent nature of blockchain. All the users on the network are aware about the transactions as they agree on consensus algorithm [6]. Safeguarding the individual privacy by using private or consortium blockchain, read or write permission can be granted. Encryption algorithm is used to provide confidentiality and integrity for records stored on blockchain.

3.3 Universal Data Sharing

There are circumstances when patients move to other country for leisure or because of any other reason. Sometimes they have to visit hospital due to any symptoms. Respective doctor of that country should be aware of patient's medical history for providing better quality of treatment. Blockchain provides a solution for sharing medical records no matter where it is located. Users have full control on their electronic health records. Better monitoring can be done when doctors are aware about patient's previous surgeries, allergies, current medications, and lab reports [6].

3.4 Data Access Control

Users are not aware which party is accessing their records and for what purpose and are they permitted to access it. Access control makes sure that only legitimate users can read and process the information stored on network. It depends on owner of data, type of data, who can control and modify the data. Access control is important for security of health organization who shares the data internally as well as with external parties such as research institute and government. Users can allow, deny, or revoke hospital request to access various parts of their medical records. It only depends on users whether they want to share their records with external party or not. Smart contract verifies the access control when a new request comes to process the records. All the requests are stored on blockchain for future reference.

3.5 Payments

Conventional system for payment is very complicated and prone to security breaches [3]. Records are not only shared with hospital but with insurance companies also for payment purpose. It consumes lots of time and resources to clear a bill. There are cases when health insurance of the patient is rejected. And they struggle to pay their bills on time. Many a times rejected claims may result due to coding errors. Healthcare solution based on blockchain technology allows insurance companies to view treatment records which are verified by the administrator and removes human errors. It reduces resources and time which result in less rejected claims and fast processing of payment [3].

3.6 Drug Traceability

Many pharmaceutical organizations are facing losses due to drug fraud [4]. According to World Health Organization, 16% drugs sold are forged. Transactions recorded on the blockchain are timestamped and immutable which means whole manufacturing chain can be tracked and tampering with records are not possible. Organizations have distributed ledger with manufactures and distributer of the drug. Based on smart contract, access can be granted to authorized parties. Information about drug will be stored on network whenever drug transmits from one organization to another. Organization can keep track of their products and even action can be taken by identifying the location when there is any defect in the drug.

4 Drawbacks in Existing Healthcare System

Healthcare industry has made astonishing advances in past decades. Many factors that affect the healthcare systems are rising cost, prevalence of large number of diseases, and high chances of survival. One of the reasons for inefficient healthcare system is due to patient's lack of knowledge for decision-making. They are unaware about the health and economic effects of their own health.

4.1 Absence of User Control

Healthcare records are under the control of several service providers, system suppliers or distributed through various medical organizations [5]. Instead, it should be managed by users itself. Users face unauthorized access of their medical data because of inability to control it [12]. Letting users free and convenient access to their personal information and allowing them to choose whether or not to share that data with third party should be up to them.

4.2 Security Concern

Users actual data is necessary for treatment in healthcare system. Conventional approaches for maintaining privacy by including noisy data are ineffective [1]. Patients are reluctant in sharing their medical records on online platform. They have a fear of loss or leakage of their medical data [4]. There are increasing cybercrimes in medical field [8]. Unauthorized users gain access to patient's insurance details and use them for malicious purposes. They do the transactions through legitimate credentials of patients which leads to false records, high premium, ruin credit history. Treatment cost will add up in victim's medical data and they will not be aware until creditors will contact them. Even it can increase patient premium. It will delay the treatment, as if the unauthorized user got treated on their identity then misdiagnosis will happen. User will face difficulty in getting appropriate medical care and health insurance benefits. Health records are essential components in ensuring appropriate treatment to patients [15]. Along with identity theft, medical records could be sold for illegal activities.

4.3 Lack of Supply Management System

Logistics are the backbone of the industry. Effective logistics practices make sure constant medical services. It causes issues such as shortage of drugs, inefficient medical resources and facilities. Doctors face difficulties in providing better quality of service. Even patients feel frustrated and it endangers their health [16]. Medical conditions are unpredictable and hospitals have to address them. It results in massive financial losses for caregivers due to defective stock, incorrect distribution of medicines, and problems faced by patients. Drug shortage enables providers to either buy even more costly substitutes or keep drugs in large quantity that have possibility to go out of stock. It also increases cost of supply chain.

4.4 Monitoring Medical Records

Doctors should have access to patients' medical records regardless of where the information or samples are collected and which technology is being used for producing it. Time is very crucial especially with critical patients [14]. Early the data will be available to caregivers, better the treatment will be provided to patients. Changes in personnel such as nurses or doctors create issues with proper access to

EHR for any patient [7]. When the person moves from one place to another, they require their prescription and pathology reports for further consultation. It becomes quite difficult to extract the previous data. When the information is on different electronic health records, then it is not accessible by doctors. They get it through fax or electronic mail. The requirement of data varies from doctor to doctor. A cardiologist will require different data as compared to a physician. The absence of structured data management and distribution system within different organization result in medical records fragmentation [4].

5 Applications of Blockchain in Other Fields

Initially blockchain was introduced in finance field. Later on, it gained popularity in various areas because it is secure, distributed, and decentralized [13]. There has been an unremitting stream of innovations in the field of blockchain technology, with new applications being forever announced, the landscape is constantly evolving. Below are the fields where its presence has enhanced the current system.

5.1 Education

Blockchain can entirely change the way of storing and data security, storage, payments through smart contracts. Data about learning content, attendance, grades, and certificates can easily be accessible without being tampered by any third party on the network. For both teachers and students, it has a very vast potential. Teachers can create the learning content and distribute it over the network without risk of any loss. Smart contract can be used in learning when some condition is fulfilled. Teacher can set up an assignment and set its conditions such as instructions, conditions, and due date. Once it is fulfilled, students can be redirected to next topic in the course. Teachers can be paid in digital tokens and can reward students with credits. It will encourage both teachers and students to contribute in learning. There are various issues with traditional system such as loss of data, fake certificates, tempering with records by third party which can be overcome by implementing blockchain. If students have lost their certificate, transcript, or degree, they have to write multiple applications to authority. Authority has to check for previous record of the student. As all the details are stored on centralized system so verification also becomes quite difficult. When students migrate from one place to another, they require their academic qualifications for new admission. At this point record sharing is not easily possible [17]. There are various features of blockchain such as transparency, immutable, distributed, encrypted hash which make it more beneficial in education sector.

5.2 Government

Blockchain being transparent through distributed ledger enables users to view and verify the data. It will build trust among citizens for independent verification of government schemes. Government can use blockchain for proper identity management of every citizen. Data management, transactions, and credentials can be stored securely [19]. Fair elections can be conducted without using any corrupt means. Budget can be allocated with transparency and effectiveness. A tax officer after scanning citizen ID could identify their name and place of residence. Even academic records, pension can be accessed and verified. Registries regarding land, marriages, or crime can be done in real time, decentralized and irreversible which simultaneously minimizes the corruption. Government administration has to manage education, healthcare, defense, and pension which involves sharing of confidential data that cannot be put on stake. Digital identity supported by blockchain where users can observe their pension from starting till end of the life cycle. Every democratic government faces challenges with counting of votes such as time delay, security, and transparency. Blockchain supports immutable and transparent voting after collation period.

5.3 Tourism

In tourism industry, lockchain allows safe and traceable transactions. First thing which comes when we decide to travel is booking of flight tickets. It encourages transparency, low cost and monitors over boarding of passengers. Fast and secure payment for the services can be achieved with blockchain solutions [18]. Using cryptocurrencies such as bitcoin eliminates the involvement of online payment which also eliminate the need for third party applications. Payment is done directly between two parties. Individual can directly move to airport without wasting their time by standing in long queues. Payments are done by automatically deducting the money from individual digital wallet. Authenticated user identification for tourists pertaining to security concerns. In traditional system users have to verify their identity in airport during check in and later on in immigration office and then in hotels which consumes lots of time. Loyalty schemes run so that users again prefer their service when they come again [35]. Users can be granted cryptocurrencies as a reward which can be used in future for travel. Many times, passenger faces issue with baggage tracking. They lose their baggage because it is handled by multiple parties. Baggage tracking issue can be overcome by using smart contract.

5.4 *Finance*

Financial services are centralized all over the world. It includes multiple middlemen. Database security completely depends on middleman and it lacks transparency. Even with maximum protection, data is prone to security breaches [20]. User is not aware about the security attack until something goes wrong. Certain level of transparency is necessary for financial institutes and for their customers as well. Blockchain is immutable which make sure that information is unaltered, secured, and authenticated. Verification of financial records can be done without disclosure through zero-knowledge proof in blockchain. Financial sector spends heavily in human labor, commission of middleman, record keeping, maintaining security and privacy of data. But it doesn't guarantee that system is safe from security attacks. Smart contract can be used to bring down the cost of third party, auditing and subsequently increasing the security of database. Blockchain is based on peer-to-peer network which removes intermediaries [34]. It makes monitoring of loan grant more reliable and efficient. It provides instant billing settlement as multiple-layer architecture is removed in blockchain. Records in the Blockchain are irreversible so auditors can ensure proper compliance of records. Any suspectable transaction is easily tracked with timestamp. It also minimizes cost with respect to cross border payment as it includes limited entities on the network.

6 Comparative Analysis of Work Carried Out by Other Researchers

Many researches have contributed in providing blockchain-based solution to enhance the healthcare system. Some have focused on providing completer user control for monitoring and sharing their own health records [21], whereas others have given importance to enhance interoperability of data globally and to secure the sensitive data of patients (Table 2).

7 Challenges of Blockchain

Blockchain will not provide complete solution for all the issues despite the advantages mentioned above. Blockchain is in its initial stage and even cooperate sector is struggling to merge it with existing system [24]. Following are the concern that needs to be addressed first before implementing blockchain in healthcare.

Table 2 Blockchain-based solution in medical field

Author	Objective	Description	Pros	Cons
Tooska Dargahi et al. [1]	Privacy preserving blockchain solution in healthcare	Smart sensors are connected to patient's body for collecting vital information. Collected data is transferred via bluetooth to mobile phones. It then sends that data to healthcare system. Collected data is stored on computer and performs asymmetric cryptographic algorithms. Access policies are specified for every user	Privacy preserving Access control Interoperability	Sensors have low storage and processing speed
Liviu Hirtan et al. [1]	Online healthcare data access while ensuring privacy	Public and private blockchains in the system are used to provide greater confidentiality. Trusted nodes are validated by medical organizations and untrusted nodes where third party want to access medical data. Every participant can send request on the network but only trusted nodes have privilege to add new block	User control High security policies Quick and lightweight consensus algorithm Digital signature	Lacks tracking medical records
Joaqun Ordieres-Mere et al. [5]	Medical records sharing system using blockchain and supported by cloud computing and machine learning	Categorized the data as continuous and instant where continuous data changes rapidly and instant data a fix measurement. It uses symmetric cryptography and uploads the encrypted data on cloud server. For decrypting the data, key is divided among various entities on network	User has complete access control	Difficulty in storing and sharing health records Prone to loss or leakage of data Less secure
Mingqiang Li [13]	Blockchain-based intelligence healthcare with privacy control	Proposed an application to provide users with full control on their own data where they can share it with whom they want to share. Data is shared securely without causing hindrance towards privacy constraints	Patients are responsible for monitoring their own data	Only design implementation is mentioned

7.1 *Storage Capacity*

Healthcare industry consists of collection of medical details, certificates, and pathology reports. Every user on the network have knowledge about the medical data of other users. Considerable amount of storage is required to manage these things [23]. The blockchain technology is still not widely acceptable among various nations.

7.2 *Security Risk*

Blockchain provides security mechanism in healthcare. Still security will remain the major concern for blockchain [3]. Legitimate users should be able to process the data in database [22]. Sharing of data should be done with uttermost safety. What if accidently something went wrong and it will pose a great risk towards personal details of the users.

7.3 *Lack of Willingness to Share*

Patients hesitate to share their sensitive information such as name, address, mobile number, blood group on electronic health record (EHR) because they have a fear of misuse of their data. Even healthcare organization is reluctant in sharing the records with other parties such as health insurance companies or research institutes [9]. Entities should have trust among them so that they can agree on a consensus which will make healthcare system more efficient [4].

7.4 *High Operational Cost*

Blockchain consumes large amount of processing power which result in high cost. Whenever there is new block added on the network, miners have to do proof-of-work which makes these numerical problems not suitable in real world.

7.5 *Low Scalability*

Blockchain faces network congestion when more users join the network. It can store patients ID in the ledger but will not be successful with storing ECG reports of all the users. It cannot handle large data records. These issues can be overcome by using permissioned blockchain [33].

8 Blockchain Healthcare Use Cases

Blockchain transformation introduced advanced cryptography and removed the control of a central authority. It provides much higher transparency for both organizations and consumers. Users have complete control on their transactions [25]. Companies like Facebook, Amazon, and Google have centralized system for their economic advantages but instead they should put the control on the hands of individuals [36]. There exist many challenges in healthcare sector such as drug delivery, sharing health records, tracking prescriptions and other medical equipment but many enterprises are trying to overcome it by implementing blockchain.

8.1 Credential Verification for Healthcare Staff

US-based firm, ProCredEx had built a credential verification system for the healthcare staff by using R3 Corda blockchain protocol. Medical organizations can learn about the professional experience of their staff by log in the credentials to deliver better user experience [26]. It is beneficial for streamlining the recruitment process. It provides transparency and trust to the patients about the past experience of their doctors.

8.2 Remote Monitoring Through IoT Devices

IoT devices are more prone to privacy and security challenges. Patient records must be kept private and in secured environment so that no tempering of data takes place. Smart devices are subject to security attack such as man in the middle and DDoS attack [27]. Blockchain cryptographic algorithm ensures that only authorized users can access confidential data. Records on the blockchain are stored as the hashing function. Users have to perform complex mining process to decode the information which is quite impossible and time consuming. IoT device can communicate with each other without being in a central system [32]. Decentralized blockchain removes the third party, thus protecting the information from security attacks. Blockchain healthcare use cases are in their early stage of development and research continues to grow in future.

8.3 Patient-Centric EHR

Blockchain-based electronic healthcare system was introduced by a leading firm, MedicalChain. Blockchain-based EHR can be interlinked with existing EHR system to provide complete visibility and accessibility to patients. New records such as lab reports, billing, and prescriptions can be converted in a unique hash id [28]. The hash ids can only be decoded when the owner will provide its consent. Patients can share their details with the research institutes and can set the time duration how long it should be visible to them. Insurance provider immediately gets the accurate data of health services from the patients without taking any middleman charges.

8.4 IBM Blockchain

Counterfeit pharmaceutical goods can cause dispute between the suppliers and consumers which can harm the health of the patients and turn into a legal case. Blockchain enables secure and auditable transactions to stakeholders with its tamperproof and immutable features. Huge volume of data is generated during clinical trials which may result in human errors unintentionally or due to poor planning of the concerned authority. Blockchain verifies the information with the participants active on the peer-to-peer network [30].

9 Conclusion and Future Scope

Enhancement in medical treatment with recent technological advancement is a need of an hour. Blockchain supports interoperability which allows doctors to give proper treatment to patient while considering their medical history. In our study we found that most of the blockchain solutions are using smart contract for proper validation of records [29]. Technology should not only be in favor of patients but doctors and other institutes who want to access the healthcare data can view it without affecting the privacy of patient. More focus should be on allowing patients to monitor their health records. They can set access permission about to whom and up to which level other users can view their data. There are various issues with existing healthcare system which can be overcome by blockchain [31]. Distributed, immutable, consensus algorithm and decentralized nature had made blockchain efficient in various fields. There are still some problems which need to be solved first before implementing this technology in healthcare. Blockchain can improve the trust, security, and privacy of individual in coming years. Unlike other healthcare systems which are centralized, blockchain is decentralized. Future of blockchain in healthcare is quite promising.

References

1. L. Hirtan, J.M. Batalla, C. Dobre, P. Krawiec, *Blockchain-Based Approach for E-health Data Access Management with Privacy Protection* (IEEE, Piscataway, NJ, 2019)
2. S.M. Idrees, M. Nowostawski, R. Jameel, Blockchain-based digital contact tracing apps for COVID-19 pandemic management: Issues, challenges, solutions, and future directions. JMIR Med. Inform. **9**(2), e25245 (2021)

3. S. M. Idrees, P. Agarwal, M. A. Alam (eds.), *Blockchain for Healthcare Systems: Challenges, Privacy, and Securing of Data* (CRC, BocaRaton, FL, 2021)
4. S. Tanwar, J. Vora, N. Kumar, S. Tyagi, M.S. Obaidat, J.J.P.C. Rodrigues, *BHEEM: A Blockchain-based Framework for Securing Electronic Health Records* (IEEE, Piscataway, NJ, 2018)
5. J. Ordieres, X. Zheng, R. Vatrapu, R.R. Mukkamala, Blockchain-based personal health data sharing system using cloud storage, in *2018 IEEE 20th International Conference on e-Health Networking, Applications and Services (Healthcom)*, (2018)
6. T. Hardin, D. Kotz, Blockchain in health data systems: A survey, in *2019 Sixth International Conference on Internet of Things: Systems, Management and Security (IOTSMS)*, (2019)
7. T. Dargahi, K.M. Hossein, M.E. Esmaeili, A. Khonsari, Blockchain-based privacy-preserving healthcare architecture, in *2019 IEEE Canadian Conference of Electrical and Computer Engineering (CCECE)*, (2019)
8. M. Li, X. Yue, W. Jiang, H. Wang, *Healthcare Data Gateways: Found Healthcare Intelligence on Blockchain with Novel Privacy Risk Control*. PMID: 27565509. https://doi.org/10.1007/s10916-016-0574-6
9. T. Malas, M. Kassab, P. Laplante, V.V.G. Neto, G. Destefanis, Exploring research in Blockchain for healthcare and a roadmap for the future, in *IEEE Transactions on Emerging Topics in Computing*, (IEEE, Piscataway, NJ)
10. S.M. Idrees, M. Nowostawski, R. Jameel, A.K. Mourya, Privacy-preserving infrastructure for health information systems, in *Data Protection and Privacy in Healthcare*, (CRC, Boca Raton, FL, 2021), pp. 109–129
11. C. Stirbu, V.M. Ionescu, N. Bizon, A. Cirstea, F.M. Enescu, Blockchain technology applied in health the study of Blockchain application in the health system (II), in *ECAI 2018—International Conference—10th Edition Electronics, Computers and Artificial Intelligence*, (2018)
12. S. Tanwar, R. Gupta, B. Sadoun, M.S. Obaidat, N. Kumar, S. Tyagi, *HaBiTs: Blockchain-based Telesurgery Framework for Healthcare 4.0* (IEEE, Piscataway, NJ, 2019)
13. Rather, I. H., & Idrees, S. M. (2021). Blockchain Technology and Its Applications in the Healthcare Sector. In *Blockchain for Healthcare Systems* (pp. 17-25). CRC Press.
14. P.N. Pathirana, D.C. Nguyen, A. Seneviratne, Blockchain for secure EHRs sharing of mobile cloud based e-health systems, in *Special Section on Healthcare Information Technology for the Extreme and Remote Environments*, vol. 7, (IEEE, Piscataway, NJ, 2019), pp. 2169–3536
15. Y. Yang, Y. Ren, K. Fan, S. Wang, H. Li, MedBlock: Efficient and secure medical data sharing via blockchain. J. Med. Syst. **42**(8), 136 (2018). https://doi.org/10.1007/s10916-018-0993-7
16. S. Tanwar, R. Gupta, A. Kumari, Fog data analytics for Iot applications, in *Studies in Big Data Book Series (SBD)*, vol. 76, (2020), pp. 3–17
17. M. Mettler, Blockchain technology in healthcare: The revolution starts here, in *2016 IEEE 18th International Conference on e-Health Networking, Applications and Services (Healthcom), Munich, Germany*, (2016), pp. 1–3. https://doi.org/10.1109/HealthCom.2016.7749510
18. E. Breeden, K. Clauson, T. Mackey, C. Davidson, Leveraging Blockchain technology to enhance supply chain management in healthcare. Blockchain Healthc. Today **1** (2018)
19. J.T. Wilson, M.N.K. Boulos, K.A. Clauson, Geospatial blockchain: Promises, challenges, and scenarios in health and healthcare. Int. J. Health Geogr. **17**, 25 (2018)
20. J. Zhao, X. Liang, J. Liu, S. Shetty, D. Li, Integrating Blockchain for data sharing and collaboration in mobile healthcare applications, in *IEEE 28th Annual International Symposium on Personal, Indoor, and Mobile Radio Communications (PIMRC), Montreal, QC*, (2017), pp. 1–5
21. P. Italiya, J. Vora, S. Tanwar, S. Tyagi, M.S. Obaidat, N. Kumar, Ensuring privacy and security in E-health records, in *2018 International Conference on Computer, Information and Telecommunication Systems (CITS), Colmar*, (2018), pp. 1–5
22. S. Tanwar, J.J. Hathaliya, N. Kumar, S. Tyagi, Securing electronics healthcare records in healthcare 4.0: A biometric-based approach. Comput. Electr. Eng. **76**, 398–410 (2019)
23. V. Balasubramanian, A. Stranieri, I. Gondal, M.A. Uddin, Continuous patient monitoring with a patient centric agent: A block architecture. IEEE Access **6**, 32700–32726 (2018)

24. J. Ainsworth, G. Leeming, J. Cunningham, A ledger of me: Personalizing healthcare using blockchain technology. Front. Med. **6**, 171 (2019)
25. R. Singh, G. Srivastava, A.D. Dwivedi, *Automated Remote Patient Monitoring: Data. Sharing and Privacy Using Blockchain*. http://arxiv.org/abs/1811.03417 (2018)
26. M. Palombini, D. Grishin, T.-T. Kuo, B. Gummadi, K.A. Clauson, G. Church, K. Obbad, R. Barkovich, T.K. Mackey, 'Fit-for-purpose?'–challenges and opportunities for applications of blockchain technology in the future of healthcare. BMC Med. **17**(1), 68 (2019)
27. K. Salah, N. Nizamuddin, A. Al-Fuqaha, M.H.U. Rehman, Blockchain for AI: Review and open research challenges. IEEE Access **7**, 10127–10149 (2019)
28. Z. Xu, A. Dubovitskaya, S. Ryu, F. Wang, M. Schumacher, Secure and trustable electronic medical records sharing using blockchain, in *AMIA Annual Symposium Proceedings*, vol. 2017, (American Medical Informatics Association, Bethesda, MD, 2017)
29. T. Hayajneh, O. Ossipova, K.N. Griggs, C.P. Kohlios, E.A. Howson, A.N. Baccarini, Healthcare blockchain system using smart contracts for secure automated remote patient monitoring. J. Med. Syst. **42**(7), 130 (2018)
30. A. Marcelo, P. Sylim, F. Liu, P. Fontelo, Blockchain technology for detecting falsified and substandard drugs in distribution: Pharmaceutical supply chain intervention. JMIR Res. Protoc. **7**(9), e10163 (2018)
31. M. Tremblay, Medicines counterfeiting is a complex problem: A review of key challenges across the supply chain. Curr. Drug Saf. **8**(1), 43–55 (2013)
32. J. Zhang, J. Ma, Y. Ji, C. Yang, X. Yao, BMPLS: blockchain-based multilevel privacy-preserving location sharing scheme for telecare medical information systems. J. Med. Syst. **42**(8), 147 (2018)
33. J. DeFranco, M. Kassab, T. Malas, V.V.G. Neto, G. Destefanis, Investigating quality requirements for blockchain-based healthcare systems, in *Proceedings of the 2nd International Workshop on Emerging Trends in Software Engineering for Blockchain*, (IEEE, Piscataway, NJ, 2019), pp. 52–55
34. R. Boodoo, A. Alsadi, J. Taylor, T. Patel, M. Diaz, M. Singh, A future for "bitbiopsy" and "cryptospecimen"? proposed use cases of blockchain technology in anatomical and clinical pathology. Mod. Pathol. **31**, 583–583 (2018)
35. I. Inayat, S. Marczak, S.S. Salim, S. Shamshirband, M. Daneva, A systematic literature review on agile requirements engineering practices and challenges. Comput. Hum. Behav. **51**, 915–929 (2015)
36. H. Li, M. Shen, L. Zhu, S. Liu, F. Gao, X. Tao, Blockchain-based data preservation system for medical data. J. Med. Syst. **42**(8), 141 (2018). https://doi.org/10.1007/s10916-018-0997-3

Blockchain Technology for Contact Tracing During COVID-19

Giuseppe Ciaburro

1 Introduction

The blockchain is an incorruptible digital register of transactions, of various types, which can be programmed to record, in a definitive and therefore un-erasable manner, data of any nature and value [1]. The main task of blockchain technology is to eliminate the need for trust between two parties wishing to carry out a transaction, as well as to eliminate centralized third parties that verify said transaction. To do this, it is therefore necessary to create an immutable register of all transactions carried out to verify if the transactions are valid. Blockchain is a technology based on a distributed ledger in which the sequence of transactions is stored in a chain of blocks and each node of the system has a complete copy of that chain. Each block in the chain contains valid transactions verified by the mining nodes [2].

To create a new transaction, the user must have an object to transact and know both his public address, created from his public key, and the public address of the recipient. In a blockchain, all the blocks are linked together through the calculation of a unique identifier, which guarantees the absence of modifications or immutability and not repudiation [3]. Each network based on this technology can use a different consensus algorithm, which defines some rules that must be satisfied to reach an agreement among all the validator nodes. Network users generate transactions that must be added to a block, which in turn must be added to the chain and chained to at least six successive blocks. This condition is essential for this transaction to be considered valid and immutable [4].

G. Ciaburro (✉)
Department of Architecture and Industrial Design, Università degli Studi della Campania Luigi Vanvitelli, Borgo San Lorenzo, Aversa (CE), Italy
e-mail: giuseppe.ciaburro@unicampania.it

S. M. Idrees, M. Nowostawski (eds.), *Transformations Through Blockchain Technology*, https://doi.org/10.1007/978-3-030-93344-9_9

When a node receives a new transaction, it must check that the sender and the recipient have valid public addresses and that the issuer has sufficient funds to carry out the transaction. In this case, a block is added to the transaction that must be filled: Once completed, the block will be submitted for approval by the rest of the miners in the network. Approval will depend on compliance with certain network-dependent conditions. In general, the content of the block should match the content of the copy that each miner owns of the same block and proof of work must be valid. The node that wants to add a new block to the network, also known as a miner, must solve some complex computational calculations which, thanks to certain characteristics of specific functions used, can only be solved by computational force. This allows you to adjust the difficulty and therefore the time required to add new blocks to the chain [5].

We are currently experiencing a complex social situation due to a pandemic caused by the uncontrolled spread of a form of coronavirus called Covid-19. There is a serious lack of contagion control and, consequently, it is necessary to establish mechanisms that give certainty to the population, so that they can understand when a person needs to make a diagnosis of Covid-19 disease [6]. All this is to prevent or detect infection early and relieve symptoms. Although there are currently applications developed for mobile devices to track contacts among the population, these applications are subject to different privacy protection policies according to the legislation of the various countries in which they are applied, as well as other restrictions [7]. These limits are due to the operating system in which they operate, to the insufficient transparency with which its operation is regulated, to the few downloads by users due to the distrust that sensitive data can be revealed, the excessive battery consumption of the devices. For these reasons and with the aim of improving existing applications, in this section we will see how it is possible to overcome these critical issues through the development of an application for mobile devices based on blockchain technology. This application will support the management of the contact tracing necessary to limit the spread of the virus [8].

In this chapter we will first introduce the general concepts underlying blockchain technology, and then move on to analyze practical cases of use of this technology in various fields. Finally, solutions based on blockchain technology for contact tracing will be analyzed with particular interest in the management of the pandemic situation that has affected the entire globe.

The chapter is organized as follows: In Sect. 2, blockchain technologies are introduced. Crucial concepts such as Transaction, Smart Contract, and Proof of Work are also covered. Section 3 analyzes the applications of technology in different fields. Section 4 explores examples of applications of blockchain technology in contact tracking. Finally, Section 5 summarizes the essential points of the work.

2 Exploring Blockchain Basic Concepts

2.1 From the Idea to Its Evolution

We are currently witnessing an authentic technological revolution that will lead to a radical change in the exchange of value, this innovation is called blockchain: Although it is a relatively recent technology, in fact it is starting to arouse interest only in recent years, it really comes from a long evolution over several decades. Humanity has always tried to minimize the uncertainty associated with any decision-making process [9]. The global spread of trade has highlighted the need for the intervention of institutional bodies that could regulate such activities: governments, banks, companies. These institutions were the precursor tools for intervention in the economy, with the aim of reducing the uncertainty generated by the lack of control that the growth of world trade entailed [10].

The bankruptcy of the Lehman Brothers bank in the United States generated a cascading economic crisis that soon brought the entire globe to a phase of recession. Some of the elements that caused the crisis were errors in economic regulation, speculation, corruption, and inappropriate practices of banking entities. These behaviors have undermined consumer confidence in regulatory bodies and controllers of the entire economic process at the root, giving rise to the need for a new intermediation system. The role of the banks in the transactions was to guarantee the correct procedure of this activity, providing guarantees to both parties on the contents of this transaction. But for this function the banks have always requested an economic consideration and given the number of transactions it represented a considerable gain for consumers [11].

We can divide the evolution of the blockchain into three generations, starting from 2008 to the present day. Satoshi Nakamoto, whose true identity is still unknown, developed the Bitcoin project in 2008 and, consequently, blockchain. Collecting previous failed studies on virtual currency, he was able to develop a decentralized electronic system that did not have to depend on third parties. In this way, there was no need to rely on a body which guarantees transactions such as banks. Blockchain is a technology capable of supporting and securely storing any transaction in an immutable history. Bitcoin was the first decentralized digital currency not to require the intermediation of a bank or an entity. At this stage, blockchain was only used to support cryptocurrency. But Bitcoin was gaining authority and hundreds of digital currencies based on the same concept began to emerge. The number of virtual payments made thus began to multiply. In this way, the beneficial features that blockchain had, capable of acting like a large ledger that instantly recorded all the monetary interactions that took place anywhere in the world, were highlighted [12].

The second generation begins with the development of Ethereum [13]. This platform allows developers to create applications and programs directly on the blockchain. At first, this platform was created simply as a database to support the blockchain, but ended up allowing financial instruments, such as loans or bonds and even decentralized social networks, to have a place in this innovation, not just digital currencies. Ethereum's most revolutionary contribution has been the so-called smart contracts. These contracts can be formalized automatically. Specifically, a smart contract is computer software that is programmed to perform tasks that are previously established between multiple parties, that is, compliance with it is not subjective. If an event is satisfied, the consequence starts automatically. This transaction does not need any intermediary body, the blockchain is enough to guarantee compliance and certainty of the contractual conditions. Blockchain has gone from being a simple ledger to a powerful economic and social platform [14].

Currently, blockchain is in its third generation. At this stage, they are trying to solve two of the biggest drawbacks of the blockchain [15]. The first problem is in cryptocurrency miners. These miners do not have a choice as a tool, but their computers. Their work is based on solving complex mathematical operations with their computer equipment to validate and process all transactions that take place on the blockchain. It could be said that they are the ones who support and update the ledger. As a recompense, they receive an amount of the cryptocurrency they are mining. This in the long run implies a high cost, so in this third-generation new platforms are being developed that allow the blockchain to self-manage. The second and biggest problem with blockchain is speed. The speed of this type of operation is measured in transactions per second. Blockchain platforms are very slow. Therefore, in this third generation, developers are placing great emphasis on drastically increasing this speed problem [16].

2.2 Features of Blockchain

Blockchain, also known as a distributed ledger, is a distributed database made up of block chains designed to avoid manipulation and modification. Once a data has been published using a reliable timestamp, it is linked to a previous block to form a chain. In other words, blockchain is an immutable digital book that records all transactions

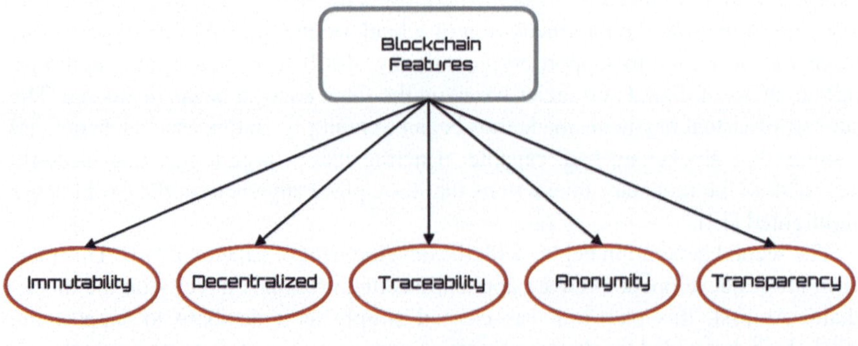

Fig. 1 Essential blockchain features

that are carried out on the network securely thanks to cryptography (Fig. 1). What makes the blockchain so secure is that all the information with all the operations in the database is fragmented and stored in different locations, thus creating a distributed ledger. All data is divided into equal sized chunks which are housed in the computers connected to the network, which act as nodes. So, to steal or manipulate information, the whole system would have to be hacked, which is impossible. Furthermore, the fact that the data is fragmented and distributed makes the information public and verifiable by anyone, even if always in an anonymous form [17].

An essential feature of a blockchain is access to data: Based on access to data, it is possible to distinguish between public or private blockchain. Access to public data is free from any barrier to reading, writing, validating, and participating in the exchange of data [18]. The properties of public access are ease of access, transparency, and less reliability since anyone can write data. Later, private access was introduced, which only allows modification, validation, and participation by trusted users. In this case it is possible to participate only by invitation, increasing confidence as all transactions are carried out in a controlled and closed environment. Although it is not usual, it is also possible to find hybrid blockchains, such as that access to data has mixed properties between public and private blockchains [19].

Another feature of blockchain is the mode of permissions granted to participants. That is, there are blockchains with permissions and others without permissions. In permission-based blockchains, the responsibility for validating and managing all transactions rests with a limited number of trusted users. While, on the contrary, in blockchain without permissions, any user can participate in the processing activities to record all the transactions that take place on the network [20].

In the current third generation of blockchains, blockchains are public and permissionless, so any user can participate and help manage the creation of new data blocks. The reason for allowing this access is that, in the long run, these features favor the growth of the blockchain and the number of participants. This is because, by helping in the consultation, validation and processing activities, these users perform the tasks of mining cryptocurrencies. Then they will be rewarded with a certain amount of any of the existing cryptocurrencies [21].

3 Elements of Blockchain

To fully understand blockchain technology, it is necessary to analyze in detail what its constituent elements are and what meaning they assume. Only in this way will it be possible to understand how the synergy between these resources makes it possible to create a complex system capable of returning an automatic system of transactions [1]. We can then list the following constituent elements of a blockchain (Fig. 2):

- *Transactions*: Transactions are structures made up of various elements, such as inputs, outputs, scripts, or addresses, which indicate the value transferred through

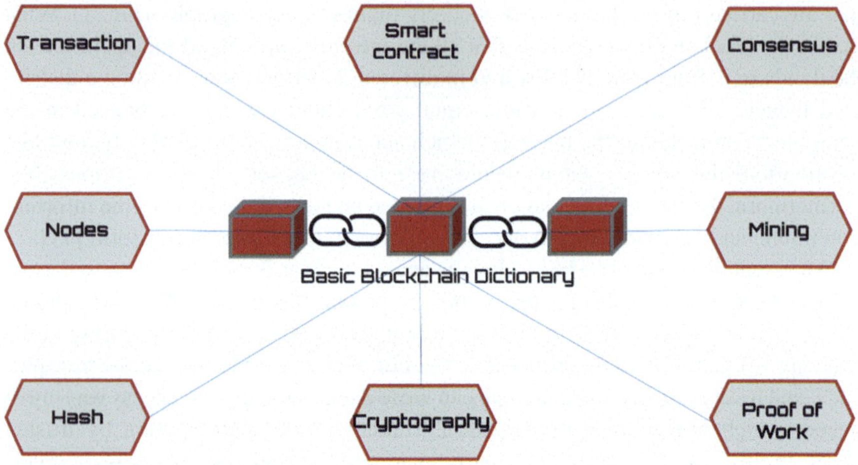

Fig. 2 Basic Blockchain dictionary

an information infrastructure that follows various protocols and algorithms. Each block contains a code that refers to the transfer value of the previous block, this is how the chain is formed and connected [22].

- *Smart contracts:* A smart contract is basically a contract that allows for automatic verification and fulfillment. These types of contracts are decentralized, so they have the intrinsic essence of the blockchain. When the first part satisfies the conditions of the contract, the second part is automatically performed to fulfill the contract, without being subject to any form of evaluation. In this way, the possibility of fraud, infringement, misappropriation, or intrusion by third parties is prevented [23].
- *Cryptography*: Cryptography is a science that investigates the algorithms used to hide information. In blockchain, this technique is fundamental since all encrypted data is shared massively, passing through immense networks of servers and computers, without following any kind of order or hierarchy [24].
- *Mining*: Each block in the blockchain contains information on transactions. The block must be sealed before moving on to the next one, but for this the so-called miners must validate the information contained in it by solving complex mathematical operations. Several users can mine the same block at the same time, although the reward they get will depend on the percentage of mining that each has obtained. Therefore, there is stiff competition among miners, who are totally dependent on the power of their computers. Once the transaction has been validated and joined to the blockchain, the fragmented information is distributed among all users or nodes of the blockchain network, obtaining a copy of everything that is happening.
- *Nodes*: Nodes are the users, or rather the computers, connected to the blockchain network. Its functions are the storage and distribution of fragmented and updated blockchain data. In computer language, it is known as peer-to-peer (p2p), which

is equal to a network of equals or a network of peers. Blockchain is a distributed peer-to-peer network, in which nodes are connected to each other. There are three types of nodes: Broadcast nodes can perform transactions and store information on the blockchain only through third parties. Full nodes can issue transactions, distribute blockchain information, and verify consent compliance. Finally, the mining nodes perform the same functions as a full node, in addition to those of mining.

- *Proof of Work*: This is a mathematical algorithm that, in blockchain, allows you to determine which block of data has priority to enter the chain. The goal is to increase security by preventing cyber-attacks. If blocks have been subjected to computer attacks in the past, they will have a lower priority in joining the chain than other blocks that have not, since this ensures the reliability of the transactions performed [25].
- *Consensus*: It is a protocol that dictates the rules that blocks must respect to be included in a chain. Furthermore, these rules establish the number of rewards that are paid to users who handle transactions so that they act honestly. The consent rules also include procedures to be followed in a multitude of circumstances. For example, if two miners validate blocks at the same time, which block will have priority to enter the chain first? In this case, according to Proof-of-Work, the block that required the most computational effort will be the first to join the blockchain, so the reward of the miner who validated that block will be greater [26].
- *Hash*: In the same way that a fingerprint synthesizes the identification of a person, a hash contains the identity of certain digital data. In blockchain, transactions contain very extensive data. Basically, a hash function is a tool that deals with summarizing these messages, thus managing to lighten the computational calculation necessary to validate and process the data [27].

3.1 How a Blockchain Works

Blockchain is an incorruptible digital book of transactions that can be programmed to record not just financial transactions, but virtually anything of value. Blockchain works like a ledger, where pages are the blocks that make up the chain. Each transaction would imply a new page in the accounting book and each user connected to the network will receive a copy, in read-only mode, on their computers, guaranteeing the reliability of the blockchain system [28] (Fig. 3).

All the blocks that will be added to the chain, in addition to all the information concerning the transactions, contain two codes, one refers to the block that precedes it and the other to the block that succeeds it, except if they are respectively the first or last block. These codes, facilitated by the hash function, are what causes the blocks to be concatenated in the blockchain. The problem is that complex mathematical algorithms must be solved to get the hashes. This is where the miners come in. These, or rather their computer equipment, must solve the voluminous number

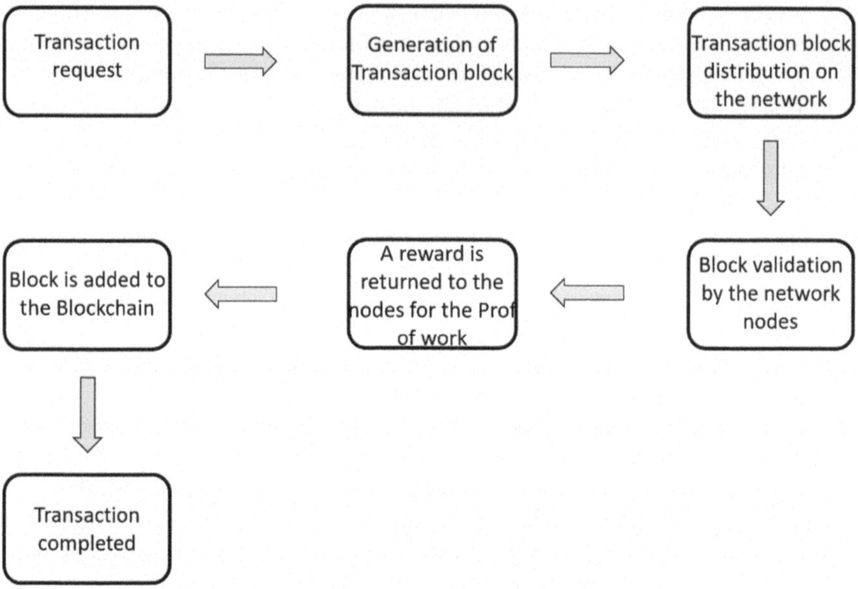

Fig. 3 Operation diagram of a transaction in a blockchain-based framework

of mathematical operations through a process of success and error, validating the information contained in the blocks, before they join the chain and are distributed through the blockchain network. This is the process known as Proof-of-Work. After this procedure, each node connected to the network will receive a copy of the operation. Thanks to the countless number of copies distributed, in addition to the property of immutability, the integrity and truthfulness of the information is guaranteed. Once the transaction has been validated and the block has joined the chain, the transaction will be officially completed, and the related operation will be performed [29].

To be part of the blockchain, not too many conditions are required: A specific blockchain software is enough to start acting as an additional node in the network, being able to perform the mining functions. In this way it will be possible to receive and validate the data so that, later, they are added to the blockchain. If you do not have a computer and specialized equipment, there is no financial benefit from data mining. This is because the necessary mathematical operations require abundant computing capacity. Otherwise, to enter the blockchain network, simply purchase products or services from a website that accepts cryptocurrencies as a form of payment. We will feel like we are buying by credit card, while in the background, miners will check that there is enough balance in the buyer's digital wallet and that the seller's digital account exists. If all data is correct, once consensus is reached, the operation will be validated, and the transaction will be executed [30].

3.2 Blockchain Pros and Cons

Among the advantages offered using blockchain technology, there is the decrease in transaction costs. Today the role of the intermediary is necessary to make any transaction more reliable, since it is the figure who intervenes between the two parties, ensuring and guaranteeing the fulfillment of the contract and the satisfaction of all. But this entails an increase in the monetary and temporal cost of any operation. Blockchain allows you to eliminate the figure of the intermediary and third parties, thanks to its solidity and immediacy. Such is the security that protects this technology that, in its 10 years of life, the blockchain network has never been hacked [31].

Another positive feature of blockchain is consistency. Having thousands of users connected as nodes from anywhere in the world, the system will be in continuous operation, so the network will never fail or suffer. Furthermore, the number of nodes and the correct functioning of the blockchain is directly proportional, that is, the more users there are, the more robust the blockchain will be.

Another benefit involves the security and transparency of transactions. Being a distributed ledger, the information on transactions is shared to all nodes connected to the network and, after its consent and validation, becomes part of a block in the chain in an immutable way, so the data will always be available for future consultation. This information will be 100% reliable, since before being added to the blockchain, it must be verified by half plus one of the nodes connected to the network. The basis of this technology is consent: If all users have the same information, it must be true. Furthermore, being immutable, the data can never be altered, so the information is more rigorous, robust, and transparent.

Finally, there is immediacy, since the figure of third parties is not necessary, there are no intermediaries that slow down the process, be they banks, companies, or any other entity. If two parties wish to enter a transaction, when the established conditions are met, the conditions are executed automatically, without the need for a third party to evaluate and interpret.

Now let's see the disadvantages in the use of this technology: One of the major drawbacks of the blockchain is linked to one of its main advantages. The blockchain is so secure because all nodes that are connected to the network receive a copy of every transaction, but this is a big waste of energy and time, as the same process must be repeated in all nodes to reach consensus. The whole procedure necessary for the correct functioning of the network is much more cumbersome than if it were carried out in a single centralized node.

The blockchain database is constantly growing, every time a block is validated and added to the chain. This entails a necessary increase in the computational requirements necessary for correct functioning since the nodes must increasingly support a greater weight than the blockchains. What causes an unconscious centralization of the process, why, over time, higher technical requirements are needed, generating barriers to entry for any user who wants to participate in the blockchain.

Another problem with blockchain is speed. Distributing the information to all nodes on the network and doing the expensive miner work takes some time.

This time slows down the number of transactions that can be performed in the unit of time, and even today the comparison with the operations performed through traditional channels is winning for the latter [32].

Another notable drawback is the power consumption involved in data mining. Miners require computationally powerful computing equipment that runs 24/7, consuming huge amounts of power. All the benefit that a virtual technology can offer to the environment is limited by the high energy consumption that the operation of the blockchain network implies.

4 Applications of Blockchain Technology

The possible uses of blockchain technology to support real-life needs are manifold. Below we will list some of them trying to group them by sectors of interest.

4.1 Supply Chain

In the field of logistics and supply chain, blockchain has great potential to improve current systems thanks to the advantages of technology such as immutability, non-repudiation or transparency of data and transactions (Fig. 4). The sale of food treated with blockchains provides advantages such as greater efficiency in terms of planning capacity and workload, reduction of food in poor condition caused by poor storage conditions and finally, allows an improvement in the accuracy of the monitoring of each ingredient to ensure compliance with health and hygiene regulations [33].

The agricultural sector can benefit greatly from this technology: The traceability of products is an obvious case of use. Blockchain could be used to create a more solid market for products derived from agriculture, tracing the products in all their

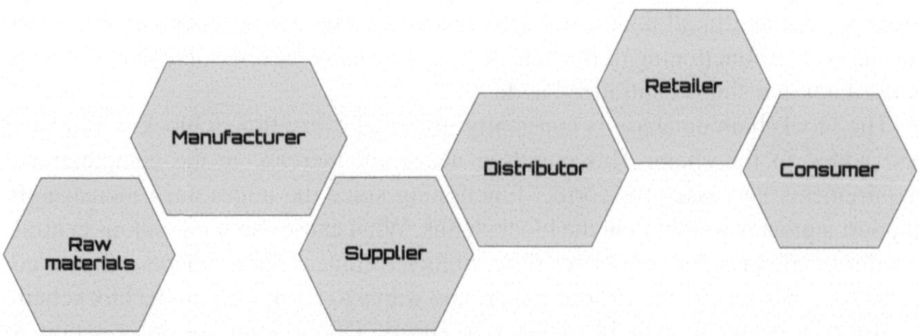

Fig. 4 Supply chain management flow

phases guaranteeing total transparency for the consumer. If we archive all the details of each product in all its phases, the efficiency of the supply chain increases, improving response times in the face of unexpected events, such as counterfeit, contaminated, or damaged products. In addition, it will be possible to obtain a complete trace and find the source of the problem more easily [34].

But this technology can also revolutionize resource monitoring operations: Storing data collected by sensors located on agricultural plantations could be another possible use case. With this data stored in the blockchain, the probability of expansion on international markets increases, respecting their standards, thanks to the transparency and security obtained with this type of solution.

Within the education sector, an educational platform to manage academic certificates can be treated with blockchain technology: Using smart contracts distributed on the public network and storing certain information on the network. The use of blockchain allows the detection of an unauthorized change in any of the platform certificates [35].

Other distributed ledger technologies allow for peer power exchange and decentralized power generation. The use of this technology can lead to a great improvement of the usual processes in this sector, improving internal processes, services to the consumer and reducing costs. It also offers a wide range of possibilities such as tracing the origin of energy, recharging vehicles with tokens or buying and selling energy produced by users of the electricity grid who have the infrastructure to generate excess energy, adding it to the network and receiving some kind of remuneration.

4.2 *Public Administration*

Blockchain technology can prove to be a valid support for public administrations by allowing them greater control and transparency of transactions between government agencies and citizens. In this way, the costs of operations can be significantly reduced, also adding an effective control on fraud and financial errors [36]. For example, the blockchain can prove extremely useful in identity management and in the keeping of personal records. A blockchain-based system that manages citizens' digital identities can be used in any government bureaucratic process, for notary documents or in any third-party service. Among the advantages over a traditional system are citizens' full control of data and the ability to share certain data with third parties. Furthermore, these third parties do not have to store private data of citizens, thus reducing the risks associated with the security of the data and increasing the compliance of the same [37].

In a traditional property register system, there must be a central body in charge of verifying all the necessary requirements. Using the blockchain, it will be possible to eliminate this central authority, replacing it with a verification model based on users who meet certain requirements. Any user identified with their ID will be able to add a resource with its associated documentation, signature, and date and time, and various authorized users will be able to check if this information is correct [38].

An important technological development could relate to voting systems. In this sense, governments could use the blockchain to improve current voting systems. This technology provides greater transparency throughout the process, as well as storing an immutable record of it. In a system of this type, various solutions could be implemented such as the delegation of votes, the vote of ideas proposed by citizens or deputies, models of random elections [39].

4.3 Financial Sector

The financial sector obviously offers more ideas for application since it is precisely from this field that it all began [40]. Cryptocurrencies constitute the first use case of the blockchain and the one that carries the most weight, so far. Today there are almost countless cryptocurrencies on the market, a figure that increases over time. Its main purpose is to be used as an electronic means of payment. But this innovation is not limited only to the use of currencies but can also be exploited for the creation of an e-commerce platform based on cryptocurrencies that connects buyers and sellers without charging a commission to make purchases or sales on those platforms. Furthermore, the platform does not collect any data from the user, and, in the event of a hacker attack, it will not affect users since their private keys are not stored [41].

Banking systems will certainly have benefits in using these technologies: Some banks have already introduced the blockchain in some of their processes, reducing the time required to complete certain transactions. While this technology was only used in a first tier of banking transactions, it could also be used to facilitate real-time exchanges and payments through a specific network, instead of having a centralized entity [42].

A platform composed of a set of smart contracts that are used to provide blockchain-based e-commerce solutions for all types of companies can represent an additional application. The main objective in this case is to facilitate the adoption of blockchain at the corporate level by minimizing the barriers to entry of new communities on the platform, low inactivity of their cryptocurrency, and high security [43].

4.4 Smart Vehicles

Blockchains can be leveraged for sharing data between intelligent vehicles, thus creating an environment of trust between vehicles using the algo-rhythm of consent and rewards [44]. Currently several ad-hoc networks are used for communication between vehicles, but these do not provide sufficient security in data transmission.

Security can be increased using blockchains, thanks to its use of different cryptographic primitives. Such a system can help improve the privacy of intelligent vehicles, provide fast and secure communications between vehicles, and allow for a detailed history of connections between vehicles to be detected. If an accident involving an intelligent vehicle occurs, all necessary data will be sent to the competent authorities [45]. Furthermore, blockchains can help the consumer in the search for economic convenience while respecting their privacy. For example, users can publish their needs and nearby charging stations offer an offer for this demand. Based on all the offers received, the user can freely choose his next charging station. This system allows the user to choose the cheapest and/or nearest charging station without revealing their current location or vehicle data, they will only be revealed when using the charging station [46].

5 Applying Blockchain for Covid-19 Contact Tracing

The health emergency linked to Covid-19 has for some time been demonstrating all the potential typical of a long-lasting global pandemic, characterized by a high mortality rate and a very high impact capacity on the health system of the affected countries [47]. Given the uncontrolled spread of the virus around the globe, states are committed to managing this emergency using a strategy aimed primarily at containing the spread of the virus and preventing further contagion. Almost all of them, therefore, are focusing on the isolation of the infected and their decontamination, on quarantine, on social distancing, on scrupulous hygiene measures and finally on the tracing of social contacts. All the studies focusing on the latter issue agree that, given the level of contagiousness of Covid-19 and the high percentage of transmission through individuals, the control of the epidemic through manual contact tracing is impossible and therefore useless [48].

5.1 Contact Tracing

Contact tracing is a fundamental tool for the prevention and control of the spread of communicable diseases from person to person: it is used daily for the control of infectious diseases [49]. The purpose of this activity is to quickly identify people exposed to existing cases and prevent further transmission of the infection. Unfortunately, at least in the early stages of the pandemic, the social contact tracking tool proved ineffective due to technological limitations and extreme distrust on the part of users. Except in some cases, which had already been able to test these tools in previous pandemics, the tracing did not bring great benefits in controlling the spread of the virus, with the consequence of having to resort to the lockdown which had heavy repercussions on the economy of these countries. The analysis of the experiences in recent months has highlighted the extreme need to develop an effective tracking system, but one that is accepted by the community especially in Western countries [50].

To identify contacts, it is necessary to conduct a detailed epidemiological investigation. In addition, it is necessary to consider the specific contexts identified where exposures to the case may have occurred, and for each context to consider the most appropriate ways to identify all potential contacts. The purpose is to try to quickly identify all the people who may have been exposed to the infection during the period of contagiousness of the case. It is therefore essential to reconstruct, hour by hour, the activities of the case throughout this period, starting from 48 h before and up to 2 weeks after the onset of symptoms or the collection of the positive sample (Fig. 5). Once the exposed persons have been identified, it is also necessary to evaluate their level of exposure and insert them in a database [51].

The contact tracing procedure requires that a series of checks be carried out:

- Establish with certainty the positivity of the person by administering specific diagnostic tests.
- Interview the positive person to identify possible contacts made in an epidemiologically significant period.
- Track down and interview reported contacts.
- Quarantine close contacts for 2 weeks.
- Diagnose subjects during quarantine.
- Repeat the procedure for contacts with positive results.

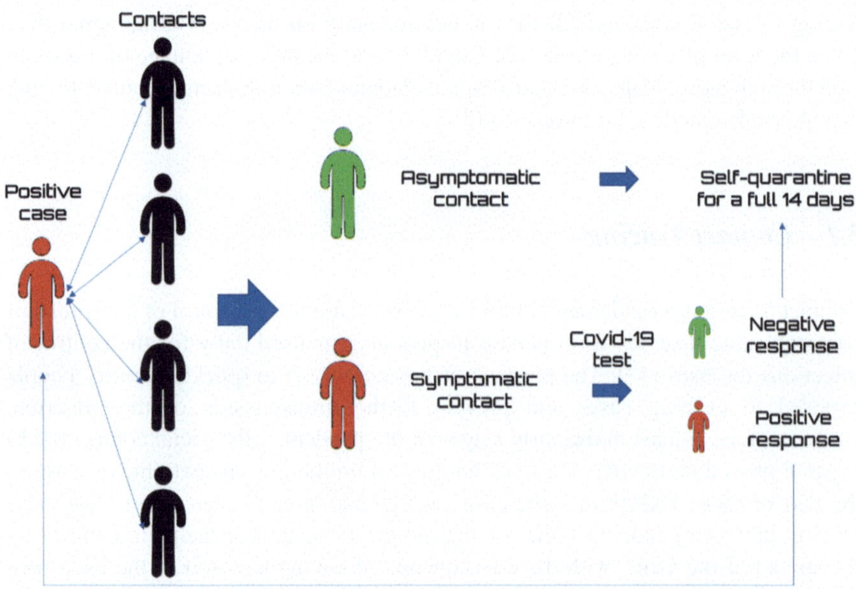

Fig. 5 Contact tracing management procedure

In all these procedures the criticality lies in the identification of the people who met the infected person. In diseases that are transmitted from person to person, the prompt identification of people who have met an infection and their isolation correspond to a clear logic of containment of the contagion. The primary indicator of the effectiveness of contact tracing is the number of secondary cases identified and isolated, that is, the number of cases of people infected by a first confirmed infected person, identified following a test. In addition to the presence of an effective surveillance system and the availability of diagnostic tests, to identify cases of COVID-19 in the population, the effectiveness of contact tracing depends on numerous system variables [52]. The main ones are listed below:

- The convinced and voluntary collaboration of the population in identifying contacts.
- The equally convinced collaboration of the contacts in carrying out the quarantine and monitoring their symptoms.

Scientific evidence on the current COVID-19 epidemic shows the importance of contact tracing, both as a method for containing the virus in the presence of a limited number of cases, and as an effective tool in the context of widespread transmission [53].

5.2 The Best Practices of Contact Tracing in the World

Among the states that have successfully applied the tracing protocols of the infection was South Korea. The Asian country has adopted contact tracing, demonstrating that it has managed to contain the infections in their territories by detecting the movements of citizens [54]. The model adopted by South Korea has shown the possibility of stemming the contagion by acting immediately, without having to resort to a total lockdown of all activities. Swabs were carried out directly from the cars with very fast response times. However, the element that represents the novelty in the fight against the virus is the use of new technologies that has given the possibility to trace and, therefore, identify the outbreaks of Covid-19 present in the country through a system that reported on a map the presence of positive subjects to the virus indicating also when the transit in that determined area had occurred. The collection of these data, anonymized, has allowed the South Korean authorities to map the virus and its spread, indicating, in the individual geographical areas, the presence in a given period of time, of people infected with the virus [55].

A similar approach was adopted by the Socialist Republic of Vietnam which managed to contain the spread of the virus through the adoption of an effective contact tracing campaign [56]. After defining the chain of contagion by identifying close contacts by degree of infection, they came up with a contact locator app called Bluezone that was downloaded more than 20 million times. Additional data, such as

Facebook or Instagram posts and cell phone location data, were also used to compare a person's movements with those reported to contact trackers. In addition, the contact detectors have adopted face-to-face interviews and used the additional surveillance data to enrich the information on the chain of contacts. This is a very invasive approach but has returned excellent results in terms of isolation of the infection. This approach was the result of studies carried out in correspondence with previous pandemics that had hit the country in the previous years [57].

In China, several applications for contact tracking have been developed, even it seems that each city has its own version [58]. These apps have been integrated into the WeChat app which is very popular with the population. Once the contacts have been identified, and a quarantine period has been prescribed, people must enter their daily temperature and health status in the app. At the end of the isolation period, a health status page is generated, which users can use to gain access to buildings and shopping centers. In addition, cell phone owners can send a message to their telecommunications companies to get a list of the locations they have visited. But as we know this option does not return usable data due to the large size covered by the cells. Finally, China's widespread facial recognition technology proved ineffective on a population that used face masks as a protective device [59].

Taiwan has also successfully exploited contact tracing policies, adopting technological solutions to trace the chain of contacts and monitor the social distancing of infected or quarantined people [60]. Cell phone tracking technology has been used to ensure that quarantined people remain in their homes. The system monitors the telephone signals to alert the authorities in the event of non-compliance with the isolation obligation or in the event of the telephone being switched off [61].

The winning contact tracing models that we have listed show that this tracking method allows you to control the virus, but on the other hand, the tracking of people's data can have negative implications within the freedoms recognized and guaranteed by a country. It has been established that tracing is a form of control, of surveillance of the population. What needs to be considered and discussed is the way in which contact tracing is implemented. It must necessarily consider the provisions on privacy and protection of personal data. The systems used by South Korea, but also by China, involved a control over the population, including through video surveillance, with the aim of imposing on infected subjects the obligation to stay at home and not to violate the measures adopted; tracing was adopted through the movements of subjects found positive to reconstruct the ties and relationships with other people, and finally the creation of a sort of census of the virus-positive population to allow the authorities to carry out targeted interventions. As early as February 2020 in South Korea, the government used the geolocation app [62]. There was a prohibition on violating the measures adopted by the government, but in the event of non-compliance with them, a drastic retaliation mechanism was initiated against the person who had violated the quarantine.

5.3 Digital Technologies for Contact Tracing

Even in the event that effective collaboration by the population is ensured, the impossibility of timely identification of the entire chain of contacts must be considered, given the difficulty of identifying people who have had occasional contact with the infected person and of whom the generalities are not known [63]. To overcome this criticality, it is necessary to adopt technological solutions that allow us to trace anyone who has had contact with the infected person in the period of contagiousness of the case. It is evident that the techniques traditionally adopted for the identification of contacts are not able to overcome these criticalities. Traditional contact tracing means that widely practiced in the past makes exclusive use of interviews with cases and contacts, carried out by telephone or with home visits, and other ways to be able to identify all the contacts of a case, depending on the context of exposure, laboratory tests, contact monitoring to verify the possible occurrence of symptoms and application of quarantine/isolation measures and other preventive measures [64].

To tackle this problem with a different approach, it is essential to make use of digital technologies. Digital contact tracing makes use of technological solutions based on digital health, with the electronic-health (e-health) and mobile-health (m-health) components, using technologies for accessing telematic databases and proximity tracing with technologies Bluetooth or GPS. In addition, contact tracing procedures require a high labor cost from operators, so digital tools can play a role in lightening the burden of these procedures by speeding up the contact tracing process [65, 66].

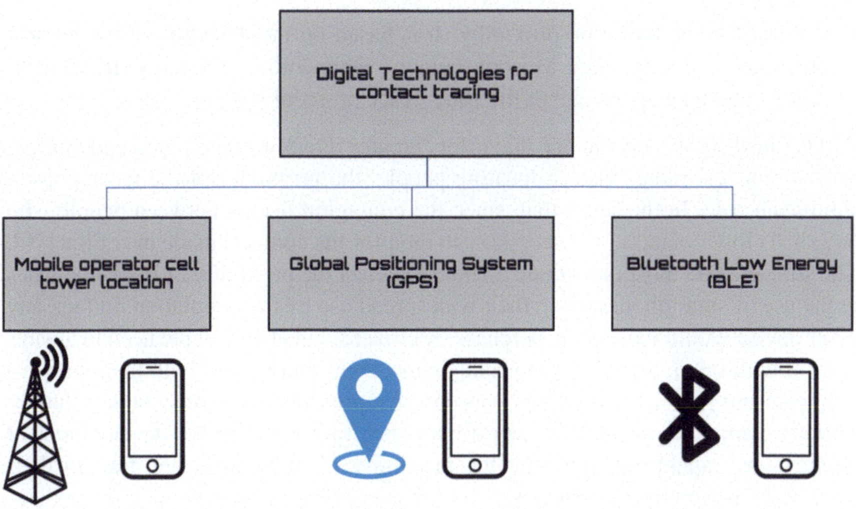

Fig. 6 Digital technologies for contact tracing

Digital technologies for contact tracing can essentially follow three different approaches (Fig. 6):

- Contact tracing based on data provided by the mobile operator: Companies that offer a mobile phone service can detect the position of a device because it constantly communicates with the towers where the signal repetition antennas are installed. These antennas constantly emit unique Tower IDs, to communicate in real time, the position of the towers close to the position of the device. This is necessary to be able to route telephone communication and internet traffic. For the correct functioning of any mobile network, the territory to be covered with the signal is divided into many cells of regular shape, usually hexagonal.

- These cells are of different sizes according to the surface to be covered, the orographic configuration, and the number of people who live in each area. At the center of each cell of the mobile network there is a repeater, called radio base station, which is entrusted with the task of receiving and redistributing the radio signal in its cell of competence [67]. The position of the device can be roughly deduced from the presence in a cell and from the strength of the signal. The operator keeps a register of this tracking which can only be consulted by the police. The main disadvantages of this technology are poor accuracy and privacy risks. For these reasons, it is not typically used to perform contact tracing, but rather to assess the impact of blocking measures and to detect potential contagion hotspots.
- Location-based contact tracing: In this case, the absolute positions of smartphones are used, recorded by the Global Positioning System (GPS) sensor or by Wi-Fi access points. The disadvantages of this technology are the low reliability of the current locating mechanisms because the nodes could easily provide false information: Furthermore, this solution can present serious privacy problems [68].
- Contact tracing based on proximity: It is based on the detection of the relative positions of smartphones. Most of them use Bluetooth Low Energy (BLE) technology and take advantage of the Blue-Trace protocol [69].

The third solution is the one that offers greater reliability, since in the identification of contacts rather than in locating people, the proximity of the latter plays a significant role. In the sense that, since the contagion occurs between people who have had close contacts, it is necessary to monitor the contacts made over time [70]. The simplest and most immediate method to detect the proximity of individuals lies in the use of smartphones given their widespread use by the population. In fact, any other device would require the purchase by citizens, but above all the need to always carry it with them. Since people usually carry their smartphone with them to communicate and for leisure purposes, this device represents the most suitable solution. Various companies around the world have therefore acted in the development of applications capable of detecting the proximity of individuals and tracing their occurrence in appropriate databases. In this way, when a case of contagion is diagnosed, it is possible to trace the chain of contacts that this person had in the period of contagiousness. With this in mind, several countries have acted in the development of efficient proximity tracking applications in compliance with the rights and fundamental freedoms of citizens, including guarantees regarding the processing of their personal data, in line with the values and international standards [71].

The development of such a system represents a great challenge: The underlying technology must provide a proximity tracing mechanism that can be applied homogeneously even beyond individual national borders. Based on it, each country developed its own local version of the application and provided its own secure infrastructure. This allowed each participating country to operationally apply the technological solution in coordination with the local health authorities for the needs of the local population. Each country then worked to inform its citizens in a transparent way to convince them, without resorting to authoritarian impositions, to participate voluntarily in this system. The basic technology, developed in constant comparison with authoritative experts from different disciplines, at least in terms of expectations, wanted to provide an important contribution to allow the tracing of proximity, even in cross-border mode, in respect of privacy, according to a scalable model. and open, which could be used by any country [72].

The selection of the solutions was therefore oriented towards full compliance with the laws and principles regarding privacy and data protection. The mechanisms and technical standards sought have been aimed at protecting privacy, transparency, and security in data management, exploiting the possibilities of digital technology to maximize the speed and real-time capacity of response to the pandemic. These mechanisms include proven proximity tracking technologies, secure and encrypted data anonymization, reliable mechanisms to enable contact between the user and health officials. All this is in a data protection compliant environment, with digital data exchange interfaces capable of providing anonymized contact chains and risk assessment to other applications [73].

The essential functionalities of these applications are:

- Send a notification to people who may have been exposed to a COVID-19 case.
- Invite these people to contact their doctor.

These applications use proximity tracking based on bluetooth low energy technology, without using geolocation [74]. When a user installs this application on his smartphone, the app begins to exchange anonymous identifiers (random codes) with other devices that have installed the same app. These are anonymous codes that do not allow to trace the corresponding device, much less the identity of the person, in full compliance with privacy laws. When a user is positive, the healthcare worker who communicated the outcome of the diagnostic test asks him if he has downloaded the app and invites him to select the option for transferring his anonymous keys to the central system on his smartphone. The app returns a numeric code that the user communicates to the healthcare provider. The code is entered by the healthcare operator within a dedicated management interface and the upload is confirmed by the user. The App notifies users with whom the case has been in contact with the risk to which they have been exposed and the directions to follow, through a message whose text is unique throughout the national territory and which invites them to contact the doctor who will make an initial assessment of the subject's actual risk exposure.

5.4 Blockchain Solutions for Contact Tracing

Despite all these premises, the use of the applications by the population has been insufficient, and many countries, in the critical periods of the various on-dates of contagion that have occurred in the last 2 years, have had to do without this precious tool containment of the contagion and rely on the lockdown. The distrust on the part of citizens lies in the fear of being traced in their movements, and that such information can be used for other purposes [75]. All this even though it has been repeatedly indicated that the contact tracking measure prepared complies with the principles of minimization and the criteria of priority by design and by default. This is because it provides for the collection of only the proximity data of the devices, their processing in pseudonymized form excluding the use of geolocation data and limiting their conservation to the time strictly necessary for the purposes of pursuing the indicated purpose, with automatic cancellation upon expiry of the term of the health emergency [76]. It can be deduced that the refusal to use this technology by citizens lies in the lack of trust in the third party which is represented by the states that must guarantee its safety. To overcome this criticality, a technology based on blockchains can be adopted which guarantees the security of the tracking data and which returns the necessary confidence in the use of this solution [77].

The applications of blockchain technology have covered various sectors, proving to be extremely effective for processing transactions regarding data in a secure way: This peculiarity makes this technology suitable for tracking digital contacts. The timestamp and immutability of the data have revealed the properties of the blockchain essential for the retrieval of accurate and transparent information, while the data encryption guarantees the integrity of the information provided. Finally, the anonymity of the user's identity solves the criticalities related to privacy and confidentiality problems [78].

Blockchain technologies have already been used in healthcare. Hylock et al. [79] have exploited blockchains for the treatment of patients' health data. This is sensitive data that must be treated with great caution, which is why the authors have proposed a new blockchain-based framework called Health Chain. The Health Level-7 Fast Healthcare Interoperability Resources was used to guarantee data exchange, which represents a framework for the electronic exchange of health information. Encryption is guaranteed thanks to the generation of a pair of public and private keys. Public keys are stored in the blockchain and are suitable for securing and verifying transactions. Chenthara et al. [80] have proposed a framework based on blockchain technology for the management of Electronic Health Records (EHR). The framework uses the distributed ledger platform Hyperledger: A distributed register is a source of digital data that is replicated, shared, and synchronized geographically distributed across multiple sites, countries, or institutions. The framework created by the authors then uses InterPlanetary File System (IPFS) to store the EHRs. For the security of transactions, the framework uses a unique cryptographic public key cryptographic algorithm. Dubovitskaya et al. [81] have developed an authorized blockchain-based EHR management system. With the use of this system, the patient will be able to manage their health data also referring to several hospitals. The protection of the patient's privacy will be guaranteed, as well as the security of management of health data, and it will be possible to control access to the patient so that the latter can know who has had access to her data. Xia et al. [82] have proposed a new system for sharing health data based on blockchain. The system was called MeDShare and ensures data auditing and control, through continuous monitoring of access to data and the adoption of smart contracts. Other applications of blockchain technology have concerned the personal health record exchange [83], medical data access and permission management [84], the validation of the patient's identity [85], and a secure supply of drugs [86].

Given the technological features offered by blockchain technology and given the limits manifested by digital health applications, with the electronic-health (e-health) and mobile-health (m-health) components, on the occasion of the pandemic event that hit the world, numerous researchers have studied the implementation of a blockchain-based contact tracing system. Xu et al. [87] have developed a blockchain-based contact tracking system. Blockchain technology ensures transactions compatible with privacy protection regulations. The strong point of this solution is a limited use of the smartphone battery. Hasan et al. [88] proposed a blockchain-based contact tracing framework. The integrated functionalities of the blockchain are exploited to protect user information during use, furthermore users choose who to share their information with and when. The authors adopted the decentralized Ethereum platform integrating it with oracles to manage off-chain data. Oracles are agents that verify and forward data external to the blockchain so that smart contracts can use it. Bandara et al. [89] have developed a blockchain-based digital contact tracking platform called Connect. This platform retrieves the data of people with a positive diagnosis and informs people who have been in contact with them in the immediate vicinity. This procedure significantly reduces the rate of spread of the infection provided that enough people use the platform and follow the recommendations given to contacts. The choice of blockchain preserves the privacy of the people involved by storing digital identities and activity tracking records on a platform that uses proof of personal identity (SSI). Klaine et al. [90] investigated a new contact tracking framework using blockchain. Named as BeepTrace-Active, it exploits the blockchain for the transmission of information and the storage of pseudo-IDs provided by confirmed patients. The user's personal information is never shared, in fact by publishing the pseudo-IDs via the blockchain, the data is protected with the maximum preservation of privacy, since the access information is not traceable if a user matches the record locally. Cases of positive COVID-19 test can be published in the chain, allowing tamper-free, public access to perform local matching. Lv et al. [91] have developed a decentralized and permissionless blockchain protocol for contact tracing. The authors developed a protocol that ensures privacy by linking an interactive zero-knowledge proof protocol and the key escrow mechanism. In this way, personal identity and ownership of location information are decoupled. The owner of the data can access it without revealing the private key. In addition, an incentive assignment mechanism based on the artificial potential field is adopted to incentivize IoT witnesses to pursue the maximum diffusion of monitoring coverage. Arifeen et al. [92] proposed a blockchain-based solution for contact tracing. The authors highlighted the problems associated with the development of contact tracing applications: security problems relating to the third-party server intended to store information and subject to possible cyber-attacks. These security concerns make such applications unpopular among users who do not trust and do not use them. In this context, a blockchain-based contact tracking framework can make contact tracking secure and privacy-aware, which will entice users to use the application. Users can verify their data and preserve their privacy through real-time monitoring. Each access is identified and added to the chain in an immutable way so that it can be identified by the same user. This procedure ensures the user on the security of the transactions.

5.5 *How Blockchain-Based Contact Tracing Work*

The success of blockchain technology is due to the security of transactions that are stored in a distributed decentralized public ledger. A blockchain-based contact tracking system can treat each contact as a transaction. But not only that, access to contact information also becomes a transaction, as does the formulation of a query to know the status of a contact, and finally the registration of a positive case [93].

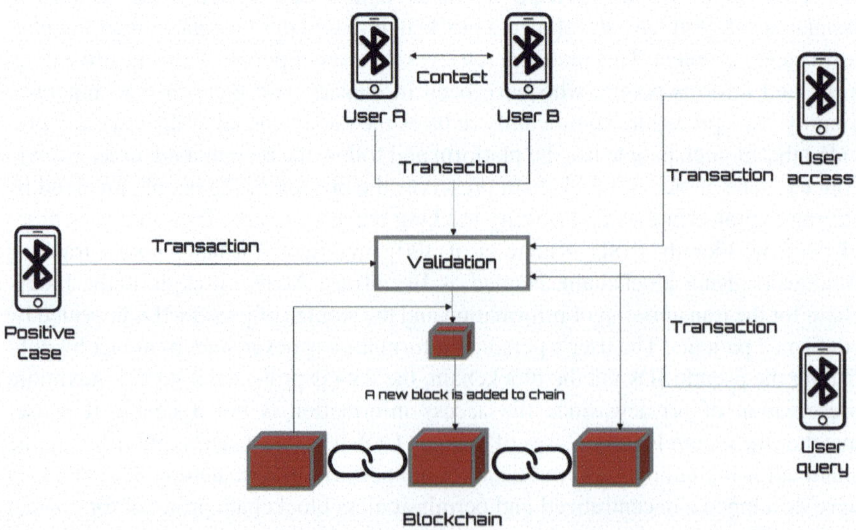

Fig. 7 Blockchain-based contact tracing functional scheme

Whenever a transaction is carried out, it will be verified by the mining nodes and transmitted to all the other nodes of the blockchain (Fig. 7).

Validation will involve adding a new block to the chain and sharing this information with all nodes. A diagnosis of positivity to a Covid-19 test will be added to the chain through a new transaction, in this way it will be shared with all the contacts that will have had in the period of contagion (last 14 days). Each user can then check the status of his contacts through a new query transaction of the distributed database.

The blockchain contact tracking solution enables user-centric data management, so each user holds the power of ownership of the data. Furthermore, the immutability of the data assures us that once a block has been inserted it can no longer be modified, guaranteeing reliability and transparency to all users. The data within the network undergoes a timestamp through the application of a timestamp that eliminates the possibility of discrepancies. The redundancy of the data then ensures its availability for each user since the data is distributed among all the nodes within the network. The origin of the data is also guaranteed through the adoption of a marking system based on the digital signature of the source, and adequate storage of encrypted data that can only be decrypted by an authorized user.

Alongside these evident potentials, it is necessary to highlight the possible limits of the blockchain technology, partly due to the novelty of this solution: In the blockchain, security is guaranteed by the distribution of a copy of each transaction to all nodes of the network they receive, but this requires a use of energy and time, having to repeat a process in all nodes to reach consensus. Furthermore, the blockchain database undergoes a progressive increase in size, which requires important computational resources from the nodes. The speed of transactions currently represents a major limit to the diffusion of blockchain, distributing information to all nodes of the network and performing the expensive work of the miner takes time. Finally, we cannot fail to mention the energy consumption involved in data mining.

6 Conclusion

Blockchain technology is based on the use of a digital transaction register, which records data in a definitive and unchangeable manner. The ledger is distributed among the nodes of the chain, and the sequence of transactions is stored in a chain of blocks and each node of the system has a complete copy of that chain. Each block in the chain contains transactions that are valid and verified by the mining nodes. The main advantage of blockchain is to eliminate the need for trust between two parties who wish to carry out a transaction.

In this chapter we have analyzed how blockchain technology can be exploited to develop applications for contact tracing that guarantee the security of users' privacy. In the pandemic situation we are going through, the need to carry out an effective contact tracing procedure to contain the spread of the contagion has been highlighted. Given the numbers of infected subjects, a traditional approach to tracking contacts immediately proved impractical, it was therefore necessary to resort to digital technologies. For this reason, numerous applications based on the use of smartphones were developed, but many problems were soon raised regarding the security of users' privacy. To overcome these limits, frameworks based on blockchains have been proposed that guarantee the security of users' privacy and encourage them to use it. This is because the success of digital contact tracing depends on the number of users who use this system.

But what future developments can this technology concern? Machine learning and artificial intelligence can expand the potential of blockchains [94–96]. Machine learning allows machines to perform specific tasks by exploiting the experience acquired from the mistakes made, without the need for continuous human contribution to programming [97, 98]. In this way, through a training process, the machines can extract knowledge from the data that can be exploited to identify structures and to predict future scenarios. We have seen that every transaction in a blockchain must be validated. This operation is the task of the mining that validates a new pending block by returning a reward for the efforts of the miners. Through the application of algorithms based on machine learning it is possible to develop more performing mining strategies. Another important contribution to the enhancement of blockchains can concern security: Deep learning lends itself effectively to the identification of malicious programs and the detection of intruders [99–101]. Finally, the ability of machine leaning-based algorithms to process so-called Big Data can help speed up transactions.

References

1. M. Nofer, P. Gomber, O. Hinz, D. Schiereck, Blockchain. Bus. Inf. Syst. Eng. **59**(3), 183–187 (2017)
2. Z. Zheng, S. Xie, H.N. Dai, X. Chen, H. Wang, Blockchain challenges and opportunities: A survey. Int. J. Web Grid Serv. **14**(4), 352–375 (2018)
3. K. Wüst, A. Gervais, Do you need a Blockchain? in *2018 Crypto Valley Conference on Blockchain Technology (CVCBT)*, (IEEE, Piscataway, NJ, 2018), pp. 45–54
4. X. Li, P. Jiang, T. Chen, X. Luo, Q. Wen, A survey on the security of blockchain systems. Futur. Gener. Comput. Syst. **107**, 841–853 (2020)
5. M. Swan, *Blockchain: Blueprint for a new economy* (O'Reilly Media, Newton, MA, 2015)
6. T.P. Velavan, C.G. Meyer, The COVID-19 epidemic. Trop. Med. Int. Health **25**(3), 278 (2020)
7. R. Weissleder, H. Lee, J. Ko, M.J. Pittet, COVID-19 diagnostics in context. Sci. Transl. Med. **12**(546), eabc1931 (2020)
8. Y. Shi, G. Wang, X.P. Cai, J.W. Deng, L. Zheng, H.H. Zhu, Z. Chen, et al., An overview of COVID-19. J. Zhejiang Univ. Sci. B **21**(5), 343–360 (2020)
9. M. Dabbagh, M. Sookhak, N.S. Safa, The evolution of blockchain: A bibliometric study. IEEE Access **7**, 19212–19221 (2019)
10. D.W. Allen, C. Berg, B. Markey-Towler, M. Novak, J. Potts, Blockchain and the evolution of institutional technologies: Implications for innovation policy. Res. Policy **49**(1), 103865 (2020)
11. N. Kube, Daniel Drescher: Blockchain basics: A non-technical introduction in 25 steps. Financ. Mark Portf. Manag. **32**, 329–331 (2018)

12. J. Angelis, E.R. da Silva, Blockchain adoption: A value driver perspective. Bus. Horiz. **62**(3), 307–314 (2019)
13. C. Dannen, *Introducing Ethereum and Solidity*, vol 318 (Apress, Berkeley, CA, 2017)
14. D. Vujičić, D. Jagodić, S. Ranđić, Blockchain technology, bitcoin, and ethereum: A brief overview, in *2018 17th International Symposium Infoteh-Jahorina (Infoteh)*, (IEEE, Piscataway, NJ, 2018), pp. 1–6
15. W. Yang, S. Garg, A. Raza, D. Herbert, B. Kang, Blockchain: Trends and future, in *Pacific Rim Knowledge Acquisition Workshop*, (Springer, Cham, 2018), pp. 201–210
16. V. Titov, M. Uandykova, O. Litvishko, T. Kalmykova, S. Prosekov, T. Senjyu, Cryptocurrency open innovation payment system: Comparative analysis of existing cryptocurrencies. J. Open Innov. Technol. Market Complex. **7**(1), 102 (2021)
17. M. Risius, K. Spohrer, A blockchain research framework. Bus. Inf. Syst. Eng. **59**(6), 385–409 (2017)
18. N. Rifi, E. Rachkidi, N. Agoulmine, N.C. Taher, Towards using blockchain technology for IoT data access protection, in *2017 IEEE 17th International Conference on Ubiquitous Wireless Broadband (ICUWB)*, (IEEE, Piscataway, NJ, 2017), pp. 1–5
19. D. Shrier, W. Wu, A. Pentland, Blockchain & infrastructure (identity, data security). MIT Connect. Sci. **1**(3), 1–19 (2016)
20. A. Ouaguid, N. Abghour, M. Ouzzif, A novel security frame-work for managing android permissions using blockchain technology. Int. J. Cloud Appl. Comput. **8**(1), 55–79 (2018)
21. C.V. Helliar, L. Crawford, L. Rocca, C. Teodori, M. Veneziani, Permissionless and permissioned blockchain diffusion. Int. J. Inf. Manag. **54**, 102136 (2020)
22. X. Xu, I. Weber, M. Staples, *Architecture for Blockchain Applications* (Springer, Cham, 2019), pp. 1–307
23. A. Singh, R.M. Parizi, Q. Zhang, K.K.R. Choo, A. Dehghantanha, Blockchain smart contracts formalization: Approaches and challenges to address vulnerabilities. Comput. Secur. **88**, 101654 (2020)
24. A. Kosba, A. Miller, E. Shi, Z. Wen, C. Papamanthou, Hawk: The Blockchain Model of cryptography and privacy-preserving smart contracts, in *2016 IEEE Symposium on Security and Privacy (SP)*, (IEEE, Piscataway, NJ, 2016), pp. 839–858
25. S.M. Idrees, M. Nowostawski, R. Jameel, A.K. Mourya, Security aspects of Blockchain technology intended for industrial applications. Electronics **10**(8), 951 (2021)
26. Z. Zheng, S. Xie, H. Dai, X. Chen, H. Wang, An overview of blockchain technology: Architecture, consensus, and future trends, in *2017 IEEE International Congress on Big Data (BigData Congress)*, (IEEE, Piscataway, NJ, 2017), pp. 557–564
27. M. Wang, M. Duan, J. Zhu, Research on the security criteria of hash functions in the blockchain, in *Proceedings of the 2nd ACM Workshop on Blockchains, Cryptocurrencies, and Contracts*, (2018), pp. 47–55
28. B. Singhal, G. Dhameja, P.S. Panda, How blockchain works, in *Beginning Blockchain*, (Apress, Berkeley, CA, 2018), pp. 31–148
29. I.H. Rather, S.M. Idrees, Blockchain technology and its applications in the healthcare sector, in *Blockchain for Healthcare Systems*, (CRC, Boca Raton, FL, 2021), pp. 17–25
30. S.S. Sarmah, Understanding blockchain technology. Comput. Sci. Eng. **8**(2), 23–29 (2018)
31. J. Golosova, A. Romanovs, The advantages and disadvantages of the blockchain technology, in *2018 IEEE 6th Workshop on Advances in Information, Electronic and Electrical Engineering (AIEEE)*, (IEEE, Piscataway, NJ, 2018), pp. 1–6
32. M. Niranjanamurthy, B.N. Nithya, S. Jagannatha, Analysis of Blockchain technology: Pros, cons and SWOT. Clust. Comput. **22**(6), 14743–14757 (2019)
33. M.M. Queiroz, R. Telles, S.H. Bonilla, Blockchain and supply chain management integration: A systematic review of the literature. Suppl. Chain Manag. Int. J. **25**(2), 241–254 (2019)
34. R. Azzi, R.K. Chamoun, M. Sokhn, The power of a block-chain-based supply chain. Comput. Ind. Eng. **135**, 582–592 (2019)

35. C.G. Schmidt, S.M. Wagner, Blockchain and supply chain relations: A transaction cost theory perspective. J. Purch. Supply Manag. **25**(4), 100552 (2019)
36. H. Hou, The application of blockchain technology in E-Government in China, in *2017 26th International Conference on Computer Communication and Networks (ICCCN)*, (IEEE, Piscataway, NJ, 2017), pp. 1–4
37. Z. Engin, P. Treleaven, Algorithmic government: Automating public services and supporting civil servants in using data science technologies. Comput. J. **62**(3), 448–460 (2019)
38. V.H. Navadkar, A. Nighot, R. Wantmure, Overview of block-chain technology in government/public sectors. Int. Res. J. Eng. Technol **5**(6), 2287–2292 (2018)
39. D. Cagigas, J. Clifton, D. Diaz-Fuentes, M. Fernández-Gutiérrez, Blockchain for public services: A systematic literature review. IEEE Access **9**, 13904–13921 (2021)
40. J.S. Cermeño, Blockchain in financial services: Regulatory landscape and future challenges for its commercial application. BBVA Res. Pap. **16**, 20 (2016)
41. Q.K. Nguyen, Blockchain—A financial technology for future sustainable development, in *2016 3rd International conference on green technology and sustainable development (GTSD)*, (IEEE, Piscataway, NJ, 2016), pp. 51–54
42. K. Fanning, D.P. Centers, Blockchain and its coming impact on financial services. J. Corp. Acc. Financ. **27**(5), 53–57 (2016)
43. P. Treleaven, R.G. Brown, D. Yang, Blockchain technology in finance. Computer **50**(9), 14–17 (2017)
44. C. Oham, R.A. Michelin, R. Jurdak, S.S. Kanhere, S. Jha, B-FERL: Blockchain based framework for securing smart vehicles. Inf. Process. Manag. **58**(1), 102426 (2021)
45. R.A. Michelin, A. Dorri, M. Steger, R.C. Lunardi, S.S. Kanhere, R. Jurdak, A.F. Zorzo, SpeedyChain: A frame-work for decoupling data from blockchain for smart cities, in *Proceedings of the 15th EAI International Conference on Mobile and Ubiquitous Systems: Computing, Networking and Services*, (2018), pp. 145–154
46. A. Dorri, M. Steger, S.S. Kanhere, R. Jurdak, Blockchain: A distributed solution to automotive security and privacy. IEEE Commun. Mag. **55**(12), 119–125 (2017)
47. COVID, G. A., & Post-Acute Care Study Group, Post-COVID-19 global health strategies: The need for an interdisciplinary approach. Aging Clin. Exp. Res. **32**, 1613–1620 (2020)
48. S. Tian, N. Hu, J. Lou, K. Chen, X. Kang, Z. Xiang, J. Zhang, et al., Characteristics of COVID-19 infection in Beijing. J. Infect. **80**(4), 401–406 (2020)
49. R. Breban, J. Riou, A. Fontanet, Interhuman transmissibility of Middle East respiratory syndrome coronavirus: Estimation of pandemic risk. Lancet **382**(9893), 694–699 (2013)
50. R.S. Dhillon, D. Srikrishna, When is contact tracing not enough to stop an outbreak? Lancet Infect. Dis. **18**(12), 1302–1304 (2018)
51. The Health Protection Agency (HPA) UK Novel Coronavirus Investigation Team, C, Evidence of person-to-person transmission within a family cluster of novel coronavirus infections, United Kingdom. Eur. Secur. **18**(11), 20427 (2013)
52. J. Abeler, M. Bäcker, U. Buermeyer, H. Zillessen, COVID-19 contact tracing and data protection can go together. JMIR Mhealth Uhealth **8**(4), e19359 (2020)
53. W.J. Bradshaw, E.C. Alley, J.H. Huggins, A.L. Lloyd, K.M. Esvelt, Bidirectional contact tracing could dramatically improve COVID-19 control. Nat. Commun. **12**(1), 1–9 (2021)
54. Y.J. Park, Y.J. Choe, O. Park, S.Y. Park, Y.M. Kim, J. Kim, COVID-19 National Emergency Response Center, Epidemiology and Case Management Team, Contact tracing during coronavirus disease outbreak, South Korea, 2020. Emerg. Infect. Dis. **26**(10), 2465–2468 (2020)
55. S. Park, G.J. Choi, H. Ko, Information technology–based tracing strategy in response to COVID-19 in South Korea—Privacy controversies. JAMA **323**(21), 2129–2130 (2020)
56. H.L. Quach, N.A. Hoang, COVID-19 in Vietnam: A lesson of pre-preparation. J. Clin. Virol. **127**, 104379 (2020)
57. T.H. Nguyen, D.C. Vu, Summary of the COVID-19 outbreak in Vietnam—Lessons and suggestions. Travel Med. Infect. Dis. **37**, 101651 (2020)

58. Q. Bi, Y. Wu, S. Mei, C. Ye, X. Zou, Z. Zhang, T. Feng, et al., Epidemiology and transmission of COVID-19 in 391 cases and 1286 of their close contacts in Shenzhen, China: A retrospective cohort study. Lancet Infect. Dis. **20**(8), 911–919 (2020)
59. J. Zhang, M. Litvinova, Y. Liang, Y. Wang, W. Wang, S. Zhao, H. Yu, et al., Changes in contact patterns shape the dynamics of the COVID-19 outbreak in China. Science **368**(6498), 1481–1486 (2020)
60. H.Y. Cheng, S.W. Jian, D.P. Liu, T.C. Ng, W.T. Huang, H.H. Lin, Contact tracing assessment of COVID-19 transmission dynamics in Taiwan and risk at different exposure periods before and after symptom onset. JAMA Intern. Med. **180**(9), 1156–1163 (2020)
61. R. Steinbrook, Contact tracing, testing, and control of COVID-19—Learning from Taiwan. JAMA Intern. Med. **180**(9), 1163–1164 (2020)
62. F. Almeida, Technological solutions for screening and geolocation of COVID-19 infected people: A country comparison approach. Front. Health Inform. **10**(1), 62 (2021)
63. P.N. Owusu, Digital technology applications for contact tracing: The new promise for COVID-19 and beyond? Glob. Health Res. Policy **5**(1), 1–3 (2020)
64. M. Leslie, COVID-19 fight enlists digital technology: Contact tracing Apps. Engineering (Beijing, China) **6**(10), 1064–1066 (2020)
65. A. Anglemyer, T.H. Moore, L. Parker, T. Chambers, A. Grady, K. Chiu, L. Bero, et al., Digital contact tracing technologies in epidemics: A rapid review. Cochrane Database Syst. Rev. **8**, CD013699 (2020)
66. P. Rodríguez, S. Graña, E.E. Alvarez-León, M. Battaglini, F.J. Darias, M.A. Hernán, L. Lacasa, et al., A population-based controlled experiment assessing the epidemiological impact of digital contact tracing. Nat. Commun. **12**(1), 1–6 (2021)
67. I. Ekong, E. Chukwu, M. Chukwu, COVID-19 mobile positioning data contact tracing and patient privacy regulations: Exploratory search of global response strategies and the use of digital tools in Nigeria. JMIR Mhealth Uhealth **8**(4), e19139 (2020)
68. T. Martin, G. Karopoulos, J.L. Hernández-Ramos, G. Kambourakis, I. Nai Fovino, Demystifying COVID-19 digital contact tracing: A survey on frameworks and mobile apps. Wirel. Commun. Mob. Comput. **2020**, 1–29 (2020)
69. G. Cencetti, G. Santin, A. Longa, E. Pigani, A. Barrat, C. Cattuto, B. Lepri, et al., Digital proximity tracing on empirical contact networks for pandemic control. Nat. Commun. **12**(1), 1–12 (2021)
70. D.J. Leith, S. Farrell, Coronavirus contact tracing: Evaluating the potential of using bluetooth received signal strength for proximity detection. ACM SIGCOMM Comput. Commun. Rev. **50**(4), 66–74 (2020)
71. V. Von Wyl, S. Bonhoeffer, E. Bugnion, M.A. Puhan, M. Salathé, T. Stadler, N. Low, et al., A research agenda for digital proximity tracing apps. Swiss Med. Wkly. **150**, w20324 (2020)
72. R. Hinch, W. Probert, A. Nurtay, M. Kendall, C. Wymant, M. Hall, C. Fraser, et al., *Effective Configurations of a Digital Contact Tracing App: A Report to NHSX* (2020). Accessed 23 Jul 2020
73. P.C. Ng, P. Spachos, K.N. Plataniotis, COVID-19 and your smartphone: BLE-based smart contact tracing. IEEE Syst. J. (2021)
74. M. Cunche, A. Boutet, C. Castelluccia, C. Lauradoux, D. Le Métayer, V. Roca, *On Using Bluetooth-Low-Energy for Contact Tracing* (Doctoral Dissertation, Inria Grenoble Rhône-Alpes; INSA de Lyon) (2020)
75. M. Platt, A. Hasselgren, J.M. Román-Belmonte, M.T. de Oliveira, H. De la Corte-Rodríguez, S.D. Olabarriaga, T.K. Mackey, et al., Test, trace, and put on the Blockchain?: A viewpoint evaluating the use of decentralized Systems for Algorithmic Contact Tracing to combat a global pandemic. JMIR Public Health Surveill. **7**(4), e26460 (2021)
76. L. Ricci, D.D.F. Maesa, A. Favenza, E. Ferro, Blockchains for covid-19 contact tracing and vaccine support: A systematic review. IEEE Access **9**, 37936–37950 (2021)

77. D. Marbouh, T. Abbasi, F. Maasmi, I.A. Omar, M.S. Debe, K. Salah, S. Ellahham, et al., Blockchain for COVID-19: Review, opportunities, and a trusted tracking system. Arab. J. Sci. Eng. **45**, 1–17 (2020)
78. A. Hasselgren, K. Kralevska, D. Gligoroski, S.A. Pedersen, A. Faxvaag, Blockchain in healthcare and health sciences—A scoping review. Int. J. Med. Inform. **134**, 104040 (2020)
79. R.H. Hylock, X. Zeng, A blockchain framework for patient-centered health records and exchange (HealthChain): Evaluation and proof-of-concept study. J. Med. Internet Res. **21**(8), e13592 (2019)
80. S. Chenthara, K. Ahmed, H. Wang, F. Whittaker, Z. Chen, Healthchain: A novel framework on privacy preservation of electronic health records using blockchain technology. PLoS One **15**(12), e0243043 (2020)
81. A. Dubovitskaya, F. Baig, Z. Xu, R. Shukla, P.S. Zambani, A. Swaminathan, F. Wang, et al., ACTION-EHR: Patient-centric block-chain-based electronic health record data management for cancer care. J. Med. Internet Res. **22**(8), e13598 (2020)
82. Q.I. Xia, E.B. Sifah, K.O. Asamoah, J. Gao, X. Du, M. Guizani, MeDShare: Trustless medical data sharing among cloud service providers via blockchain. IEEE Access **5**, 14757–14767 (2017)
83. H.A. Lee, H.H. Kung, J.G. Udayasankaran, B. Kijsanayotin, A.B. Marcelo, L.R. Chao, C.Y. Hsu, An architecture and management platform for blockchain-based personal health record exchange: Development and usability study. J. Med. Internet Res. **22**(6), e16748 (2020)
84. A. Azaria, A. Ekblaw, T. Vieira, A. Lippman, Medrec: Using blockchain for medical data access and permission management, in *2016 2nd International Conference on Open and Big Data (OBD)*, (IEEE, Piscataway, NJ, 2016), pp. 25–30
85. M. Mettler, Blockchain technology in healthcare: The revolution starts here, in *2016 IEEE 18th International Conference on E-health Networking, Applications and Services (Healthcom)*, (IEEE, Piscataway, NJ, 2016), pp. 1–3
86. S. Baunm, Health IT startups working to secure pharma supply chains?. Medcity News. [Online] (2017). https://medcitynews.com/2017/01/drug-supply-chain-security-and-technology/. Accessed 12 May 2021
87. H. Xu, L. Zhang, O. Onireti, Y. Fang, W.J. Buchanan, M.A. Imran, BeepTrace: Blockchain-enabled privacy-preserving contact tracing for COVID-19 pandemic and beyond. IEEE Internet Things J. **8**, 3915–3929 (2020)
88. H.R. Hasan, K. Salah, R. Jayaraman, I. Yaqoob, M. Omar, S. El-lahham, COVID-19 contact tracing using Blockchain. IEEE Access **9**, 62956–62971 (2021)
89. E. Bandara, X. Liang, P. Foytik, S. Shetty, C. Hall, D. Bowden, K. De Zoysa, et al., A blockchain empowered and privacy preserving digital contact tracing platform. Inf. Process. Manag. **58**(4), 102572 (2021)
90. P.V. Klaine, L. Zhang, B. Zhou, Y. Sun, H. Xu, M. Imran, Privacy-preserving contact tracing and public risk assessment using Blockchain for COVID-19 pandemic. IEEE Internet Things Mag. **3**(3), 58–63 (2020)
91. W. Lv, S. Wu, C. Jiang, Y. Cui, X. Qiu, Y. Zhang, Towards large-scale and privacy-preserving contact tracing in COVID-19 pandemic: A Blockchain perspective, in *IEEE Transactions on Network Science and Engineering*, (IEEE, Piscataway, NJ, 2020)
92. M.M. Arifeen, A. Al Mamun, M.S. Kaiser, M. Mahmud, *Blockchain-Enable Contact Tracing for Preserving User Privacy During Covid-19 Outbreak* (2020)
93. S.M. Idrees, M. Nowostawski, R. Jameel, Blockchain-based digital contact tracing apps for COVID-19 pandemic management: Issues, challenges, solutions, and future directions. JMIR Med. Inform. **9**(2), e25245 (2021)
94. G. Ciaburro, G. Iannace, J. Passaro, A. Bifulco, D. Marano, M. Guida, F. Branda, et al., Artificial neural network-based models for predicting the sound absorption coefficient of electrospun poly (vinyl pyrrolidone)/silica composite. Appl. Acoust. **169**, 107472 (2020)
95. F. Hutter, L. Kotthoff, J. Vanschoren, *Automated Machine Learning: Methods, Systems, Challenges* (Springer Nature, Cham, 2019), p. 219

96. G. Ciaburro, G. Iannace, M. Ali, A. Alabdulkarem, A. Nuhait, An artificial neural network approach to modelling absorbent asphalts acoustic properties. J. King Saud Univ. Eng. Sci. **33**(4), 213–220 (2021)
97. A. Rajkomar, J. Dean, I. Kohane, Machine learning in medicine. N. Engl. J. Med. **380**(14), 1347–1358 (2019)
98. G. Iannace, G. Ciaburro, A. Trematerra, Fault diagnosis for UAV blades using artificial neural network. Robotics **8**(3), 59 (2019)
99. G. Ciaburro, Sound event detection in underground parking garage using convolutional neural network. Big Data Cognit. Comput. **4**(3), 20 (2020)
100. G. Ciaburro, G. Iannace, Improving smart cities safety using sound events detection based on deep neural network algorithms, in *Informatics*, vol. 7, (Multidisciplinary Digital Publishing Institute, Basel, 2020), p. 23
101. G. Ciaburro, Security Systems for Smart Cities Based on acoustic sensors and machine learning applications, in *Machine Intelligence and Data Analytics for Sustainable Future Smart Cities*, (Springer, Cham, 2021), pp. 369–393

Transforming Healthcare Sector in India Through Blockchain Technology: Challenges and Opportunities from Legal Perspectives

Shambhu Prasad Chakrabarty and Souvik Mukherjee

1 Introduction

Information technology (IT) today has simplified and automated various aspects of work once done through a slow and complicated mechanism. Errors can now be mitigated at the origin, providing adequate assurance of quality output. Professionals from all sectors now rely upon IT, whether on hardware or software. Various industries currently use sophisticated data manipulation techniques. A wide variety of industries have also included data manipulation techniques to facilitate their functioning. Even law enforcement agencies are using technology to refrain from illegal activities on the one hand and identify wrongdoers on the other. Complicated challenges like detecting credit card fraud are now being mitigated using statistical data mining techniques [1]. The IT revolution has changed the world order.

Information technology is growing at a meteoric speed; often, society, in general, cannot keep up with the pace of development. The development of law in the sector is also hindered by the mismatched pacing of the technology-making entities and law-making entities. Hence, it is not unexpected that laws are mostly playing catch up with technological development. Thus, the pace factor is becoming one of the significant challenges for innovation in information technology and its applicability at the outset (Fig. 1).

S. P. Chakrabarty (✉)
Centre for Regulatory Studies, Governance and Public Policy, The West Bengal National University of Juridical Sciences, Kolkata, West Bengal, India
e-mail: spc@nujs.edu

S. Mukherjee
The West Bengal National University of Juridical Sciences, Kolkata, West Bengal, India

S. M. Idrees, M. Nowostawski (eds.), *Transformations Through Blockchain Technology*, https://doi.org/10.1007/978-3-030-93344-9_10

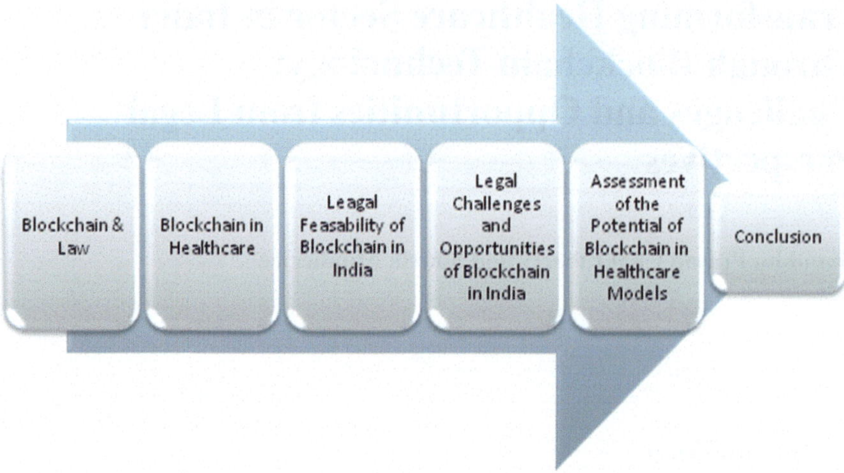

Fig. 1 Structure of the chapter

The significant contribution of this chapter is as follows:

- It reviews the interaction of blockchain technology with law.
- It reviews the legal impediments and opportunities in the application of blockchain technology in healthcare in India. It also highlights the potential design and deployment challenges that the technology needs to mitigate to be effectively acceptable.

The methodology adopted to justify the objectives of this chapter is doctrinal. The researchers substantially explored various aspects of the method. They critically assessed the existing legal positions of blockchain in healthcare and related challenges thereto, in the Indian context, to identify strategic models that may be adopted to make blockchain technology in healthcare viable in developing economies.

2 Blockchain Technology and Law

Discoveries and inventions we enjoy today often appear insignificant and behave spontaneously. The technological marvels are given much lesser acknowledgement by society than they deserve. Inventions or discoveries cannot exist in a vacuum and must be relevant from a social perspective to be a part of human discourse. In addition to the social utility, the invention or discovery also needs to adapt itself to the existing legal norms and principles that govern human life. If there lies a vacuum, then technological development needs to be crafted in compliance with potential legal issues and future challenges.

Blockchain is such a species of IT that is at the receiving end of much attention and trend. However, blockchain technology struggles with legal feasibility and is indeed on a cross-road with law, whether domestic or international.

2.1 The Blockchain Technology

By its name, blockchain technology speaks for itself, that is a combination of blocks and chains. A kind of web technology that relies on a growing network of data containers or chunks of data, each called a block, linked together or chained together, forming a network of the decentralized database system, which securely, immutably and most important privately preserves data, which are illegible by any unauthorized person or user [2]. Linking of 'block to block' is done using hash function. Each block contains the previous block's cryptographic hash, which ties together a timestamp of data creation, modification and transaction data. No central authority, but blockchain data are managed in a decentralized manner [2]. The transaction entered in the public ledger is subsequently verified by the consensus of most participants who had provided inputs. Once recorded, a Cryptographic Hash Function is created, altering the data becomes highly challenging, furthermore changing the information in the previous blocks in the chain is virtually impossible. Hence, a blockchain is considered the most secure, unalterable, and immutable so far its decentralized structure is concerned. Cryptographic applications make the data in the blockchain impossible to be read by an unauthorized participant or user. Since there is no centralized authority, who manages a central database and thus supplies specific relevant data to a particular user, who accesses the system of the database with the help of user ID and password security-based portal, and a completely decentralized network of a database is managed collectively by each of the participants, it is also called a DLT, i.e. Distributed Ledger Technology. It is managed by a peer-to-peer network [3]. The ingenuity of the security feature offered by blockchain technology potentially can not only change the landscape of supply chain management, rather it could have its presence reflected in general management of establishments in the digital age [4]. The environment and the platform upon which the blockchain technology operates make it apt for its adaptation in several sectors such as data management, healthcare, governance, banking, manufacturing industry, etc. [5].

The blockchain technology resurgence is undergoing an elaborate evolutionary process, starting with the Bitcoins at 1.0, Smart Contracts at 2.0, Decentralized Applications (DApps) at 3.0, presently it is being considered for Industrial Applications at 4.0 [6]. Once smart contracts are implemented by blockchain technology, their execution was decentralized and distributed within the network of nodes [7].

Irrespective of tons of confusion and loads of overstatement on the capacity of blockchain technology, it is applied in almost every field of commercial and noncommercial operations [8]. Blockchain network, with its inherent 'Distributed

Ledger Technology', can be introduced reliably. Cryptocurrencies, financial services, energy trading, food trading, video games, healthcare are the primary fields where the application of blockchain network has high potential to deliver good results. There has been a plethora of literature addressing the technological aspect of blockchain technology. In contrast, limited literature explored the concerns with the technology's interaction with the law, especially developing states such as India; the following segments would explore this very aspect of blockchain.

2.2 *Blockchain Technology in Crossroads with Law*

Blockchain technology (public) is an evolved technology for the maintenance of data. It is a distributed ledger that is immutable and tamper proof. On paper, the technology is a leap forward in terms of maintenance of record. Does the question arise what possible challenges it may face from the legal or policy positions? Ideally, a fool-proof record-keeping mechanism should not be subjected to debate. It only brings transparency, security, certainty, immutability in the frame; these are desired qualities in any system. However, often laws of several countries refused to look eye to eye with the technology. This hesitancy could be apportioned (to the existing process) for framing the law for potential challenges technology poses towards the modern concept of state, governance, the legal system, policy goals, social structure and individual's rights.

Cryptocurrencies, the application of blockchain technology 1.0 has been the buzz word over the last decade; often, the blockchain and bitcoins are used interchangeably by simpletons. With cryptocurrency, a peer-peer electronic cash system was established, which did not need trusted third parties' intermediaries. Still, two parties who did not need to trust each other could engage in the transaction [9]. Such an environment was the gift from blockchain technology due to its novel contribution of 'distributive/decentralized', 'immutability' and 'incorruptibility' characteristics.

In light of the above, let us deliberate upon the potential reason for reluctance towards the technology by the states and, thereby, framing laws that is illegal to deal with specific blockchain applications, e.g. cryptocurrency. The technology makes it a potent tool in the hands of the people who are technologically educated and in a specialized field; it is a distributed system where the participants or 'miner' could be anyone with the technological expertise and act. The technology, albeit independent, is managed by a fraction of the world's population due to its operational nature. These individuals can act beyond the man-made geographical boundaries; thus, they may or may not be governed by specific laws of any particular state. Information is considered the most potent weapon in the twenty-first century; therefore, it cannot be overstated that states and citizens may feel vulnerable to such technology. The 'independent' characteristics of the ledger maintenance could be considered the most susceptible characteristics of the technology from the legal and policy perspective of particular states. As the novel element of the technology is decentralization, it could often be potentially participated by such individuals and groups with nefarious intentions towards certain sections of people or a community of states. The nation states are wary of technocrat terrorists, who may use the blockchain technology to acquire information and alter the information by mode of majority consensus; what we know of it is that the possibilities are endless. These security concerns are worthy of paying attention to before accepting the technology in critical spheres of governance and administration. The fears regarding decentralization also bring up accountability; the decentralized and indiscriminate geographical participation raises the issue of eventual accountability and responsibility regarding the misuse of the information.

Blockchain 1.0 application through cryptocurrency took the world by storm. There was this new form of currency in existence that one would not feel physically or tangible; however, it could be used to exchange goods and services and even convert them into the recognized currency of a particular nation state. The question arises that could cryptocurrency be considered as money. In the modern world, defining money could be challenging as it is; could it be a medium for exchange, or means of payment, wealth ascertainment tool? The definition could be any of the above, but it could not be identified without being dependent on its institutional certainty [10]. There exist debates on the valuation of money, as it is considered as that historically it was valued upon the scarcity of gold [11]. However, the connections of money with gold in terms of transformability are historical accounts of acceptability of the form of money [11], just as the bartering mechanism. Money represents a promise; assuming that it is the accurate representation of money's character, anyone can create money theoretically. But mere promise without another person accepting the promise, money cannot be created. Hence, there need to be two sides in the structure, one making the promise, another accepting the promise [10]. The concept of state came with the uniform exchange tokens. The work of Keynes and Kanpp's elaborates on the relation between state and money. Keynes states that:

> *The State money was reached when the state claimed the right to declare what should answer as money to the current money of account- when it claimed the right not only to enforce the dictionary but also to write the dictionary. Today all civilized money is, beyond the possibility of dispute, chartalist* [10].

With modern economies, the state is responsible for establishing a stabilized economy by maintaining money and preventing economic depression [11]. It is well accepted that money is a creature of the state. The most critical aspect of money, i.e. its acceptability, is determined by the state, which brings us to the initial challenge posed by acceptance of blockchain 1.0, i.e. cryptocurrencies as money. The technology may have created an exchangeable entity. However, its acceptance as money lies with each state. With no role in its creation and limited connection with that state's economy, it is a rather challenging endeavour to accept the state's acceptability in general. Hence, the reluctance and hesitant towards cryptocurrency may provoke the state to declare them illegal for all practical purpose within their jurisdiction.

Blockchain technology, especially the public one, raises critical questions towards the security interest of the state. The security concern may not limit itself to an armed attack. Instead of an economic crisis, health crisis and information crisis, it can also be apportioned to the general security threats that the state may feel regarding the application of blockchain technology in critical areas. One may argue that other than military or armed threat, the rest does not amount to security threat; however, the world has changed over the years, we have witnessed a significant rise in respect towards the political independence and territorial integrity of a particular state when it comes to direct military action; however, there has been a spike in terrorism through non-state actors, such activities have also moved from armed events and have engaged in the use of technology to inflict injuries to states. Furthermore, it would be rather not so prudent to assume that economic crisis could not form security concerns, as the difficult financial situation can create significant havoc and raise questions on the lives of the population and the ability of the government to administer, akin to military attack [12]. The fact that the miner's indeterminate geographical location can impact the creation of the blockchain also acts as a deterrence towards accepting the technology by the lawmakers, as the miner is beyond the territorial jurisdiction of the home state. Governance of the state will potentially be weakened when interfered with by individuals beyond a particular state's jurisdiction.

The potential resistance which the technology might face from the law could also be apportioned to the mechanism laws that are made in a state. Apart from the inherent challenges the technology poses towards the state, as discussed in brief in paragraphs described above, it is imperative to note the law-making process and consider the factors that influence the framing of law in a particular jurisdiction. Irrespective of the schools of thought, whether it is natural, positivist/analytical, historical, sociological, or realist school, it cannot be denied that laws are connected to the subjects it intends to govern. The modern legal system is not influenced exclusively by a particular school of thought; instead, it is an amalgamation of the views propounded by the schools. The creation of law is predominantly a sovereign act, and such design will only occur when a state feels the need to regulate, permit, or restrict a particular phenomenon. The critical question that arises here is whether the state is keen to enact laws regulating or allowing or restricting blockchain technology or not. To determine the problem mentioned earlier, states need to inquire on the following aspect: what technological aspect and the extent of implementation? What is the life cycle of technology? Is society ready to induct such technology? How far has preceding technology been inducted? Is the present system where the technology is intended to be implemented a failed system? Would the technology bring in additional financial burden on the state and its subject? Is the introduction of technology going to run contrary to constitutional principles? Would it adversely impact the fundamental rights of individuals? What is the extent of potential abuse of the technology? Would people denying the technology amount to injustice? Who is going to benefit from the incorporation of the technology? How would the international relation or obligations to international law be impacted by acceptance, rejection or regulating the technology? and so on. While determining the legal position of blockchain, some of the questions are essential to realizing that while some are legal dilemmas, others could be policy questions, sociological concerns, economic issues—many of such concerns from the Indian perspective in a later segment.

One may argue that the legal concerns involving privacy, accessibility, security concerns can be reduced significantly by using the private blockchain instead of public; however, the private blockchain technology continues to be plagued by specific problems. The most critical of those is that private blockchain could be potentially tampered with. The ingenuity of the public blockchain being immutable is not precisely true for private blockchain. The potential higher cost of operation and technology akin to existing database management systems makes the technology less attractive altogether [13]. Thus, we are re-routed to the legal challenges and security concerns as discussed earlier.

However, there are certain areas where it is widely believed to be a worthy application of technology. One such area is the healthcare sector. Blockchain 4.0 is specifically designed to have its application in the industrial sector. In the following segment, we would engage in this very application of technology.

3 Blockchain Technology 4.0: Potential Application in Healthcare

Healthcare and blockchain technology are one of the trending news across the globe. Blockchain technology may successfully be used in healthcare, as shown in pilot projects conducted by MIT [14]. These promising signs have encouraged researchers to explore the possibility of invoking this healthcare technology on a larger scale [15]. With such potential to provide solutions to some of the inherent challenges of modern healthcare, like data security, safety and accessibility, blockchain in healthcare can be the next big thing. This context invokes the relevance of testing the existing healthcare models to determine the most suitable model for blockchain technology.

3.1 Scope of Blockchain Technology 4.0 in the Existing Healthcare Models

Blockchain can be applied within healthcare, e.g. clinical trials, health insurance, pharmaceutical tracing and device tracking. However, certain major issues could be settled within the modern healthcare industry in India by facilitating blockchain.

- Healthcare data interchange: Once the blockchain is introduced in healthcare formally, all the service providers would be able to access relevant medical data when and where required, albeit under the authority of the individual who is the owner of the record.

- Nationwide interoperability and beyond: The existing medical data management varies from almost one service provider to another. Uniformity in data management would be introduced, prompting to standardizing the system. Incentivizing the contribution to data generation could certainly make the difference.
- Device tracking: There would be a revolution in the supply chain management of medical devices when blockchain technology would be introduced. This would facilitate hassle-free management of such devices.
- Drug tracking: Blockchain would effectively track medicines and would mitigate the challenge of counterfeiting in drugs. Supply chain management is one of the critical areas where blockchain technology can contribute immensely.
- Clinical trials: Technology like blockchain would reintroduce and reassure trust in clinical trials, as the information would be immutable and un-tampered.

The reliability of data, the transparency in the system, the ability not to tamper with existing data, etc., are the key to any successful business, and healthcare is not an exception. What is essential is the commitment to the data, and it is expected that blockchain could bring in the much-required commitment in storing and managing such data.

The blockchain technology is fast finding its place in the healthcare sector across the world, especially, in the management of remote patient monitoring, supply-chain management, medical record and insurance [16]. As the world continue to grapple with COVID-19 pandemic, the health sector was confronted with unprecedented challenge, blockchain technology came to aid in identity management, contact tracing, supply chain in South Korea [17] (Opportunities and Challenges of Blockchain Technologies in Healthcare), the associated benefit of the technology is expected to find place across the world, with Estonia inducted the technology in their healthcare sector as early as 2012, albeit the low population of Estonia has assisted such induction [18]. However, the diversity present in the world can bring forth much challenge and opportunities, as discussed in the forthcoming segments.

3.2 Blockchain Technology 4.0 and Healthcare Insurance Model

The other key area in the healthcare sector where blockchain technology can contribute is the insurance models. There are some healthcare models of which this study would broadly consider three, viz. the 'Beveridge model', the 'Bismarck model' and 'the national health insurance model'.

The Beveridge model is named after Sir W. Beveridge, who promoted the government as the sole player providing standardized healthcare benefits to everyone from funds accumulated from tax [19]. Some of this model's challenges include overutilization and situations arising out of a national emergency or a pandemic like COVID-19. Blockchain technology can be adapted and applied in a set up like this where inter alia, the regulatory mechanism is uniform. Needless to say, technology, including blockchain, has stumbled upon legal and regulatory hurdles. European countries like the UK and Spain have used this model in their jurisdictions. Blockchain may most certainly complement and supplement the system by securing medical data and making it accessible on demand.

The Bismarck model, which is named after Otto Von Bismarck, is based on a decentralized healthcare model. Similar to Beveridge, it is non-profit seeking, and the state regulates the price. But the members covered by this model must hold an employer-employee relationship. They create a fund commonly known as the 'sickness fund', which would fund their insurance [19]. Blockchain can be effectively used to enrich this model to secure the claims in time. The distributed ledger technology would assure transparency in the system. Blockchain technology may be used in this platform as the caregivers and service providers are private while the insurer is public. The multiplicity of hospitals and service providers can be observed in this model, which blockchain technology would best fit. European countries like Switzerland, Germany and Belgium have adopted this model in their healthcare plans. The model has also been used in the US in some employer-based healthcare plans.

The national health insurance model proposes private providers, whereas the government is a single payer. It is based on a non-profit, non-deny claim model. This model has some classical challenges like delay in treatment due to long waiting lists [19]. Perhaps, blockchain is most certainly required in this model to assure trust and transparency in the system. Once implemented, the system would be able to mitigate unnecessary complications like incomplete information in the claim leading to litigation and expenses related to such disputes. This model is in use in some advanced jurisdictions like South Korea and Canada [19].

4 Legal Opportunities and Challenges Involved in the Application of Blockchain Technology in the Healthcare Sector in India

To explore the possible opportunities and challenges in the application of blockchain technology in healthcare in India, it is imperative to have an overview of the legal framework pertaining to healthcare in India. It is pertinent to state in this regard that sanctions can play a pivotal role in case of violation of technology law in general and blockchain technology law in particular. This position would be in line with the EU's cyber sanctions regime which is effective since 2019. The USA initiated to impose sanctions since 2014 in similar situations [20]. It would be fair for India to promote this remedy in tune with the EU and USA.

4.1 Laws Regulating Healthcare in India

Blockchain technology 1.0's most famous application, i.e. cryptocurrency, was not readily acceptable by the central bank of the country, i.e. Reserve Bank of India (RBI). RBI has frequently raised concerns about the financial stability of cryptocurrency, money laundering, security, the overall impact on the economy of the country [21]. Irrespective of the fact that cryptocurrency is widely gaining popularity, RBI is not very optimistic about it. The law-making procedure in India is not the simplest of mechanisms. The fact that the first popular application of the technology is yet to be accepted legally sanctioned application; in fact, the efforts are in the opposite direction. Such an environment cannot be considered conducive towards the application of technology in other sectors. Healthcare is a booming sector, and numerous stakeholders are at play. Due to the sheer numbers of stakeholders, data protection and privacy remain the chief concern. While the European States have regularly worked collectively together, the Indian counterpart is playing a constant catch-up in terms of data security [22]. The diversity in terms of infrastructure, treatment methods, awareness, socio-economic conditions and culture is tremendously high, and if one adds the factor of population, socio-legal challenges turn manifolds. At the same time, blockchain technology can be a very useful technological development in the non-essential sectors or with information which are not sensitive. However, regulating and protecting data related to the health and ancillary fields can be challenging in the present set of laws in force in India. A selective list of legal provisions of related laws regulating healthcare in India is provided in Table 1.

Blockchain technology has primarily found its popularity in cryptocurrency and not in the healthcare sector. The laws, rules and regulations developed for

Table 1. Legislations to protect data privacy in India (selected)

Sl. no.	Legislations (selective)	Year	Provisions (selective)
1	Indian Telegraph Act and Indian Telegraph Rules	1885	Sections 4 and 5 Rule 419A of the IT Rules
2	Indian Post Office Act	1898	Section 26
3	The Indian Wireless Telegraphy Act	1933	Sections 3 and 4 vesting the power on the government to regulate data
4	Information Technology Act and Information Technology Rules	2000	Section 69
5	Unlawful Activities Prevention Act	1967	Section 4
6	Code of Criminal Procedure	1973	Section 91
7	Consumer Protection Act	2019	Unfair Trade Practice to disclose the data of the consumers
8	Right to Information Act	2005	Section 8 (including related provisions)
9	Personal Data Protection Bill	2019	Yet to be enacted

cryptocurrencies are used for blockchain in healthcare. This absence can be felt as more and more issues of healthcare are emerging.

It is pertinent to state in this regard that law enforcement as well as intelligence agencies must comply with the law while accessing or processing personal data in furtherance of criminal investigation [23].

4.2 Personal Data Protection Bill (PDPB)

The proposed law on data protection in India in the current socio-economic scenario is quite an endeavour to adopt on the one hand and a significant challenge to implement, on the other. The Bill comes with a lot of promise; once enacted, it would be vesting clear responsibilities on people in the business of handling and processing sensitive private data.

Even though the proposed law with its 98 sections divided into 14 chapters, appears to be detailed, it bears significant resemblance with the popular European Union General Data Protection (GDPR), but some classical differences could be noticed on a closer look. On the aspect of scope, the proposed Indian law is broader as it includes any entity irrespective of its affiliation and is involved in the mere processing of personal data in India. This may discourage potential investors in India and narrowing down on the scope has been proposed to exclude such broader application of the law [24]. The Bill also allows the government compelling information to be disclosed, which is outside the purview of personal data, making its application wider than private data.

Indian law on data privacy also overlaps with GDPR on aspects relating to sensitive data with a broader definition, including financial data and allowing additional categories by the government within its ambit. On a different note, however, the GDPR provides clearly a set of additional rules for processing matters relating to criminal convictions, which is not the case in PDPB.

Heath data under the PDPB has been incorporated as sensitive data vide Sec. 3(36)(*ii*) and is given a comprehensive interpretation which inter alia assures considerable protection under Sec. 34(b)(*i*).

4.3 Blockchain Technology 4.0 in Indian Healthcare Sector

Blockchain technology, like any other technology, would face some strategic challenges when it is to be applied in an existing traditional ecosystem. Blockchain 4.0, when it is being applied to healthcare platforms in India, the experience would be somewhat similar when it comes to challenges. In most countries, the healthcare industry is complex and interconnected with various entities. All entities involved store patient data and records in their own unique information system. For blockchain to be successful, all the stakeholders would require to accept this new entrant within its fold of technical infrastructure. This transformation towards integrating healthcare entities would be the key in reaping benefits that blockchain technology is capable of producing, which includes inter alia, portability, care delivery management, interoperability and cost-effective administration.

All the stakeholders would be immensely benefited from the introduction of blockchain technology in India. The patients would be able to access and share their prior medical records without much complication or efforts from their individual end. This availability of medical history simplifies the complications in choosing proper medical care by the doctors and healthcare professionals. From the perspective of healthcare professionals, this technology can revolutionize the interoperability of medical databases. Access to archived medical databases would prevent in improper diagnosis and/or experimentation in medication, furthermore, it wouls also be of assistance during medical emergencies.

The market is adapting to changes brought by evolution in the last few decades. Laws prevalent in this domain are mostly conventional, with the ADR being the new change. In India, consumer protection laws, which are modern versions of the law of torts, largely follow compensatory jurisprudence. The big question arises if this technology could be integrated on a large scale in India.

5 Can Blockchain Fit in with the Existing Healthcare Models in India?

In order to address this question, two areas are needed to be explored, viz. the design and the deployment challenges.

5.1 The Design Challenge

The designs in use in healthcare are based on distributed ledger technology. The technology works across jurisdictions in near real-time, and the stakeholders are updated accordingly [25]. Like most of the other online technologies, this design brings with it the challenge of extraterritorial application. Even if theoretically, it does in some instances, as it does through the Information Technology Act 2000, but practically bringing perpetrators outside the territorial jurisdiction of India has always remained a distant dream in non-high-profile cases.

Thus, the design of blockchain technology inherently poses challenges of deployment of the said technology, practically within the territory of India. The silver lining, however, is the fact that blockchain brings with it trust, which is also encouraged and promoted by her legal system. Thus, under the Indian legal system, the technology can, mutatis mutandis, act as a supplementary layer towards protecting and securing data. It can also very well provide relevant trust infrastructure in the years to come to be adopted formally by the legal system.

It may be used at this point where the technology is solely functioning within the territory of India and where no existing legal impediment prevents using the technology. Rather, it would be interesting to see its possible pilot application in the smart cities to gauge how the technology pans out in a uniform and controlled environment. The technology can be of significant value when used for model contact tracing, warning measure in an emergent health crisis akin to the one we witnessed during the COVID-19 pandemic [26].

Blockchain can be more relevant in India than the outdated traditional methods prevalent in the healthcare sector today. The challenges involved in the absence of previous health records with the patients usually create scope for improper disease detection and treatment. Starting from practical challenges to the authenticity of healthcare diagnostic equipment and data generated there from, blockchain 4.0 can bring in a revolution in personalized healthcare across the globe. Various court orders have considered distributed ledger technologies similar to that of partnerships or joint ventures between participants [27].

The application of this technology can revolutionize the deployment of law. Once in use and at an advanced stage, law enforcement mechanisms can rely on this technology and assist the justice system considerably. The technology, which is in a nascent stage, can actually see the light of the day once it adapts to the requirements of the law. The changes in technology are required to objectively assist the law and not contradict it. It must be capable of performing its intended functioning along with supplementing and complementing the legal system.

- Healthcare information: Privacy, confidentiality and security are three major concepts surrounding healthcare data.
- Privacy is the right conferred to an individual with regard to the control and disclosure of health data.
- Confidentiality, on the other hand, 'is the controlled release of personal health information to a care provider or information custodian under an agreement that limits the extent and conditions under which that information may be used or released further' [28].
- Security may be construed to be 'a collection of policies, procedures, and safeguards that help maintain the integrity and availability of information systems and control access to their contents' [28].
- Right to be forgotten: Laws in many jurisdictions, including that of India, protects the privacy rights of an individual very strongly. The European Union General Data Protection Regulation (GDPR) is a strong example of this claim. The right to privacy includes the right to be forgotten. A system based on blockchain runs on DLT, which enables the history of the digital asset unalterable, transparently. Thus, making the ordinary practice of 'right to be forgotten' technically in-alienable.

Transparency of confidential information contradicts the law of privacy. Healthcare information is considered confidential, and the patient may even have the reservation to expose them to the doctors [29].

5.2 The Deployment Challenges

India does not have an adequate infrastructure where blockchain technology may be deployed. The digital divide has plagued various parts of the country where the people living in remote stretches and economically challenged are deprived considerably. With the vast population to cater to, the government has to face innumerable challenges, including skilled human resources and century-old laws. This divide challenges the viability of technology-driven governance and healthcare protection, including that of blockchain. Similarly, a separate but uniform healthcare IT system is not available in India [30], which doubles the challenge.

> *Blockchain is going to be the driver of convergence between government, hospitals, pharmaceuticals, medical device manufacturers, and insurance providers* [30].

However, the country is moving forward towards an electronic superpower with the passage of every financial year more, and more budget allocation can be seen to bridge this divide. Blockchain can at this moment be used in pilot projects as well as in various independent and favourable platforms.

A classic example of the use of blockchain technology can be seen when the UN used it for tracking food aid to refugees [31]. This paper focuses on three main challenges, viz. the apoptosis challenge, the lifespan challenge and multiple medical schools challenge, with regard to application of blockchain technology in healthcare in India.

5.2.1 The Apoptosis Challenge

With age, there can be an alternation of human health conditions. For instance, a non-diabetic can become diabetic. Thus, the doctors relying on the previous health history maintained through blockchain technology may proceed with the wrong treatment causing more harm than good. Therefore, real-time diagnostic innovations may sound like a better alternative in certain situations.

5.2.2 Technological Lifespan

Every technology is replaced by another, and the former's huge investment may sound misspending in many cases. The life of blockchain technology is also not eternal, irrespective of its being claimed as a revolutionary innovation.

According to Gregory Tassey, '*no matter what the final outcome with respect to the distribution of value-added across national economies in one technology life cycle, global markets will increasingly experience shifts in leadership in the following life cycle. This greater competition is due to the fact that a larger number of economies are acquiring the requisite innovation in infrastructure to become competitive in technology-based markets*' [32].

5.2.3 Multiple Medical Schools

India recognizes almost all types of medical practices within its jurisdiction. Allopathy, Homeopathy, Ayurveda, Unani and other indigenous methods of healing are practiced parallelly. More so, healthcare systems have their specific preferences primarily based on the availability of healthcare professionals. For example, if a person residing in Kolkata goes to Sundarbans, around 70 km south, he or she would have to rely upon indigenous healthcare support in a case of emergency or else would have to return to Kolkata to initiate the treatment. This preferential practice is also dependent upon the economic parameters. As most rural populations do not have health insurance benefits, they are not able to support themselves with evidence-based healthcare practices. Homoeopathy, ayurvedic and other inexpensive methods have gained popularity primarily because of the affordability factors. The popularity of Homeopathy in India can be assessed by using it to prevent COVID-19 [33]. A contrast was noticed with the rise in popularity and use of Homeopathy in the USA, according to a recent report published in 2016 [34]. Similar trends are visible with other forms of alternative medicines in the USA [35]. It is imperative to state that purely technology-driven healthcare propositions are improbable in such a diversified healthcare environment.

On the contrary, the system can be applied only to the existing platform where technology is used. Thus, economic inequality in healthcare system stands as an impediment in the practical adaptability of blockchain in healthcare in the rural areas covering almost half of India. It is not merely the digital divide but the multiple medical schools that need to be mitigated before formally adopting blockchain technology in healthcare in India.

6 Conclusion

After a comprehensive discussion of the challenges and opportunities that blockchain technology in healthcare is surrounded with, some critical findings are unravelled. The gap between law and technology is too big to cross as of now. Influence of technology on law is not overwhelming as is the position with other professional discourse. Irrespective of the fact that IT has been adopted in all professional sectors, including that of law, these tools are far lesser in terms of sophistication than is commonly found in other industries. Blockchain works well in a homogenous environment whereas it faces too rough a mountain to climb. This can be seen in successful selective use of blockchain technology in the healthcare industry in some jurisdictions. There is a clear gap between the extent to which it may be adopted and the extent to which it has been adopted. This gap has manifested in exploring avenues to make it acceptable in various sectors, including that of healthcare. One of the significant hindrances identified in this chapter is the gap that exists in various technologies on the one hand and the typical nature of law on the other. The gap would persist unless the legal profession adopts technological tools as a whole. Digital divide has plagued India as it does in similar heterogeneous economies.

Various new avenues must be explored to bridge this gap to make blockchain technology adaptable in various jurisdictions, including India. The typical practice of technology to infringe the rights of the people have also contributed to the conservative approach of judicial decision-making, whether in the European Union or in India. Inappropriate use of technology in privacy, data protection and surveillance has set wrong precedents for the technology to be acceptable on face value, making the slope slippery for technology to stabilize. Collaborative interdisciplinary research in this domain which is in tune with UNSDG G-17 is what is desirable to make blockchain technology a reality in India and other developing economies. The gap of digital divide is also gradually decreasing with noticeable political intent in the last decade in India.

Although it is hard to predict when blockchain technology will emerge to be legally compliant, one can be confident that once adopted, blockchain technology will almost certainly bring revolutionary changes to the existing practice of healthcare for good.

References

1. P.K. Chan, W. Fan, A.L. Prodromidis, S.J. Stolfo, Distributed data mining in credit card fraud detection. IEEE Intell. Syst. Appl. **14**(6), 67–74 (1999)
2. D. Yaga, P. Mell, N. Roby, K. Scarfone, Blockchain technology overview (2019), arXiv preprint arXiv:1906.11078, https://arxiv.org/pdf/1906.11078.pdf
3. Blockchain Training Alliance, What is Blockchain? (2017), https://www.buffalo.edu/content/dam/www/ubblockchain/files/basics/001%20What%20is%20Blockchain.pdf
4. S.M. Idrees, M. Nowostawski, R. Jameel, A.K. Mourya, Security aspects of blockchain technology intended for industrial application. Electronics **10**(8), 951 (2021)
5. S.M. Idrees, I. Aijaz, R. Jameel, M. Nowostawski, Exploring the blockchain technology: Issues, applications and research potential. iJOE **17**(7), 49 (2021)
6. A.G. Khan, A.H. Zahid, M. Hussain, M. Farooq, U. Riaz, T.M. Alam, A journey of WEB and blockchain towards the industry 4.0: An overview, in *2019 International Conference on Innovative Computing (ICIC)*, (IEEE, Piscataway, 2019), pp. 1–7
7. P. De Filippi, S. Hassan, Blockchain technology as a regulatory technology: From code is law to law is code (2018), arXiv preprint arXiv:1801.02507
8. A. Walch, Blockchain applications to international affairs: Reasons for skepticism. Georgetown J. Int. Aff. **19**, 27–35 (2018)
9. G. Gabison, Policy considerations for the blockchain technology public and private applications. SMU Sci. Tech. L. Rev. **19**, 327 (2016). https://scholar.smu.edu/cgi/viewcontent.cgi?article=1043&context=scitech

10. S. Bell, The role of the state and the hierarchy of money. Camb. J. Econ. **25**(2), 149–163 (2001)
11. A.P. Lerner, Money as a creature of the state. Am. Econ. Rev. **37**(2), 312–317 (1947)
12. LG&E Energy Corp., LG&E Capital Corp., and LG&E International, Inc. v. Argentine Republic, ICSID Case No. ARB/02/1, https://www.italaw.com/cases/621
13. M. Profitt, Blockchain 101: Fundamentals of new economics (2018). Create Space Independent Publishing Platform
14. A. Ekblaw, A. Azaria, J.D. Halamka, A. Lippman, A case study for blockchain in healthcare: "MedRec" prototype for electronic health records and medical research data, in *Proceedings of IEEE Open & Big Data Conference,* vol. 13 (2016), p. 13, https://static1.squarespace.com/static/59aae5e9a803bb10bedeb03e/t/5a6fd2f7e2c48387dff12676/1517278000381/blockchain-medical-records-patient-data-medrec-eckblaw.pdf
15. P. Pandey, R. Litoriya, Implementing healthcare services on a large scale: Challenges and remedies based on blockchain technology. Health Policy Technol. **9**(1), 69–78 (2020). https://reader.elsevier.com/reader/sd/pii/S2211883720300046?token=66B15F4C2E6B54A644F34E8B2777A82A8BD91B96853DDF00633A4B955E14538CD0353A8AB83F3D4E32AD00786863BFBA
16. R.B. Fekih, M. Lahami, Application of blockchain technology in healthcare: A comprehensive study, in *International Conference on Smart Homes and Health Telematics*, (Springer, Cham, 2020)
17. Idrees, S. M., Nowostawski, M., & Jameel, R. (2021). Blockchain-based digital contact tracing apps for COVID-19 pandemic management: Issues, challenges, solutions, and future directions. JMIR medical informatics, 9(2), e25245.
18. S. Daley, How using blockchain in healthcare is reviving the industry's capabilities. Builtin.com (2021), https://builtin.com/blockchain/blockchain-healthcare-applications-companies
19. D. McCane, Health care systems-four basic models. Physicians for a National Health Program (PNHP) 6, (2010), https://www.pnhp.org/single_payer_resources/health_care_systems_four_basic_models.php
20. K. Bannelier, N. Bozhkov, F. Delerue, F. Giumelli, E. Moret, M. Van Horenbeeck, in *GUARDIAN OF THE GALAXY: EU Cyber Sanctions and Norms in Cyberspace*, ed. by P. Pawlak, T. Biersteker, (European Union Institute for Security Studies (EUISS), Paris, 2019), pp. 79–86. https://doi.org/10.2307/resrep21136.12
21. P. Mulye, Why India's Central Bank has no faith in cryptocurrency. Quartz India (2021), https://qz.com/india/1977698/why-rbi-is-concerned-about-bitcoin-other-cryptocurrencies/
22. S.P. Chakrabarty, S. Mukherjee, A. Rodricks, The role of law in protecting medical data in India, in *Data Protection and Privacy in Healthcare Research and Innovations*, ed. by A. Elngar, A. Pawar, P. Churi, (CRC Press, Milton Park, 2021), pp. 229–243
23. E. De Busser, E.J. Briscoe, B.C. Dean, T. Tropina, M.B. Aparicio, *Big Data: A Twenty-First Century Arms Race* (Atlantic Council, Washington, DC, 2017)
24. K. Wimmer, G. Maldoff, D. Lee, Comparison: Indian personal data protection bill 2019 vs. GDPR. iapp (2019), https://iapp.org/media/pdf/resource_center/india_pdpb2019_vs_gdpr_iapp_chart.pdf
25. J. Cheng, B. Geva, Understanding block chain and distributed financial technology: New rails for payments and an analysis of Article 4A of the UCC. Bus. L. Today (2016), p. 1
26. U. Chakraborty, A. Banerjee, J.K. Saha, N. Sarkar, C. Chakraborty, *Artificial Intelligence and the Fourth Industrial Revolution* (Jenny Stanford Publishing Pte. Ltd, Singapore, 2021). ISBN 978-981-4800-79-2 (Hardcover), 978-1-003-00000-0 (eBook)
27. D.A. Zetzsche, R.P. Buckley, D.W. Arner, The distributed liability of distributed ledgers: Legal risks of blockchain. U. Ill. L. Rev. **4**, 1361 (2018)
28. T.C. Rindfleisch, Privacy, information technology, and health care. Commun. ACM **40**(8), 92–100 (1997)
29. C. Davis, Patient rights (2020), https://www.emedicinehealth.com/patient_rights/article_em.htm

30. R. Garg, Is blockchain in Indian healthcare worth the challenges? (2018), https://health.economictimes.indiatimes.com/news/health-it/is-blockchain-in-indian-healthcare-worth-the-challenges/64095898
31. L. Cuen, UN using blockchain technology to help refugees fight world hunger. Int'l Bus. Times (2017)
32. G. Tassey, *Technology Life Cycle, Encyclopedia of Creativity, Invention, Innovation and Entrepreneurship* (Springer Science + Business Media BV, Berlin, 2013). https://www.researchgate.net/publication/341115374_Technology_Life_Cycles/stats
33. PTI, Covid-19: Homeopathic drug is given to half of Gujarat population since March, (2020), https://www.livemint.com/news/india/covid-19-homeopathic-drug-given-to-half-of-gujarat-population-since-march-11598179862862.html
34. M.L. Dossett, R.B. Davis, T.J. Kaptchuk, G.Y. Yeh, Homoeopathy use by US adults: Results of a national survey. Am. J. Public Health **106**(4), 743–745 (2016). https://www.ncbi.nlm.nih.gov/pmc/articles/PMC4816083/; Also see, 2017, March. Harvard Study has good news for Homeopathic Medicine, https://ochm.ca/harvard-study-has-good-news-for-homeopathic-medicine/
35. F. Kronenberg, L.F. Cushman, C.M. Wade, D. Kalmuss, M.T. Chao, Race/ethnicity and women's use of complementary and alternative medicine in the United States: Results of a national survey. Am. J. Public Health **96**(7), 1236–1242 (2006)

Smart Contracts in Blockchain Application: Review Chain

Swati Srivastava, Roheet Bhatnagar, Gaurav Srivastava, and Urvashi Srivastava

Abbreviations

DAO	Decentralized Autonomous Organization
EA	Ethereum Addresses
IPFS	Inter Planetary File System

1 Introduction

1.1 Background

A smart contract is a self-executing contract in which the terms of an arrangement between two parties are directly converted into lines of code. A network of distributed, autonomous blockchains underpins the code and the agreements it contains. The code governs the execution of transactions that are both trackable and irreversible.

Smart contracts enable separate, anonymous parties to conduct trustworthy transactions and agreements without the need for a central authority, legal structure, or external enforcement.

S. Srivastava (✉) · R. Bhatnagar
Department of Computer Science and Engineering, Manipal University Jaipur, Jaipur, India

G. Srivastava
Sunrise institute of Engineering Technology and Management, Unnao, India

U. Srivastava
Maulana Azad National Institute of Technology, Bhopal, India

S. M. Idrees, M. Nowostawski (eds.), *Transformations Through Blockchain Technology*, https://doi.org/10.1007/978-3-030-93344-9_11

1.2 Motivation

Recently, online reviews saw huge utilization and growth. Many digital platforms like the tour and travel agencies, e-commerce websites like Amazon, Flipkart, etc. use famous customer review systems for their services and products they sell. The reliability of these reviews was (and is) doubtful, because the central authority can alter, exploit, or remove the reviews.

Lots of customers may be looking at an online review or ranking for a product while browsing online for anything to purchase. It seems like a perfect way to get an unfiltered view of quality, but research shows that most reviews online are too simplistic and can misguide consumers.

As of late, an Australian lodging network was fined $2.2 million for evacuating negative criticism on the well-known Trip Planner site, as expressed in. Notwithstanding TripAdvisor, there are visit instances of vote fixing, missed or erased basic surveys, and different questionable strategies that have been found to utilize different audit sites, for example, Yelp and Airbnb.

According to one survey carried out by the United States, 78.5% of the American customers were searching for online information about a product or service, and 34% had written an online review. A global survey by Nielsen showed 70% of customers trust and use online product feedback when making decisions.

Such exploitation can occur in many ways, such as false reviews, negative review omissions, and promotion of advertised reviews. It is due to the fact that, to date, all evaluation systems are largely centralized and vulnerable to abuse, hacking, and misuse of their data by insiders or outsiders, as well as malfunction and external attacks.

1.3 Aim and Objective

The main objective of this research work is to understand the application of smart contracts in blockchain application.

- Our first aim is to understand the concept of smart contracts in blockchain. We will be able to understand what is a blockchain and its application called smart contract and its implementation details.
- Our next objective is to understand how smart contracts works, how it is useful in many modern-day applications.
- Also, we aim to modify the smart contracts in blockchain to make it suitable for our any transactional based application like in supply chain management, e-commerce.
- Finally, our main goal is to study the application ReviewChain to make tamper proof review system integrated with the e-commerce websites.

1.4 Gaps in Research

- Most of the blockchain-based review systems are using third-party application for reviews.
- Whenever a third-party application is involved in review system, there is a chance of review tampering.
- Most of the review platforms were not considering purchase time for calculating hash value, which will make blockchain even more secure.
- There is no e-commerce platform with in-built blockchain-based review system.

2 Literature Review

2.1 Review Process Adopted

Daniel Martens University of Hamburg, Hamburg, Germany [1, 2] Online stages give more client contribution to the type of appraisals and surveys. Late research has featured the estimation of this criticism and checked that positive input helps the marketing projections of the merchandise and subsequently their prominence. Administrators of the online gateways fill in as focal experts in the whole audit process.

Administrators may, in the most pessimistic scenario, preclude clients from posting audits, changing existing surveys and including counterfeit surveys by imaginary clients. This chapter proposes a decentralized examination approach, ReviewChain. Our answer by utilizing blockchain innovation, decentralized applications, and capacity forestall focal specialists. It causes clients to present untampered criticism and on get them.

We represent the troubles looked in applying our way to deal with the open blockchain Ethereum. At that point, we are examining potential plan choices and their cost, security, and reliability exchange offs. At long last, we take a gander at the structure choice to acknowledge diverse exchange offs, and the subsequent blends of decentralized blockchain advancements ought to be picked, additionally with conventional concentrated innovations.

- ReviewChain means to fabricate and give nonstop access to untampered evaluations and criticism of the merchandise. The fundamental objective is to stop focal specialists. Our methodology thusly utilizes innovation, for example, blockchain as decentralized and unchangeable information stockpiling, and decentralized applications to give access to the input.
- A complexity between the customary brought together survey approach and our decentralized ReviewChain approach. Nearby specialists will absolve singular clients from posting criticism in the upper piece of the figure. Furthermore, specialists hold remarks in a brought together database. This causes them to change

the existing input and include counterfeit audits of imaginary clients who have not bought the explored item.

- That is not, at this point conceivable in the decentralized strategy, found in the base piece of the figure. At the point when setting on the blockchain, criticism cannot be changed, nor would it be able to block clients from giving input. We allude to such remarks as mimicked remarks (Fig. 1).

Ethereum Platform: It is a digital, open money network, and new types of applications. On Ethereum, you can write code that controls the money and create open applications anywhere in the world.

Transaction: Exchanges are messages marked carefully that live on the blockchain. That exchange is related to an activity, for example, cryptographic money unit moves. Exchanges incorporate a location for sender and beneficiary, speaking to open keys that have a place with various clients.

K. Salah, A. Alfalasi, M. Alfalasi [2] talk about:

- A significant test is the presentation of a decentralized, reliable, fair-minded, and open arrangement of assessment. In the present chapter, we propose an answer utilizing Ethereum Blockchain, Smart Contracts, and IPFS to give a sheltered, open, and dependable discussion for high trustworthiness and vigorous online audit system.
- Principle design engineering highlights the availability of gadget components, calculations, and rationale stream. We likewise show how we applied and tried

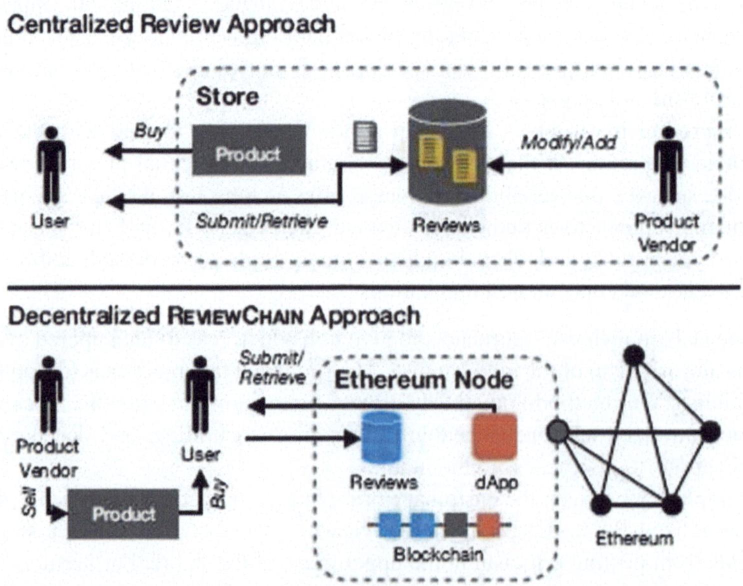

Fig. 1 Centralized reviewing/ReviewChain approach

the functionalities of the general program. Likewise, we have the keen agreement innovation powerlessness audit.

- Here we discussed a blockchain-based survey structure which furnishes the first commentators with a decentralized, reliable online audit, with the capacity to screen and follow criticism. The chronicle of input on the Blockchain implies that no gathering can ever alter, abuse, or delete this record.
- Our inventive blockchain-based rating program utilizes this as an encouragement to rating analysts. Specialist organizations are permitted to allow tokens to legitimate clients. Tokens will be utilized to check analysts and expect corrections to be transferred. After positive accommodation of a study, commentators get charged in Ether.
- This usefulness can be presented using Ethereum brilliant agreements. In short, a specialist organization, for example, TripAdvisor, transfers to the open Ethereum arrange the implicit rules trapped in a Smart Contract code and stores Ether for lawful audits to be administered. For a front-end program, the client would then have the option to demand a rundown utilizing the IPFS.
- After the examination has been checked, the Smart Contract application would repay the client with Ether for their commitments, and the check will be for all time kept in the IPFS system to be downloaded by anybody with solid trustworthiness and unwavering quality.
- The key point of this chapter is to exhibit how brilliant blockchain agreements can be utilized with IPFS to execute a straightforward, dependable, reliable online assessment organize apace.

This present chapter's central commitments can be depicted as follows:

- We are pitching a blockchain-based methodology using Ethereum shrewd agreements and IPFS for a profoundly proficient, steady, open, and reliable online audit arrange. The program expects tokens to be given to qualified analysts and charge them in Ether after online assessment has been effectively finished.
- They provide information on device description, configuration, layout, and the system's relationships with the different actors.
- We offer a full usage of the smart contract structure from Ethereum and address center issues pertinent to the frontend and backend systems that entertainers use [3].
- We have prepared to exhibit the correct gadget highlights for the Ethereum shrewd agreement application, including insurance and bug survey (Fig. 2).

Ke Wang, Zhizhe Zhang, and Hyong S. Kim [4] suggest:

- Current organized gatherings are probably going to examine misuse and abuse. A blockchain organize that is straightforward, lasting, and auditable is an ideal match to stay away from phony and contorted criticism.
- We assemble and implement ReviewChain, a structure of decentralized survey using savvy contract Ethereum. ReviewChain approves ongoing appraisals by questioning buyers' characters from a production network.

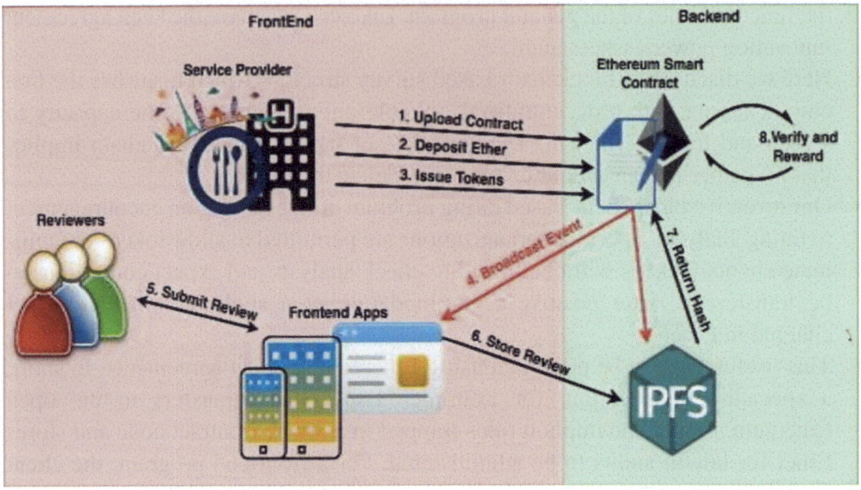

Fig. 2 System architecture

- The inventory network organizes, we believe, is centered around blockchain innovation. We are structuring and actualizing a convention to interface two separate blockchain systems focused on the Ethereum.
- Ethereum's savvy contract innovation is confined and cannot uphold outer calling techniques. So, we include public accountants called multiblock chain portal hubs.
- Our proposed public accountants and keen agreements permit us to impart between two blockchains.
- Through Ethereum, smart agreement grows the exchange information standard in the blockchain. Smart agreement requires exchanges in the blockchain to work computationally.
- An exchange in Bitcoin implies a change starting with one record then onto the next of any total of unspent exchange creation.
- This state is checked in a content language which is deficient in Turing. Smart agreement extends the idea to an idea that is Turing finished. The smart agreement greatly improves blockchain capacities.
- The blockchain stage might be utilized to address unscrupulous issues in assessment forms. To stop the system misusing investigation, we abuse the blockchain's changelessness and auditability properties.
- We are making up a blockchain-based assessment structure that ensures that each assessment is available to all members until verified and never erased or changed.
- We require the blockchain to assemble off-chain information and confirm if a client has purchased the item and forestall the rating extortion.
- The dissemination and assembling reports are typically held in an alternate production network arrangement.

- ReviewChain approves late evaluations by questioning a production network organize about the buyer's name.
- We accept store network organize depends on blockchain innovations also. In the smart agreement, assembling such as off-chain subtleties is troublesome.
- A couple of prior examinations recommend methodologies for interfacing different blockchains. Such works characterize their elevated level methodologies and give different points of interest.

We are structuring and conveying a system to tie together two separate blockchain systems dependent on Ethereum. Ethereum's smart agreement code is compelled and cannot demand for synchronization through outer procedure. There are two diverse blockchains dependent on Ethereum interfacing and communicate. We include public accountants, called door hubs.

Recently distributed remarks are checked by each Notary. The legal official examines the validity of the inquiry and votes its affirmation, as part of conjuring a clever agreement dubbed new Vote. The excavators apply the examination to the concurred agenda until the votes are through. Different legal officials that vote that the possibility of undermining a solitary public accountant [4].

3 Theoretical Aspects

The strategy utilized in writing audit, writing a survey of research papers, correlation table, qualities and impediments, and contrast examination of them have been tended to up until this point. Presently, we will investigate further the hypothetical parts of the strategies and procedures utilized in the exploration papers right now. Here we will discuss about the blockchain, its application, smart contracts in blockchain and its application, and other implementation details.

3.1 Blockchain

Blockchain is a decentralized, straightforward record that empowers exchange documentation and resource following procedures inside a business organize. A benefit can be unmistakable (a house, a vehicle, money, land) or impalpable (protected innovation, licenses, copyright, marking); virtually anything of enthusiasm on a blockchain system can be followed and traded, diminishing the hazard and limiting expenses for those included [5, 6, 7].

Google Doc is an easy analogy for understanding the blockchain technology. When we build and share a document with a group of people, the document is circulated instead of being copied or transferred. It establishes a decentralized distribution chain, which also allows everybody to access the document.

3.1.1 Applications

- Asset Management: Trade Processing and Settlement.
- Insurance Sector.
- Supply Chain Management.
- Banking System.
- Healthcare Sector.

3.2 *Smart Contracts*

The historical backdrop of savvy agreements can be followed back to the 1990s when Wei Dai, a PC engineer, made a post on unknown credits, delineating a mysterious advance plan with repayable securities and single amount expenses to be gotten at development. Szabo et al. along these lines investigated the potential kind of keen agreements and proposed to utilize cryptographic components to improve security.

These days shrewd agreements are being built with the headway of blockchain innovation as PC programs running on blockchain hubs and can be conveyed by untrusted, mysterious gatherings without the mediation of any outsider. The primary effective usage of a blockchain-based keen agreement was Bitcoin Script, a purposefully not-Turing-complete language with a fundamental assortment of precharacterized orders [7, 8].

Standard kinds of Bitcoin exchanges, including pay-to-open key-hash (P2PKH) and pay-to-content hash (P2SH), are completely determined with Bitcoin Script as basic types of savvy contract. Moreover, there are structures which take into consideration increasingly complex legally binding functionalities and capacities.

For instance, Ethereum, which receives a Turing-complete shrewd agreement language. More up-to-date blockchain frameworks, for example, Neo and Hyperledger Fabric grant the composition of brilliant agreements in different significant level dialects (Fig. 3).

3.2.1 Need of Smart Contracts

Savvy contracts acquire the properties of the basic blockchains, which incorporate an unchangeable information record, and the capacity to alleviate single disappointment focuses. Keen agreements can likewise have the option to speak with one another through calls. Not at all like conventional paper contracts which depend on go-betweens and outsider middle people for execution, brilliant agreements robotize authoritative methods, limit among parties, and diminish organization cost.

Brilliant agreements on open blockchains or "open smart agreements" have pulled in a wide scope of business applications because of simplicity of execution. While brilliant agreements on endorsed blockchains or "affirmed smart agreements"

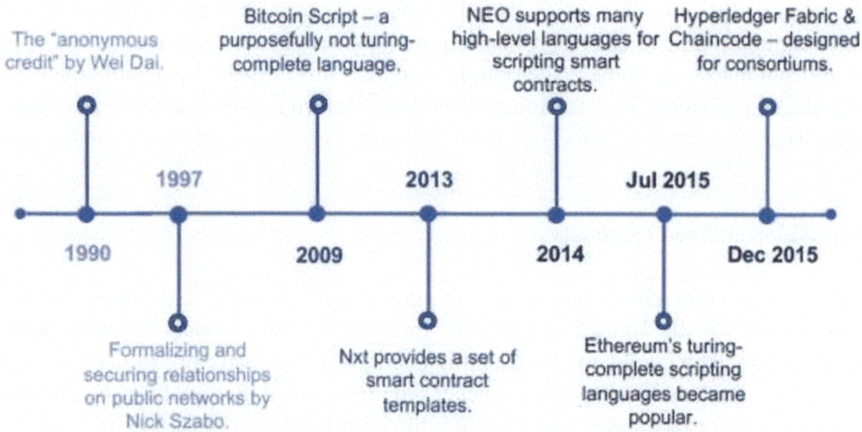

Fig. 3 Evolution of smart contracts

Table 1 Characteristics of public and permissioned smart contracts

	Public smart contracts	Permissioned smart contracts
Common	Immutable record	
	Proper encryption on data and pseudonymity	
	Interoperability among different platforms	
	Traceable modifications	
Unique	Easy to deploy	Faster settlement
	Accessible for the public	Lower operational cost
		Permissioned access

are all the more generally utilized in shared business forms, they can evade unintended changes, support biodiversity, and spare expenses [8, 9]. (Table 1).

3.2.2 Types of Smart Contracts

We divide smart contracts according to the blockchain networks on which they run to public smart contracts and approved smart contracts. Because the standards and criteria for smart contracts are often the same for both the categories, we address them separately below. All smart contracts on approved, cooperative, or private blockchains are considered approved smart contracts.

Public Smart Contracts

A large portion of the present blockchain systems can execute open brilliant agreements, implying that as found in the image beneath, any piece of information found right now is uninhibitedly accessible to the whole system.

In spite of the fact that this is incredible news for some, extraordinary use cases that require clearness, use cases that require a particular type of security can be a significant deterrent. To have the option to address this critical absence of connecting the components of exchanges with both security and straightforwardness (Fig. 4).

Permissioned Smart Contracts

Private smart contracts are a generally obscure option in contrast to the open shrewd agreements. Private shrewd agreements are content-wise comparative with open savvy contracts, with one major exemption being that the conditions of private brilliant agreements cannot be gotten to, checked, or executed by somebody other than the proprietor of that agreement, except if the proprietor awards get to [10, 11].

3.2.3 Applications of Smart Contracts

Some other uses of Smart Contracts are:

- Healthcare and medical records.
- Identity management.
- Scaling blockchains.

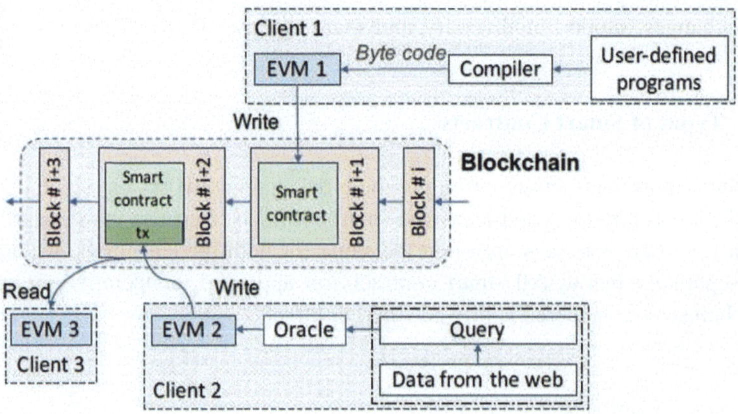

Fig. 4 Mechanism of Ethereum smart contracts

4 Design and Implementation

This segment contains portrayals of the plan and strategy utilized by different essayists for executing the particular calculation proposed. The conversation of the issue explanation is an initial phase in any investigation. It incorporates thought of all the utilization cases associated with the examination, just as what will be the strategy stream.

All stream charts were talked about in further sub-areas with the explanation of the different structures and techniques utilized by various creators.

4.1 Architecture of the Process

- All through this section, we recommend and clarify our proposition for an online audit structure that utilizes smart agreements from Ethereum and IPFS to give precise input, with open doors for analysts.
- To forestall the high cost of handling subtleties on the open Ethereum arrange, the program stores all remarks in IPFS [12, 13].
- This is an autonomous strategy for advancing remark composed by a substance supplier and analysts.
- It cannot be changed until the examination is checked, in light of the fact that it is kept up in the IPFS with full privacy, trustworthiness, and sturdiness.
- The proposed approach centered around using smart agreements to fabricate an electronic appraisal structure worked around blockchain.
- The system configuration shows various gatherings and associations who can speak with the smart agreement.
- This methodology mechanizes the prize instrument for clients to distribute their remarks in the IPFS.
- Each program customer has an Ethereum account with an Ethereum address (Fig. 5).

1. **Service Provider:** This is the association providing the client with the office, for example, a lodging, bar, or online shop. The specialist organization must apply assets (in Ether) to the smart agreement to encourage remunerating for clients. Indeed, even the specialist organization must make a token to create a synopsis for the client.
2. **Reviewers:** Analysts are the clients or shoppers who send the survey to IPFS by means of a frontend gadget. Tokens would be dispersed by the specialist organizations to the inspectors' EA (Ethereum addresses). The smart agreement would pay evaluators who have done positive appraisals to Ether to EA.
3. **IPFS:** IPFS ought to be the document organize for presenting remarks on. The survey's IPFS hash would be held in the smart agreement, alongside the Ethereum address of the creator. The smart agreement would utilize both the token and the IPFS hash to confirm the client's examination and legitimacy.

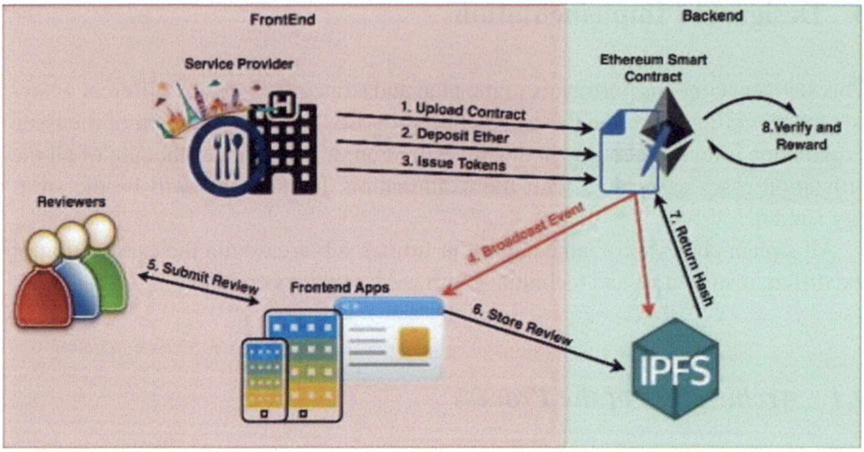

Fig. 5 System architecture

4.2 *Smart Contract Vulnerability Analysis*

- Security has been one of the more basic worries since the presentation of smart agreements than any time in recent memory.
- Keeping up the application and related information safe is critical, on the grounds that the agreement itself can be vulnerable to assaults because of the generally brought together presence of blockchain systems.
- Blames in projects aggressors despite everything stay a test to smart agreement assurance.
- Reentrancy imperfection, Transaction-Ordering Dependence, Request Stack Depth Flaw, Parity Multisig Error, and Assertion Issue are the most broadly perceived defects in smart contracting.
- In this way, there is a solid need to utilize mechanized programming to secure and review the smart agreement application [14, 15].

4.3 *Experimental Setup Used in Research Paper*

- We send ReviewChain to our datacenter. All excavators run Ethereum rendition 1.8.2.
- Each machine in the bunch has a 4-center i3700 CPU and 8G memory.
- We send audit blockchain with five excavators. We likewise convey store network blockchain with five diggers.
- There are three public accountants. A solitary server goes about as analysts to submit audits.

- The customer submits audits as per a Poisson dispersion with configurable accommodation rate.
- We creep the Amazon's site for our engineered audit information. We gather 500,000 audits from Amazon. 99.5% audits have size under 3 KB.
- The length of the gathered audits can be demonstrated utilizing a moved exponential dispersion.

5 Experimental Results and Analysis

5.1 Introduction

We tended to the different center segments of Hadoop in the past part, and we will talk about the test discoveries and breaks down right now. Therefore, the discoveries are:

- Blockchain is a new domain in modern tech world with lots of potential to be discovered for appropriate applications.
- Smart Contract is one of the most promising application of blockchain which can be used in myriad of applications.
- Smart Contract has the capabilities to make a revolution in financial sector, eradicating the need of middleman completely.
- Most of the works are based on third-party application for reviews.
- There is a lack of an e-commerce website with integrated blockchain-based review system.
- In the exploration papers, we additionally considered that the issue of overseeing stockpiling on web scale has likewise risen.
- We also studied that the system does not consider modifying the existing implementation of the smart contracts.
- Some research papers said that large computational power is required to implement such a robust system.
- Some research papers said that the data to be processed was been generated at a very high speed that the CPU was not able to cooperate with it.
- Since there is online business and services, so is the issue of authenticity and true information.

5.2 Discussion

To solid insurance, the main choice is custom fitted. Exchanges will be sent legitimately from inside the program, an ERC-20 token will be utilized to empower clients to post remarks, and information will be put away decentrally. Also, this decision is centered around smart agreements for compensating excavators who process the investigation exchanges. This elective uses a dApp which gets to a neighborhood Ethereum hub for survey recovery.

Enhanced with the subsequent choice reliability. Contrasted with the principle, it bundles the product to enroll exchanges with a private pool card, and in this way absolutely expels focal specialists. As a drawback, this component diminishes the secrecy of the code, since the private key might be recovered from the gadget and its passphrase. The last choice depends on cost improvement [16, 17, 18].

Put something aside for using a solitary document to hold the full remarks (not hashes), it is equivalent to the main choice. This diminishes the expense of capacity, with the inconvenience of decreasing the dependability of audits, since the substance of the database can be altered by its administrator.

Distinctive engineering options arrangements should be chosen dependent on the exchange offs to be reached, for example, security advancement, reliability, or cost.

These would apply either to a completely decentralized system based on creativity in blockchain or to a combination of decentralized and hierarchical architecture options, such as a blockchain to carry work hashes, and a traditional database to carry content.

6 Conclusion and Future Scope

6.1 Introduction

This part includes the conversation about the significance of work for future references. Additionally, it incorporates the upgrades that could be made to improve the existing outcomes and making a superior workplace. The section advances towards end by closing the exploration work. In the later segment, all the data in regard to reference is given.

6.2 Future Scope

- Blockchain's smart contracts can be integrated with e-commerce websites/applications to make untampered product review system.
- Cloud computing can also be used to handle computing of blockchain-based review system in e-commerce platform.
- The integrated review platform will be robust, and the reviews will not be able to be modified and faked.

The motive of this research is to study about applying blockchain in the review platform using the smart contract transactional approaches with the help of research-based modified models of smart contracts. Smart contract is a wide area for research purposes. There are many methods which can improve the performance and security of the system. Various researchers have use different methods to study about using blockchain-based review systems.

Future work includes studying the possibilities of using cloud-based services and computing power to impower these heavy blockchain transactional computations. Smart contracts are well known for their high accuracy and security in data and transactions.

6.3 Conclusions

- Blockchain is a new domain in modern tech world with lots of potential to be discovered for appropriate applications.
- Smart Contract is one of the most promising application of blockchain which can be used in myriad of applications.
- Smart Contract has the capabilities to make a revolution in financial sector, eradicating the need of middleman completely.

References

1. D. Martens, W. Maalej, Reviewchain: Untampered product reviews on the blockchain, in *Proceedings of the 1st International Workshop on Emerging Trends in Software Engineering for Blockchain*, (2018), pp. 40–43
2. K. Salah, A. Alfalasi, M. Alfalasi, A blockchain-based system for online consumer reviews, in *IEEE INFOCOM 2019-IEEE Conference on Computer Communications Workshops (INFOCOM WKSHPS)*, (IEEE, Piscataway, 2019), pp. 853–858
3. Y. Hu, M. Liyanage, A. Mansoor, K. Thilakarathna, G. Jourjon, A. Seneviratne, Blockchain-based smart contracts-applications and challenges (2018), arXiv preprint arXiv:1810.04699
4. C. Molina-Jimenez, I. Sfyrakis, E. Solaiman, I. Ng, M.W. Wong, A. Chun, J. Crowcroft, Implementation of smart contracts using hybrid architectures with on and off–blockchain components, in *2018 IEEE 8th International Symposium on Cloud and Service Computing (SC2)*, (IEEE, Piscataway, 2018), pp. 83–90
5. S. Srivastava, R. Bhatnagar, A study about process mining, in *Proceedings of International Conference on Sustainable Computing in Science, Technology and Management (SUSCOM)*, (Amity University Rajasthan, Jaipur, 2019)
6. S. Srivastava, G. Srivastava, R. Bhatnagar, Analysis of process mining in audit trails of organization, in *International Conference on Information Management & Machine Intelligence*, (Springer, Singapore, 2019), pp. 611–618
7. S. Srivastava, G. Srivastava, R. Bhatnagar, Green cloud, in *Smart Agricultural Services Using Deep Learning, Big Data, and IoT*, (IGI Global, Hershey, 2021), pp. 69–80
8. X. Xu, I. Weber, M. Staples, *Architecture for Blockchain Applications* (Springer, Cham, 2019)
9. S. Srivastava, Process mining techniques for detecting fraud in banks: A study. Turk. J. Comput. Math. Educ. (TURCOMAT) **12**(12), 3358–3375 (2021)
10. K. Hansrajani, M.B. Johnson, S. Srivastava, Cyber security. J. Adv. Robot. **6**(3), 25–28 (2019)
11. S. Srivastava, G. Srivastava, Tools and techniques for network forensics
12. S.M. Idrees, M. Nowostawski, R. Jameel, A.K. Mourya, Security aspects of blockchain technology intended for industrial applications. Electronics **10**(8), 951 (2021)

13. S.M. Idrees, I. Aijaz, R. Jameel, M. Nowostawski, Exploring the blockchain technology: Issues, applications and research potential. iJOE **17**(07), 49 (2021)
14. S.M. Idrees, M. Nowostawski, R. Jameel, Blockchain-based digital contact tracing apps for COVID-19 pandemic management: Issues, challenges, solutions, and future directions. JMIR Med. Inform. **9**(2), e25245 (2021)
15. P. Tasatanattakool, C. Techapanupreeda, Blockchain: Challenges and applications, in *2018 International Conference on Information Networking (ICOIN)*, (IEEE, Piscataway, 2018), pp. 473–475
16. O. Labazova, T. Dehling, A. Sunyaev, From hype to reality: A taxonomy of blockchain applications, in *Proceedings of the 52nd Hawaii International Conference on System Sciences (HICSS 2019)*, (2019)
17. V. Sangtani, Investigations on performance of AI based system for highways. Turk. J. Comput. Math. Educ. (TURCOMAT) **12**(13), 2526–2531 (2021)
18. G. Srivastava, Infrared plastic solar cell. PalArch's J. Archaeol. Egypt **18**(4), 1473–1480 (2021)

Blockchain Technology Transforms Digital Marketing by Growing Consumer Trust

Fazla Rabby, Ranga Chimhundu, and Rumman Hassan

1 Introduction

Companies who want to compete in highly competitive business to consumer (B2C) industries must have a customer-centric approach [1]. The communication methods used by a company can vary depending on the industry; however, the basic goals and challenges of customer interactions have not changed [2]. Emerging technology has a democratising effect on both businesses and customers, moving beyond the reach and size of large corporations; it is more open to smaller businesses. Although unpredictable return on investment, small businesses continue to invest in fee-based services and networks that they believe are necessary to remain competitive in their markets [3]. For example, the introduction of 'Martech'—a combination of marketing automation and technology—has improved how businesses communicate and connect with their customers [4]. Martech alters customer–brand partnerships and their dynamics by changing a company's outreach strategy, and by changing and boosting customers' expectations [5].

In this recent economic growth, companies are no longer reliant on launching one project after another. As an alternative, companies are focused on innovative ways of communicating with consumers to broaden their company and incorporate more synergistic, well-coordinated marketing relation strategies [6]. Companies are developing a portfolio of technologies using different social media networks and advertising campaigns to label and sell their goods, services and concepts [7]. Digital marketing helps businesses to use evolving social networking tools to engage with customers in modern, innovative, cost-effective and powerful ways [8]. As a

F. Rabby (✉) · R. Chimhundu · R. Hassan
School of Business, Faculty of Business, Education, Law & Arts, University of Southern Queensland, Toowoomba, QLD, Australia
e-mail: fazla.rabby@usq.edu.au; ranga.chimhundu@usq.edu.au; rumman.hassan@usq.edu.au

S. M. Idrees, M. Nowostawski (eds.), *Transformations Through Blockchain Technology*, https://doi.org/10.1007/978-3-030-93344-9_12

result, consumers are becoming a vital part of the conversation, and consumers influence on the marketing system is growing [9].

The Internet and other evolving technologies such as Blockchain have had a huge influence on the traditional marketing mix. Today's companies have integrated advanced technologies, large and intricate data sets within their business strategies, such as Big Data analytics, to win further customer perceptions [10]. Similarly, to outperform their competitors, stores and Internet firms increasingly incorporate social networking into their marketing messaging strategies. According to Goldstein et al. [11], the trend towards cyber shopping will continue, with consumers' willingness to engage in online shopping experiences increasing in the future [11]. As a result, many customers choose to shop online rather than in brick-and-mortar stores [12]. However, there are risks associated with digital technologies, and customers are on the frontline of security, protection, trust and transparency issues.

According to Ingram et al. [13], any time a person makes an online purchase, they leave a digital footprint of personal information about themselves—their shopping habits, spending preferences, credit cards and other personal information [13]. As data processing methods become more versatile and popular, the issues around privacy deteriorate. Data breaches are common because online businesses often fail to comply with regulatory requirements, and they have a long-term negative impact on customer trust [13]. Furthermore, many developing countries disregard privacy laws altogether. As a result, customers' awareness grows, their concerns grow, and they become more cautious when making online transactions, because their personally identifiable information can be accessed or shared without permission for monetary gain [14].

Since physical retailers also allow consumers to use rewards cards and keep track of their transactions, a centralised database that can be abused or misused, preventing Internet purchases will not solve the problem. Therefore, businesses must stay current on new privacy regulations, consumer requests, technological advancements and best practices. Proponents of consumer rights argue that systems should be built with a 'privacy-by-design' system in mind. Peer-to-peer networking is a key feature of Blockchain technology, which was originally developed to underpin cryptocurrency, Bitcoin. It eliminates the need for centralised controlling of the flow of transactions between third parties [15]. According to Salha et al. [16], Blockchain is a distributed, free and digital ledger, and the purpose of Blockchain is to establish permanent and tamperproof records by recording and adding transactions in a sequential order [16]. A Blockchain design is a series of technology, tools, and methods that work together to resolve trust and uncertainty issues in business application [17]. Thus, marketing executives must be aware of the benefits of Blockchain technology as a communication protocol, which transmits messages from the informational Internet to the value and trustful Internet [18]. Despite the increasing use of digital marketing, there is still much to be discovered about how Blockchain technology increases transparency and trust in digital marketing. This study aims to provide an overview of Blockchain technology and its applications in digital marketing to develop trust and transparency using big data.

2 Literature Review

The literature review in this chapter is a theoretical review. The aim of theoretical analysis is to look at the most recent theories about the phenomenon under study [19]. Theoretical studies are often used to demonstrate that a previous study is insufficient or incomplete in describing current or ongoing research issues [19]. The literature review for this study aimed to get a comprehensive overview of the current topic and determine that there are insufficient hypotheses about the relationship between Blockchain and digital marketing. As a result, the literature review was used in this analysis to determine research weaknesses, and, as a result, a suitable field area was identified. We set the foundations for future academic research throughout this chapter by identifying many key research areas.

2.1 Blockchain Technology

Following the global financial crisis of 2007–2008, the public's trust in traditional banking plummeted [7]. In addition to believing in technological advancement, both enthusiasts and digital developers created distinct, independent banking industries based on the belief that third-party relationships had become corrupted. The idea of digital money Bitcoin, a peer-to-peer electronic cash system proposed by Satoshi Nakamoto (pseudonym for the person or persons who invented it), was to solve this major problem in today's world [20]. Bitcoin employs a protocol based on cryptographic rules and techniques for transaction processing [21]. They include hashing, time stamping, consensus processes, and asymmetric encryption with public and private keys. The proposed cryptocurrency model ingeniously addresses the issue of double-spending [16] and develops an innovative framework for conducting transactions and quantifiable terms in a digital world [22]. To ensure transaction legitimacy, every transaction activated on the Blockchain must conform to a set of predetermined guidelines centred on protection, verifiability and peer consensus [23]. All transactions are transcoded, recorded in blocks and linked together with a public transaction record in a sequential manner. Changing the cryptographic hash of previous Blockchain is technically difficult and considered infeasible due to the large amount of computing power needed [24].

Blockchain is built on distributed ledger technology (DLT) [18]. Blockchain is a distributed database system that continuously adds new records to a growing list maintained by all of the system's nodes [25]. All data, including each transaction, are recorded in a public ledger. In other words, the Blockchain network is decentralised, meaning there is no central authority and the system is inherently democratic [25]. Information in the ledger is open to everyone because it is shared and immutable. As a result, anything built on the Blockchain must necessarily be transparent, and everyone who has a hand in its construction is accountable for their actions. Decentralised Blockchain solutions do not require third-party mediators [15]. All the information about every transaction that has ever occurred on the Blockchain is available to every network member. A decentralised transaction will have greater transparency than one that uses a third party [18]. Another great feature of Blockchain is that all the nodes are anonymous, which gives other nodes more confidence when confirming transactions. Bitcoin was the first use of Blockchain technology in a digital application. In creating a decentralised environment for digital currency, Bitcoin gave rise to cryptocurrencies to buy and sell goods [15]. On large chains, mining a block is difficult because every block has its unique nonce and hash but also points to the hash of the previous block in the chain, making it a difficult task. For the most part, Blockchains are distributed databases that are immutable, hosted by a network of computers that no single entity controls. Data is held together by cryptographic principles, with each block connected to every other block (i.e. chain) [26].

By utilising a network of computers that all have the ability to validate transactions, DLT offers a consensus validation mechanism. Every transaction is verified and, together with a number of verified transactions, is appended to an existing block of transactions, which is what leads to the term 'Blockchain' [27]. Once a transaction has been recorded on the Blockchain, it cannot be edited or removed. The process of re-mining the entire Blockchain is necessary when making any change to any block earlier in the chain [28]. That is why Blockchain technology is so difficult to manipulate. Since finding golden nonces requires much time and computing power, think of it as 'security in math' [29]. According to cryptocurrency terminology, a Blockchain is a combination of technology [30]. The vast majority of academics and professionals see Blockchain as a branch of the distributed ledger framework [26]. The ledger is a virtual book or a one-of-a-kind list of all Blockchain exchange parties' transactions. The platform can be conceived as a modern way of encrypting assets in a transaction and can also be used in several different business operations and processes [28]. In the next topics, we will go through how Blockchain technology's main characteristics enable functions and applications that will drastically alter future marketing strategies.

2.2 Marketing with Blockchain

The benefits of Blockchain technology have become quickly apparent and should not be limited to cryptocurrency applications [31]. According to Antoniadis et al. [32], Blockchain adoption will influence over 50 industries' business models and operations in the coming years. Its ability to perform digital asset transactions in near real time has made it a promising tool for executing contracts without the use of middlemen, and for storing data with unrivalled levels of confidentiality and reliability in a variety of economic activities [33].

There are limited detailed research reports on Blockchain technology as a marketing support tool [33]. As a result, the benefits of Blockchain are commonly recognised as unquestionable in practitioner literature [34]. To begin, Blockchain technology focuses on peer-to-peer networking, which disrupts business processes by encouraging direct marketing, or the elimination of mediators who deal with clean data sources and add to cost. By establishing permanent and shared data archives, Blockchain technology can increase system reliability and accelerate data access. From the users' point of view, Blockchain technology has the ability to significantly enhance customer experiences by increasing data and information transparency while also improving security and privacy [35]. It enables the development of various types of consumer loyalty items, which can add value. We will go through each of these features in greater detail in the sections below.

This chapter focuses on three main areas: management concerns, the digital market environment, and its effect on companies and sectors, and technological architecture [27]. According to the literature, Blockchain implementations have three major characteristics [36]. Since most of the analysis focuses on cryptocurrencies and fintech, the first characteristic divides applications into financial and non-financial categories [35]. Portmann [37] considers the second characteristic to be the three waves of Blockchain evolution: cryptocurrency wave 1.0 (Bitcoin), Smart Contracts (SC) systems and Blockchain 3.0, which is characterised as a programmable group [37]. According to Zheng et al. [38], the third and final characteristic is the five major areas of Blockchain applications: finance, Internet of Things (IoT), public and social networks, reputation systems, and security and privacy. Casino et al. [33] extend this strategy by providing a more extensive collection of Blockchain applications focusing on specific fields of economic operation like wellness, education, regulation, data processing, and specific business processes and activities like supply chain management [33].

When it was first introduced in the 1990s, digital marketing was mostly used in customer-facing commercials. However, as digital marketing evolved over the past two decades, the concept of customer engagement became increasingly important [15]. Digital marketing focuses on how to communicate with customers and form long-term relationships, as well as how to fulfil their ever-changing needs. Today, digital marketing is an agile, technology-enabled process in which companies collaborate with customers and partners to create, communicate, generate and sustain value for all stakeholders [39]. In addition, digital marketing encompasses a wide range of topics, including social media marketing, online marketing, analytics, e-commerce and customer data mining [40]. Companies are now concentrating on digital marketing for Internet-based search marketing, digital advertising and banner ads [41].

Emerging technologies such as the Internet, telephone, digital television, email and databases are also used in digital marketing. These tools help both interactive and non-interactive marketing strategies to attract and keep profitable clients [41]. Digital conversation interfaces, which companies use to offer meaningful content to consumers, are connected to customers' digital touchpoints. Social networking, search, e-commerce and mobile are only a few examples of customer digital touchpoints. These differ from firm conversation interfaces, such as marketing analysis, email, Internet marketing and Search Engine Optimisation platforms used by marketers [41].

Additionally, some of these studies also focus on the impact of social media on digital marketing, which is relevant to this research [10, 40, 42]. Social media marketing includes all social media channels, such as Snapchat, YouTube, Twitter, Facebook, and WordPress, because it incorporates all of these [42]. Marketers who use social media to expand their network have two options: listen and respond to their audience or listen and gather feedback from their audience. Social networking tends to have a greater effect on word-of-mouth than traditional marketing platforms [42]. Companies use social media marketing to increase brand awareness, improve brand image, draw visitors to web pages and promote customer engagement. They do this in one of three ways: by creating, fostering or boosting content [43]. Influencer marketing is one of the many ways that social networking can be used. In their recent interview, Fehrer et al. [44] share an in-depth discussion on the subject of influencer marketing, which is considered part of the larger digital marketing spectrum [44].

Blockchain is similar to the Internet in the 1980s, and social media in the last decade; it is a foundational technology that can revolutionise the marketing landscape [33]. This project aims to use DLT to revolutionise marketing. DLT, which intelligent contracts enable, can reform marketing [15]. Currently, many companies make using Blockchain in marketing campaigns feasible. When someone hears the term 'Blockchain', they immediately think of a continuously growing list of records, called blocks that are linked together without the need for a third-party verification. Blockchain was well known by the general public because it was the technology that allowed Bitcoin to exist [19]. Every transaction could be verified and recorded with Blockchain. Technology's positive attributes that customers are searching for today, like transparency and data protection, can be leveraged by marketers.

Companies that want to survive in fiercely contested B2C markets must focus on customer-centric marketing [41]. The value customers derive from a product or service is influenced by marketing. Businesses employ a variety of communication methods that differ from industry to industry. However, the goals and hurdles surrounding consumer engagement remain unchanged. Growth in new technologies also democratises businesses and consumers by making new technologies more accessible and affordable for smaller companies. Despite their lack of guarantee, small businesses are now putting their money into fee-based technologies and platforms that they believe are vital for retaining their market position [41]. In light of this emerging trend, firms that use 'Mar-tech' have seen improved results due to integrating both marketing automation and technology solutions [1]. This also raises customers' expectations and changes the way a company does its outreach [1]. Due to Blockchain, marketing and data transparency may be on the march. Many industries are undergoing rapid change as a result of Blockchain technology. The term 'Blockchain' gets frequently associated with the financial and banking world, but its implications are broader. Because Blockchain is known to be closely linked to finance, many people wonder how marketing could use it. Blockchain technology may have a role to play in marketing. Because of Blockchain, marketers can now collect and use data in different ways, answer customer needs and handle advertisements.

Data, like reputation and credibility, should be owned by the individual [16]. In other words, consumers should be able to transfer the data they obtain to other platforms, lowering the switching costs. Using Blockchain, it is possible to implement a concept like a global credibility score, which can interact with various parties who have a business together. This type of risk is known as counterparty risk and concerns the possibility that the counterparty's declared identity is fake [5]. The pseudonymity of Blockchain avoids this risk in marketing environments by implementing proof-of-identity on the protocol level [13]. Another threat to e-commerce websites occurs when many forum sites are viewed as untrustworthy. Many claim that a large portion of the content is false because some parties purposely try to influence consumer perception about businesses positively or negatively. The Blockchain technology can trace and validate the real identity of the content creator [5]. In the next section, we will discuss how Blockchain will impact digital marketing [16].

2.3 *Blockchain Applications in Digital Marketing*

Although Lu [30] acknowledged that Blockchain-based governance has useful marketing applications, no systematic documentation of such applications exists. As stated by Min [45] 'Blockchain pulls marketing into uncharted territories'. Publications on the subject are limited to websites and blogs that point to potential fields of use for Blockchain in digital marketing, as well as listings of start-ups and companies that provide solutions to marketers through Blockchain platforms.

The most obvious use of cryptocurrencies is electronic payments. They can be used to conduct transactions between customers and business (B2C) and business to business (B2B). They can be combined with smart contracts to preserve transaction speed and security, establishing trust between the parties involved [46]. Furthermore, eliminating intermediaries reduces operating and financial costs, resulting in higher profit margins [47]. We should also take advantage of the public's enthusiasm for Bitcoin and Blockchain, and use this to attract more customers—just as Malta is doing to improve its image as a tourist destination [47]. Bitcoin and Blockchain also have implications in supply chain management. Smart contracts can help track suppliers and retailers in a sustainable and resilient manner, preventing fraud and other issues that damage brand image and reputation, and increasing customer trust and loyalty [45]. Walmart and IBM are already using such solutions in the grocery industry, and companies like Origin provide similar solutions for verifying product origin, journey and effect.

Bitcoin and Blockchain are also useful for handling different elements of the marketing mix. Sales management can be automated and tracked in real time in a straightforward and cost-effective manner. Digital advertisement management can be aided by the Ad Ledger developed by IBM and Salon, and Blockchain-based markets for advertising and media planning, such as the NYIAX, introduced by NASDAQ [48]. Apart from the benefits of disintermediation, Blockchain will be able to provide improved targeting for IMC (Integrated Marketing communications) digital marketing strategies and accurate performance monitoring of marketing and advertising initiatives. This will add significant benefits to digital and email marketing because tracking can prevent manipulation, saving the industry significant capital [32]. In the digital advertising industry, consumers' trust issues have emerged around transparency and privacy. Blockchain will build trust, offering anonymity and immutable transactions [29]. Such trust will encourage consumers to consent to providing their personal data safely via a distributed ledger, where businesses can 'buy' them without the need for middlemen or centralised processes [49].

Blockchain identification apps can help companies and customers verify feedback and make it more accurate. Users can be rewarded in tokens for their contributions, creating incentives for creating quality user content, especially among social media influencers. Users can be paid in tokens for watching advertisements, as well as monitoring conversion rates, validating purchases during the campaign, and measuring the campaign's efficacy [34]. Mobile apps and Blockchain will also revolutionise the use of tokens and cryptocurrencies in loyalty programs and services in digital marketing industries [50].

Tokens would benefit the incorporation of all rewards systems (including gift cards) into a single network that would include purchases and a customer's relationship with a brand or shop–integrating digital marketing and social networking platforms [51]. Tokens would lead to a more inclusive loyalty scheme, taking into account all customer interactions with the brand in a transparent and easy-to-monitor manner. Several companies are now using Blockchain for this purpose. For example, Cathay Pacific and Air Asia have switched their air miles benefit scheme to run on Blockchain and mobile devices, merging Blockchain and gamification to provide a better consumer experience. The Blockchain platform automates data fulfilment processes, offering a detailed and transparent transaction history between the airline and participating partners while increasing operational performance and lowering back-office administration.

Full transparency and traceability of brand promises are also possible with Blockchain in digital marketing industries [10]. The brands in digital marketing claim to be socially responsible, regardless of whether their claims are genuine or not [46]. Because of this, Blockchain enables consumers to see if a brand donated to charities instantly, and check how socially or environmentally responsible the brand is. In addition, consumers will learn whether the brand they prefer uses employees in a way that is respectful of their well-being (i.e., no child workers). With Blockchain technology, customers will also be able to view, unambiguously, examples of highly transparent measures, such as complaint rates, satisfaction scores, defect rates and timely delivery [10]. Instead of vague generalisations, brands should provide proof for their claims. Blockchain also serves as a method for detecting and dealing with counterfeit goods and protecting trademarked brands [13].

The disintermediation of traditional trading methods with alternative trading facilitated by electronic channels is called disintermediation [25]. People became interested in disintermediation once the Internet became widely accessible. An electronic brokerage effect has emerged because search costs are low on the Internet, reducing the need for intermediaries [8]. But a new class of intermediaries, bypassing the trusted third party, has recently emerged to meet the need for a secure middleman. E-commerce features, such as information brokering, trust provision and search capabilities, are added to the value chain by re-intermediation [8]. As has been previously stated in the discussion of how to establish trust in situations where uncertainties are present, Blockchain does away with the need for centralised institutions to enforce trust in digital marketing industries [15]. This definition also covers the transacting parties' reputation scores, credibility and the like. As a result, we should expect true disintermediation in digital marketing instead of redistribution of intermediaries [15]. It will also be possible for customers to take advantage of a single service from one brand and use other available services from other brands. For instance, a customer who was a customer of one bank might choose to use another bank's digital wallet service while also using the bank's other services. Mass migration of consumers could take place in a matter of minutes. The only way for brands to differentiate themselves in a post-switching world is to provide seamless customer experiences.

All intermediaries will need to alter their business models as Blockchain gains traction. A strategic shift in how information is distributed, and the amount of time individuals spend interacting with advertisers will be implemented [41]. Their inherent structure can stop phishing and spam—the more spammers attempt to phish, the more financially unviable their spam becomes [50]. A better understanding of customers' behaviour and enhanced control over inbound traffic for all digital marketing efforts are both possibilities [29]. On the other hand, advertising would not be imposed on people unless they paid a fee to all of the people who were exposed to it. Consumers will also be incentivised to post a truthful online profile in order to earn money. More and more companies will begin to pay consumers directly, without the middleman of social media. When setting customer's incentives, the value of the customers must be taken into consideration [52]. Blockchain technology in digital marketing has the potential to improve societies by increasing trustworthiness and empowering individuals [52]. Changes in digital marketing and advertising are relevant because they impact marketing and advertising fundamentals. Chief Marketing Officers and other executives with strategic, financial, and technological decision-making roles must consider ways to make Blockchain-related transformations. The current levels of trust between businesses and individuals can be augmented using Blockchain technology, and it may be possible to reach a scale of connection between their products and services and their customers not possible without Blockchain [10]. Customer relationships could be reinvented using Blockchain with digital marketing and technology leaders. By taking action on this far-reaching technology as early as possible, we can ensure that companies are well-positioned to take advantage of widespread usage. Because of this vast amount of digital data used in digital marketing industries, in the next section we will explore the relationship between Big Data and digital marketing.

2.4 Big Data and Digital Marketing

Since the volume of information and data obtained from various sources like digital and social media channels and the IoT remains unchanged, it is not sensible to say that digital information and data are increasing or decreasing. Particularly since data is collected constantly though different digital marketing platforms, and consumers are now more aware of how much competition there is for their attention and time [28]. As a result, the Big Data gains in value and, as it is brimming with customer insights, it continues to be valuable to digital marketing professionals, allowing them to adjust products and services and develop compelling offers for customers that nobody can refuse [28]. Many technology behemoths have leveraged Big Data to help businesses grow. It is still too early to turn data collection, mining and analysing into a business, but the future is there for those who master the skills of data collection, sifting and processing to gather actionable insights on consumer behaviour.

This attention economy is thanks to all the different digital technologies and techniques available to process data, and as a result, we now have an economy where everyone is engaged in vying for attention [28]. Ertemel [28] contends that this spurs customers into a damaging cycle because of the different data gathering platforms, which accumulate due to the industry's competitiveness. According to Ertemel [28], Blockchain has the potential to counteract this damaging system. People gain control over their personal data through the pseudonymity attribute of the digital technology and can store it either in a safe cryptographic wallet or on the phone. Giving the data generator the option to either share their data or keep it to themselves empowers them and allows them to be entirely in control. Now, notable companies like Google and Facebook will not be able to gather user data in the same way they could in the past [28]. Personal information and behavioural data will no longer need to be shared with third parties; instead, individuals will be able to monetise that data themselves [28]. However, using these tokens instead of traditional currencies presents numerous problems due to their inherent impracticality. The performance of the digital technology remains a big concern because it has not yet been perfected [43]. Strategic marketing decisions are more accurate and advanced when applied to Big Data. This will help digital marketers gain a better understanding of the organisation and enable them to find new insights and information.

Today's digital environment consists of a large quantity of data that contains data on consumer ratings, reviews, blogs, tags, likes, shares, retweets and other actions [47]. Digital marketing relies on a combination of digital media and devices, including computers, to target consumers [10]. Big Data has become a necessity in digital marketing as a result of the extensive use of these technologies. Big Data can be utilised to identify new target customers, adversaries and the competitive market [47]. Research conducted by several organisations underscores the importance of Big Data in digital marketing. Enormous amounts of data allow businesses to track and identify online customers' preferences, thus building more targeted digital marketing strategies. They can also track the results of their marketing activities and measure their overall success [46]. By optimising digital marketing strategies, marketers can also learn a lot about customers by tracking and predicting their reactions to marketing [41].

Additionally, Big Data can help to track a business' performance [41]. Because there is so much data available for collecting and processing, it is viewed as a challenge for many companies [16]. Marketers face a challenge because they have too much data from multiple sources instead of a small amount of data [1]. This helps to illustrate the digital marketing environment's growing dependence on numerous intermediaries and their businesses. All of them use user data to offer targeted advertising, and they make full use of advertising budgets.

The application of Blockchain could alter how online businesses operate, removing uncertainty and boosting the volume of online transactions [28]. Although customers are unsure about doing business with different companies in the online marketplace, uncertainty has caused them to shy away from changing whom they do business with. Well-known brands need to build trust over time by convincing more and more customers with their reviews and ratings. Because of this, customers are less likely to alter the company they purchase from [28]. According to Ertemel [28], companies own valuable data on customers, but customers should control the data themselves. In addition, customers should be able to move the data around and do with it as they please. For example, transfer their data to another platform to make using other companies more accessible. An additional benefit of Blockchain is that it enables implementing a global reputation score that could be utilised by anyone who is about to enter into a financial agreement with another party [28]. Big Data has the potential to yield fresh and relevant insights for digital advertisers. Real-time analysis and informed marketing decisions and strategies are made possible by the fast processing abilities of today's digital marketing industries.

3 Theoretical Framework and Hypothesis

This section on theoretical framework begins with an overview of digital marketing, analytics, Big Data and digital marketing technology. Following that is a detailed summary of Blockchain and a study of its possible impact across digital marketing industries (Fig. 1).

This study aims to determine the scope of Blockchain in digital marketing by using Big Data to boost digital marketing and revamp business strategy. The following hypotheses were formulated based on previous studies.

H1: Blockchain increases transparency and trust in the digital world.
H2: Digital marketing and Blockchain are inextricably linked.
H3: The use of Big Data in conjunction with Blockchain improves digital marketing and marketing strategies.
H4: Regulatory uncertainty is a barrier to Blockchain adoption.

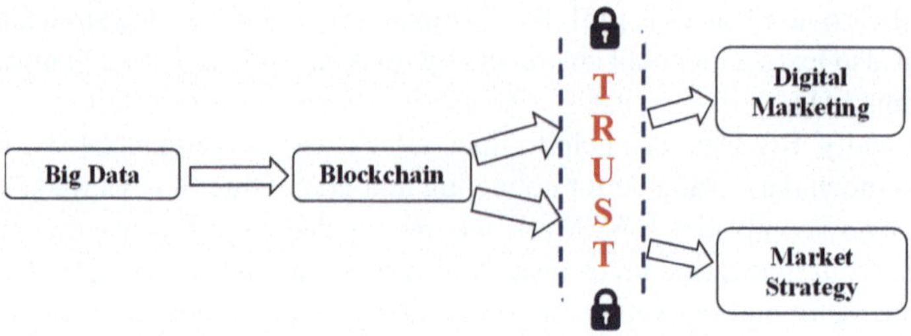

Fig. 1 Blockchain can influence the use of Big Data in digital marketing

4 Research Methodology

To discover how Blockchain will influence digital marketing in the future, the authors need to focus on a few important topics. The application and implementation of the Blockchain are highly dependent on the purpose of the technology. For instance, the researchers from the previous chapter cited findings pertaining to Blockchain's inability to handle Big Data. Previous chapters covered the context of the analysis, the problem statement, the objectives and a literature review. We used previously accepted indicators from the literature for this analysis. The investigation began with a Google search for Blockchain in digital marketing, using terms such as 'Blockchain technology', 'digital marketing', 'Blockchain trend', and 'technology in marketing'. We used an extensive scholarly peer-reviewed article and PwC's Global Blockchain Survey result to address the study questions and identify solutions to the proposed problem.

5 Findings and Analysis

5.1 Part A: Theoretical Contribution and Findings from the Literature

Big Data, digital marketing and Blockchain are being studied together to address how the relationship works. Blockchain adoption has, to date, been focused primarily on digital marketing. Although none of these previous studies has focused on the specific connection between digital marketing, Blockchain and Big Data, this study will take this concept to the next level. This study also makes a notable contribution to theoretical knowledge, with the development of a new model to illustrate how the various concepts are interrelated. Blockchain is predicted to change how the industry makes use of Big Data, and this thesis contributes to our understanding of that.

5.1.1 Transparency and Trust in Digital Marketing

'Click fraud' is today's most visible digital marketing phenomenon, and numerous theories have been proposed to explain its origins. Blockchain technology is believed to be able to displace this phenomenon in the digital marketing industry [28]. Blockchain is also anticipated to assist in the growth of digital marketing trust and transparency. It has the ability to tackle a lot of today's marketing problems. According to Ertemel [28], Internet shopping has trust issues as it is done through the use of computers and servers. Trust issues can be traced back to factors, such as a lack of clarification on who the counterparty is, a lack of insight into the asset's journey through the supply chain, a lack of understanding about brand commitments and a lack of understanding about what happens when things go wrong [28]. This lack of Internet mediums means that online businesses cannot work as well as they would like.

Blockchain is a decentralised technology, and this improvement in cybersecurity ensures that the digital marketers' transactions are secure [25]. To put it another way, it is not under centralised control; nobody controls or owns the Blockchain [18]. Few rigorous academic studies provide empirical evidence of Blockchain's application in support of marketing activities. The benefits of Blockchain are certain even though this information was difficult to uncover [25]. Most importantly, Blockchain technology is based on peer-to-peer communication, which allows for market structures to be transformed by having intermediaries removed. This is why the Blockchain economy will favour disintermediation, the process of removing intermediaries who handle and filter data streams [38]. Blockchain technology is also capable of improving data quality and increasing data access by creating immutable shared data records. Consumers have the potential to significantly transform their relationship with their data by enhancing information and data transparency, as well as improving privacy and security [50].

Ensuring Trust and Fraud Prevention

A relatively overlooked aspect of data analysis is the ability to control so-called dirty data (i.e., erroneous information). This can be a boon for the Blockchain industry. The security of the data in Blockchain transactions is assured because it is maintained in a decentralised ledger [29]. Since the data is kept on the Blockchain and is verifiable, the trustworthiness of that data is assured. When one specific data block is tagged as coming from an unconfirmed source, details of the origin and the other blocks it has interacted with are recorded on the Blockchain and automatically verified (or validated) before the block can be acted upon. It also promotes transparency, as all Blockchain network activities and transactions are recorded. Big Data has not yet solved fraud detection and risk assessment difficulties in the financial services industry [47]. Because existing detection and assessment methods are dependent on historical data. With the use of Blockchain technology, financial institutions can access real-time data from all transactions [47]. Having done that, rather than analysing past frauds, the banks can instantly identify risky or fraudulent transactions and stop them before they occur. With the use of Blockchain, financial institutions can finally monitor risk and flag unusual patterns instantly [29]. By adopting this measure, banks and their customers will be protected from fraud.

Managing Data Sharing and Data Quality

A Blockchain-based Big Data system would enable all sectors to exchange records with each other with significantly reduced risks [47]. A Blockchain is a powerful tool for data storage because it maintains data in a distributed, permanent and secure manner. Teams use this approach to prevent project teams from doing data analysis previously completed by other teams or reusing data that has already been used [47]. Another benefit to using a Blockchain platform is that data scientists can trade the analysis outcomes they store on the platform. In this way, the technology can bring about fresh ideas in data storage [16]. Businesses can improve data quality by implementing Blockchain's new storage methods. Another benefit of integration with Big Data analytics is improving the technology's foundation by eliminating its weaknesses [16]. The accuracy and full scope of analysis are greatly improved as a result. When big business' data is more secure, trusted and accessible, making decisions based on insights will be easier.

Predictive Analysis

The data in the Blockchain (and any other type of data) can be used to learn about user behaviour, marketplace trends and so on. In the context of businesses, these questions have good predictive value for future outcomes, such as customer preference, customer lifetime value, dynamic pricing, and churn rate [20]. Real-time analysis of data becomes possible in large-scale organisations thanks to Blockchain, making quick decisions possible. The world of accessing vast amounts of data, and using the results to generate insights, is still in its infancy. Using Blockchain and Big Data, businesses will reliably and effectively collect and analyse real-time data [27].

Facilitating Big Data and Streamlining Data Access

Big Data companies could process data much faster and more efficiently by storing it on a decentralised ledger rather than using a centralised solution [18]. By doing this, transaction times will be significantly reduced, and so will the fees be associated with transferring money. In addition to enabling Big Data and analytics, Blockchain could empower these emerging fields by making data easier to access [33]. Using Blockchain, different departments in organisations can gain access to the relevant data to carry out analysis work. This makes data access and analysis smoother at the same times as shortening the work cycle [47]. With the Blockchain database storing the information, a single, unalterable database is being built, which anyone authorised to view can access [36]. In the context of Blockchain, it is possible to have multiple sign-offs or permissions from network members to access a record. All employees will have all the accurate and up-to-date information they need to perform their jobs.

Summary of Findings

Using Blockchain, companies and businesses will have access to numerous advantages. Using Blockchain technology, transparency, trust and efficiency could all be better managed in the digital advertising ecosystem. It would add transparency and trust to the entire marketing process. Blockchain allows any business to build trust quickly, regardless of how long it has been in operation. Companies, regular users and governments in the cybersecurity community all benefit from Blockchain's transparency.

5.1.2 Impact of Blockchain in Digital Marketing

Blockchain technology is a distributed ledger that includes mutually private or public records of all transactions. These records contain information extracted from various digital activities carried out and shared within the Blockchain network [53]. There are several distinctions between public and private Blockchain networks. Everyone has access to public Blockchains, allowing them to participate in the scheme. Bitcoin, for example, is a cryptocurrency built on a public Blockchain framework [45]. On the other hand, a private Blockchain is a closed network that primarily consists of members or nodes who have been selected and approved by the party or group of parties running the Blockchain [45]. Brauer et al. [41] point out that the main difference between decentralised and centralised Blockchains is the degree of decentralisation, based on their findings.

Blockchain has a distributed ledger structure in which customers maintain their own private or public record of all transactions. As a result, the documentation incorporates all of the most up-to-date information extracted from a number of digital activities carried out and exchanged across the Blockchain network [32]. Bitcoin is a cryptocurrency founded on the public Blockchain system [32]. By contrast, a private Blockchain is a closed network primarily of participants or nodes identified and accepted by the party or group of parties that run the Blockchain [54]. It is anticipated that Blockchain implementation would provide both opportunities and repercussions in the digital marketing sector [28]. With a detailed understanding of consumer needs in relation to goods, it is expected that this new technology will be able to return personal information to customers, allowing them to better understand their habits and behaviours in digital marketing industries.

The Blockchain technology revolution will have a significant impact on marketing professionals and advertising. Digital marketing represent the most effective and exciting opportunities for reaching and engaging consumers, in the history of marketing [41]. Most brands nowadays use third-party data sharing, such as Facebook, to acquire customer information. On the other hand, with Blockchain, retailers could incentivise customers to share personal information by utilising micropayments instead of an intermediary [50]. In today's complex marketing environment, trust is the greatest challenge [8]. In an e-commerce setting, it is crucial. This is because a lack of total trust between the parties affects the overall amount of business that can be done online [8]. Despite extraordinary efforts, the adoption of e-commerce has been minimal. As trust is an important factor in the online platform, Blockchain technology is prepared to tackle distrust at various levels [15]. There is always some uncertainty in business dealings with others, regardless of whether the business transaction takes place online or offline. As a result of prior negative experiences, consumers will try to avoid being victimised by not-acquainted businesses. This limits the opportunity to do business over the Internet because of the large number of uncertainties that are associated with it. eBay and Amazon, which are well-known brands, serve as trusted third parties for consumers to turn to online platforms [8]. Once they have earned consumers' trust, they will have the power to trap them within their platforms for the long term [15]. One starts to gain trust with eBay by building their reputation score over time. Because of this, consumers will likely become reluctant to change platforms, increasing the overall power of the platform [10].

With the advancement of Blockchain technology, digital marketing is changing because companies can no longer steal information from customers while also offering to return the value [41]. An example of this is the Brave browser, which influences how users interact with online advertising. Instead of having advertisements constantly thrown at them, Brave users elect to have ads shown to them and get Basic Attention Tokens (BATs) in return. It is a whole new way of advertising, since online attention is now a valuable currency rather than space for potential ad sales. As far as marketing is concerned, there are both opportunities and problems associated with implementing the Blockchain [28]. Companies are thinking of returning data ownership to customers while at the same time giving customers a deeper understanding of their buying habits and behaviour in relation to brands [28]. Today, digital marketing's biggest challenge is argued to be click fraud [41]. Pilkington [46] argues that the issues associated with online advertising can be addressed with the help of technology and Blockchain fosters trust and openness in marketing. Marketers are dealing with many challenges [28]. Ertemel [28] posits that e-commerce has a distinct lack of trust. Online businesses cannot conduct as much business because there is a limited amount of trust in the online mediums.

Blockstack is a new type of network for decentralised apps, and it also uses Blockchain to 'protect your digital rights' [25]. Previously, consumers would supply data to specific applications, and the data would remain completely in the hands of the developers. This is simple, but it represented a radical change in the realm of digital marketing. Marketers will see both opportunities and risks with Blockchain technology [28]. These would have a major impact on the advertising industry and online businesses as a whole [28]. It is believed that the technology is not yet ready to solve the problems in digital marketing, even though platforms utilising Blockchain already exist [1]. Using Blockchain, companies can certify that an item was made or grown in the exact location where it was stated, the type of soil where it was grown and the wage paid to the workers employed there. This is significant, especially in the current era where customers place increasing importance on the integrity of their purchase and the company and process producing them [1].

Summary of Findings

Digital marketing is changed by the use of Blockchain technology. Blockchain solutions demand that companies produce relevant content and information in all of their channels. The technology enables consumers to retain ownership of their personal data, forcing businesses to have more respectful, in-depth dialogues with their customers to meet needs. If this is done right, the customer's knowledge will grow. However, if the customers' demands are not met, the technology will impact digital marketers' activities negatively. Customers will not allow companies to process their personal data, which will limit marketers' ability to utilise this information in digital marketing campaigns and to understand and target customers. Overall it is clear that Blockchain is set to influence digital marketing activities and companies' ability to process and collect data for marketing purposes.

5.1.3 Combining Blockchain and Big Data

A Blockchain can be applied to various applications, including energy trading, real estate, and many others. Data analytics in the field of Big Data has advanced because of this fact. For example, to deal with fraud prevention, financial institutions can perform real-time verification of every transaction with Blockchain technology [16]. In the context of digital marketing, Big Data is defined as historical statistics, statistics and information about a person [41]. Big Data is sometimes referred to as the 'raw material' of the digital economy, and it is often compared to 'digital oil' [15]. The data itself is unimportant, but the insights it provides are crucial [41]. Businesses can use problem-solving methods more effectively and achieve better results if they use these concepts. Implementing Big Data allows companies to improve their operations, creating a huge opportunity. While many organisations are using Big Data analytics to better understand their customers and increase customer loyalty, others continue to use other approaches [41].

The importance of data in digital marketing has increased. There are large volumes of data available to marketers, which allow them to track and better understand online customers' interests, develop more targeted marketing strategies and measure the effectiveness of marketing campaigns [41]. Marketing professionals use Big Data to optimise digital marketing campaigns and predict customer behaviour as a result of marketing activities [41]. Furthermore, data collected from the digital environment can help evaluate a company's performance [41]. The wide range of data available benefits the different touchpoints that businesses have with their customers, making it easier to collect data directly related to the point of communication. This knowledge is critical for analysing and optimising various online marketing actions, such as attracting and retaining clients and calculating the overall costs of customer acquisition and retention [41].

The use of Blockchain and Big Data are increasingly in demand among companies today [33]. Soon, businesses and organisations will be run in new and entirely different ways. Over the years, technologies such as these have developed in different ways [33]. By opting for Ripple, the consortium has reduced the time and associated costs required to process a transaction and nearly eliminate the risk of double-spending by eliminating numerous risks [29]. For enterprises currently dealing with Big Data issues, many increasing opportunities that distributed ledgers will finally assist them tackle the challenge. They can both handle the tasks independently, but they could bring in many opportunities when they are both present. Others assert that Big Data and Blockchain are made for each other. Big Data has presented numerous problems for both big companies and the general public. As the amount of data continues to increase, good analytics becomes increasingly difficult. When connected devices increase in number and the size of Big Data increases, it exposes companies' information to additional security risks [50].

Big Data raises the stakes on data quality management. To ensure the data's authenticity, cleanliness, and security, it is important that the data are not altered [16]. These data should also be made available to the public. For most companies, data silos remain a considerable problem, and an all-company digital transformation is simply a concept that has not become a reality [27]. Several exciting potentials can be developed thanks to the blend of Blockchain technology and Big Data. For Big Data management, there are several ways that Blockchain might be helpful. For example, it has the most significant potential in terms of data quality. Blockchain data acquired and validated by large companies will become valuable [47], and Blockchains will help achieve more realistic and accurate expectations about the data set. When Blockchain technology matures, we should see improvements in immutability, general agreement timestamping and data integrity over time [47].

Summary of Findings

The findings presented in this thesis suggest that Big Data will impact Blockchain, which will significantly affect digital marketing. Blockchain technologies will raise the value, truthfulness and speed of Big Data. The integrity of the data on the Blockchain is helping to enhance the collection and processing of data more securely. However, the findings also show that if Blockchain is implemented, the amount of data will decrease because customers will gain control over their own data, preventing companies from collecting and processing customer data without permission.

5.2 *Part B: PwC's Global Blockchain Survey Result Analysis*

In this section, the core findings of PwC's (2018) Blockchain research are discussed [55]. According to PwC's survey (2018) of 600 executives from 15 countries in Table 1, 84% of participants said their organisation is working on a Blockchain project because nobody wants to be left behind [55]. According to a PwC survey, Initial Coin Offerings (ICOs) raised US$13.7 billion in the first 5 months of 2018 and will create more than $3 trillion in annual market value by 2030 [55]. By the end of 2030, 10–20% of the world's economic infrastructure may be running on Blockchain-based systems.

Table 2 shows that 84% of respondents are actively engaged with Blockchain. At the same time, 45% believe that confidence could hinder adoption, and 28% of respondents believe that Blockchain integration is the key to success. According to the results of PwC's study, Blockchain is becoming more popular because companies can streamline digital marketing processes, allow data exchange and enhance data security by combining Blockchain with enterprise resource planning platforms [55]. According to survey participants, Blockchain can deliver trust and transparency for digital marketing strategies, and companies should set themselves on the road to successful execution in their Blockchain efforts [55]

Technological progress is more rapid than it has ever been before, and experts consider it essential to future fiscal growth. We are just at the beginning of seeing how big a disruption Blockchain technology will have on the way we work and live in the twenty-first century. Blockchain has already become a key component in ensuring the secure connectivity of the future. This advanced technology implemented an ingenious and powerful idea to offer a comprehensive security solution, and as a result, it will have profound implications on our working and personal lives in the coming century. Today, people are as excited about Blockchain technology as they were about the Internet in the 1990s. As for which country is best when it

Table 1 Companies with Blockchain.

How far along are companies with Blockchain?					
None	Research	Development	Pilot	Live	Paused
14%	20%	32%	10%	15%	7%

Table 2 Engagement of companies with Blockchain

Assessment of population involvement with Blockchain	
84%	Respondents are actively involved with Blockchain
45%	Respondents believe trust could delay adoption
30%	China as a rising Blockchain leader
28%	Say interoperability of systems is a key for success

Table 3 Trust and regulatory uncertainty.

Trust amongst users		Regulatory uncertainty	
Country	Percentage (%)	Country	Percentage (%)
Singapore	37	Germany	38
UAE	34	Australia	37
Hong Kong	35	UK	32

comes to developing Blockchain initiatives, the United States is in the lead (29%), with China, second (18%) and Australia, third (12%) (Table 3).

There will be some genuine trust in the underlying Blockchain technology, but to achieve full trust on an application level, IT artefacts must be trusted [8]. Over half (45%) believe regulatory uncertainty is a barrier to Blockchain adoption. According to the survey, Singapore (37%) is the most concerned about the trustworthiness of its users. The UAE (34%) and Hong Kong (35%) also report similar levels of concern, which, in part, reflect the importance of financial services in Blockchain development. Other notable features of the current era are social impact and user behaviour. Worry about uncertain regulations was most prevalent in Germany (38%), Australia (37%) and the UK (32%).

Summary of Findings
According to the study participants, one of the biggest barriers to widespread Blockchain adoption is a lack of trust. Majors issues, like regulatory challenges, exist predominantly on permissionless public Blockchains, where anyone can join thus reducing their value. Better user experience is possible due to increasing fragmentation of user needs and competing vendor products as time passes. The greatest benefits of Blockchain will come from open platforms that span the industry.

6 Conclusions and Future Directions

The findings of this report show that Blockchain is influencing digital marketing techniques and that these strategies are needed by industry. The data suggests that Blockchain technology will help to strengthen values, provide more reliable information and distribute Big Data faster. This is because Blockchain's characteristics make data more reliable and higher in quality than traditional data. The technology's ability to increase the speed with which data is collected and processed also ensures that it is more accurate and relevant. Furthermore, the study revealed that Blockchain has a significant effect on digital markets and provides access to new information that is more realistic than the data we currently have. Blockchain is expected to affect digital marketing practices and help customers fulfil their needs. The deduction that Blockchain encourages the holistic understanding of customers when it is introduced and integrated into industry to be preserved, is focused on the deduction that Blockchain encourages holistic understanding of customers both when it is applied and incorporated to refresh the strategic business sector, and when implemented separately.

The data presented here demonstrates that a Blockchain-based solution will increase pressure on companies to deliver useful content and information across all of their networks. This is based on the idea that people will be able to access their personal information through the technology. Businesses will be obligated to have more respectful and thorough interactions with their customers. They will be forced to follow their customers' demands if they want to use customer details. This can be done effectively, and it will enable companies to obtain a greater understanding of their customers. While an organisation's failure to fulfil customer demand can cause negative digital marketing outcomes, this technology will affect whether or not the organisation can meet those demands. Customers will be able to refuse to allow businesses to process their personal information, limiting businesses' ability to use this information. This would leave digital marketers with less information about their clients, and they would have to refine and evaluate the effectiveness of their digital marketing campaigns. In short, Blockchain will affect digital marketing activities and businesses' ability to process and collect data for marketing purposes.

There are many benefits to Blockchain for business. A well-designed Blockchain not only removes intermediaries and reduces prices but also improves speed and reach to better function as a distributed, tamper-proof database. It has the added benefit of enhancing transparency and traceability through some business processes. Blockchain will have created more than $3 trillion in annual market value by 2030 [55]. It is reasonable to anticipate that 10–20% of the global economic infrastructure will be using Blockchain-based systems by the same year [55]. Given the growing body of literature on the potential applications of Blockchain technologies, further in-depth academic research is required to show how this evolving technology will deliver a fundamental layer for improved marketing transparency and trust.

Blockchain gives marketers new opportunities and threats. In the modern market, customers are no longer only on the consumption side. Instead, they can be considered value creators. And, furthermore, because of network effects, they can also serve as brand ambassadors. It is imperative that brands adapt and implement new approaches to improve the overall ecosystem. Additionally, with Blockchain, marketers face new obstacles. Growth is impossible without a seamless customer experience. Ultimately, all brands must have an intriguing story and cause to stand out. This is the only way to galvanise and empower consumers to participate in their communities. Blockchain facilitates concepts such as these to happen. Blockchain is still in its infancy and mainstream adoption of Blockchain-based projects is still years away. With this new Blockchain technology, marketers need to keep an eye on these ramifications.

Blockchain also promises to enhance transparency and credibility in digital marketing. Blockchain's aspects, such as security, decentralisation, smart contracts, transparency, traceability and digital identity, eliminate fraudulent activities within digital marketing. Therefore, implementing Blockchain guarantees the reliability and trustworthiness of the data, and fake content and inaccurate data will be eliminated. Let's take a recent example: because of Blockchain, it will be harder for consumers to be tricked by dubious content, such as deepfakes. Blockchain even eliminates fraudulent activities, such as illegitimate clicks, reviews and followers.

This chapter concludes that the Blockchain can only handle MB of data and that consensus algorithms are not scalable. Currently, Blockchain should be designed in a way where all data is not processed and stored in the Blockchain. In other words, we need an off-chain solution. Here, the Blockchain is used to store references and keys so that off-chain data can be linked to it. Intermediaries are another area where Blockchain can help digital marketing. This follows from the current distribution of intermediaries in the digital marketing environment. Fraudulent activities are still taking place in the context of digital marketing. However, in future, there will be fewer intermediaries due to data being decentralised, and due to customers acquiring control of their personal data. Thus, the ability for huge organisations like Google and Facebook to collect personal data will be impeded. By removing middlemen, the data will be more transparent and dependable. Finally, not all actors have the incentives to use this technology. Also, some ad-space platforms do not profit unless they have the number of clicks and views they can provide. They may not have the incentive to adopt the Blockchain because of fraud, especially click fraud. Facebook and Google are disinterested in any solutions that undercut their market position. As a result, they are likely to counter Blockchain solutions they perceive as a threat. But they also suggest that Blockchain may be good for digital advertisers as well. For instance, Blockchain can guarantee that data are not created by fraudulent activity; therefore, data are likely to be of more value. Advertisers who buy that type of data can get more money. It results in a larger customer base for digital advertisers, as they can be certain that no fraudulent activities are occurring.

References

1. J.N. Sheth, The future of relationship marketing. J. Serv. Mark. **16**, 590–592 (2002). https://doi.org/10.1108/08876040210447324
2. B. Larivière, H. Joosten, E.C. Malthouse, et al., Value fusion: The blending of consumer and firm value in the distinct context of mobile technologies and social media. J. Serv. Manag. **24**, 268–293 (2013). https://doi.org/10.1108/09564231311326996
3. M. Sosna, R.N. Trevinyo-Rodríguez, S.R. Velamuri, Business model innovation through trial-and-error learning. Long Range Plann. **43**, 383–407 (2010). https://doi.org/10.1016/j.lrp.2010.02.003
4. I. Fazey, N. Schäpke, G. Caniglia, et al., Ten essentials for action-oriented and second order energy transitions, transformations and climate change research. Energy Res. Soc. Sci. **40**, 54–70 (2018). https://doi.org/10.1016/j.erss.2017.11.026
5. H. Treiblmaier, A. Strebinger, The effect of e-commerce on the integration of IT structure and brand architecture. Inform. Syst. J. **18**, 479–498 (2008). https://doi.org/10.1111/j.1365-2575.2007.00288.x
6. J. Santomier, New media, branding and global sports sponsorship. Int. J. Sports Mark. Spons. **10**, 15 (2008). https://doi.org/10.1108/IJSMS-10-01-2008-B005

7. M.P. McAllister, J. Turow, New media and the commercial sphere: Two intersecting trends, five categories of concern. J. Broadcast. Electron. Media **46**, 505–514 (2002). https://doi.org/10.1207/s15506878jobem4604_1
8. T.C. Melewar, P. Foroudi, S. Gupta, et al., Integrating identity, strategy and communications for trust, loyalty and commitment. Eur. J. Mark. **51**, 572–604 (2017). https://doi.org/10.1108/ejm-08-2015-0616
9. S.J. Berman, B.E. McClellan, Ten strategies for survival in the attention economy. Strategy Leadersh. **30**, 28–33 (2002). https://doi.org/10.1108/10878570210697513
10. M.D. Stone, N.D. Woodcock, Interactive, direct and digital marketing: A future that depends on better use of business intelligence. J. Res. Interact. Mark. **8**, 4–17 (2014). https://doi.org/10.1108/JRIM-07-2013-0046
11. D. Goldstein, A.J. Moskowitz, A.C. Gelijns, G. Ailawadi, M.K. Parides, L.P. Perrault, J.W. Hung, P. Voisine, F. Dagenais, A.M. Gillinov, V. Thourani, M. Argenziano, J.S. Gammie, M. Mack, P. Demers, P. Atluri, E.A. Rose, K. O'Sullivan, D.L. Williams, E. Bagiella, R.E. Michler, R.D. Weisel, M.A. Miller, N.L. Geller, Two-year outcomes of surgical treatment of severe ischemic mitral regurgitation. N. Engl. J. Med. **374**(4), 344–353 (2016). https://pubmed.ncbi.nlm.nih.gov/26550689/. Accessed 13 Jan 2021
12. M. Petrescu, A.S. Krishen, Marketing analytics: From practice to academia. J. Mark. Anal. **5**, 45–46 (2017). https://doi.org/10.1057/s41270-017-0019-1
13. D.J. Ingram, D.T. Cronin, D.W.S. Challender, et al., Characterising trafficking and trade of pangolins in the Gulf of Guinea. Global Ecol. Conserv. **17**, e00576 (2019). https://doi.org/10.1016/j.gecco.2019.e00576
14. M.K. Crosley, R.A. Osten, J.W. Broderick, et al., The search for signatures of transient mass loss in active stars. Astrophys. J. **830**, 24 (2016). https://doi.org/10.3847/0004-637x/830/1/24
15. P.C. Verhoef, W.J. Reinartz, M. Krafft, Customer engagement as a new perspective in customer management. J. Serv. Res. **13**, 247–252 (2010). https://doi.org/10.1177/1094670510375461
16. R.A. Salha, M.A. El-Hallaq, A.I. Alastal, Blockchain in smart cities: Exploring possibilities in terms of opportunities and challenges. J. Data Anal. Inf. Process. **7**, 118–139 (2019). https://doi.org/10.4236/jdaip.2019.73008
17. A. Rejeb, Blockchain potential in tilapia supply chain in Ghana. Acta Technica Jaurinensis **11**, 104–118 (2018). https://doi.org/10.14513/actatechjaur.v11.n2.462
18. E.D. Zamani, G.M. Giaglis, With a little help from the miners: Distributed ledger technology and market disintermediation. Ind. Manag. Data Syst. **118**, 637–652 (2018). https://doi.org/10.1108/imds-05-2017-0231
19. M. Pournader, Y. Shi, S. Seuring, S.C.L. Koh, Blockchain applications in supply chains, transport and logistics: A systematic review of the literature. Int. J. Prod. Res. **58**, 2063–2081 (2019). https://doi.org/10.1080/00207543.2019.1650976
20. G. Chen, B. Xu, M. Lu, N.-S. Chen, Exploring Blockchain technology and its potential applications for education. Smart Learn. Environ. **5**, 1 (2018). https://doi.org/10.1186/s40561-017-0050-x
21. G. Papadopoulos, Blockchain and digital payments, in *Handbook of Digital Currency*, (Elsevier, Amsterdam, 2015), pp. 153–172. https://doi.org/10.1016/b978-0-12-802,117-0.00007-2
22. T. Clohessy, T. Acton, N. Rogers, Blockchain adoption: Technological, organisational and environmental considerations, in *Business Transformation Through Blockchain*, (Palgrave Macmillan, Cham, 2018), pp. 47–76. https://doi.org/10.1007/978-3-319-98,911-2_2
23. H.M. Hussien, S.M. Yasin, S.N. Udzir, et al., A systematic review for enabling of develop a Blockchain technology in healthcare application: Taxonomy, substantially analysis, motivations, challenges, recommendations and future direction. J. Med. Syst. **43**(10), 320 (2019). https://doi.org/10.1007/s10916-019-1445-8
24. J.J. Sikorski, J. Haughton, M. Kraft, Blockchain technology in the chemical industry: Machine-to-machine electricity market. Appl. Energy **195**, 234–246 (2017). https://doi.org/10.1016/j.apenergy.2017.03.039

25. Y. Chen, C. Bellavitis, Blockchain disruption and decentralized finance: The rise of decentralized business models. J. Bus. Ventur. Insights **13**, e00151 (2020). https://doi.org/10.1016/j.jbvi.2019.e00151
26. S. Fosso Wamba, J.R. Kala Kamdjoug, R. Epie Bawack, J.G. Keogh, Bitcoin, Blockchain and Fintech: A systematic review and case studies in the supply chain. Prod. Plan. Contr. **31**, 115–142 (2019). https://doi.org/10.1080/09537287.2019.1631460
27. R. Beck, M. Avital, M. Rossi, J.B. Thatcher, Blockchain technology in business and information systems research. Bus. Inf. Syst. Eng. **59**, 381–384 (2017). https://doi.org/10.1007/s12599-017-0505-1
28. A.V. Ertemel, Implications of Blockchain technology on marketing. J. Int. Trade Logist. Law **4**, 35–44 (2018)
29. D. Berdik, S. Otoum, N. Schmidt, et al., A survey on Blockchain for information systems management and security. Inf. Process. Manag. **58**, 102397 (2021). https://doi.org/10.1016/j.ipm.2020.102397
30. Y. Lu, Blockchain: A survey on functions, applications and open issues. J. Ind. Integr. Manag. **03**, 1850015 (2018). https://doi.org/10.1142/s242486221850015x
31. P. Helo, Y. Hao, Blockchains in operations and supply chains: A model and reference implementation. Comput. Ind. Eng. **136**, 242–251 (2019). https://doi.org/10.1016/j.cie.2019.07.023
32. I. Antoniadis, K. Spinthiropoulos, S. Kontsas, Blockchain applications in tourism and tourism marketing: A short review, in *Strategic Innovative Marketing and Tourism*, (Springer, Cham, 2020), pp. 375–384. https://doi.org/10.1007/978-3-030-36,126-6_41
33. F. Casino, T.K. Dasaklis, C. Patsakis, A systematic literature review of Blockchain-based applications: Current status, classification and open issues. Telemat. Inform. **36**, 55–81 (2019). https://doi.org/10.1016/j.tele.2018.11.006
34. A. Farouk, A. Alahmadi, S. Ghose, A. Mashatan, Blockchain platform for industrial healthcare: Vision and future opportunities. Comput. Commun. **154**, 223–235 (2020). https://doi.org/10.1016/j.comcom.2020.02.058
35. S.M. Idrees, M. Nowostawski, R. Jameel, A.K. Mourya, Security aspects of Blockchain technology intended for industrial applications. Electronics **10**, 951 (2021). https://doi.org/10.3390/electronic-s10080951
36. F. Casino, V. Kanakaris, T.K. Dasaklis, et al., Modeling food supply chain traceability based on Blockchain technology. IFAC-PapersOnLine **52**, 2728–2733 (2019). https://doi.org/10.1016/j.ifacol.2019.11.620
37. E. Portmann, Rezension "Blockchain: Blueprint for a new economy". HMD Praxis der Wirtschaftsinformatik **55**, 1362–1364 (2018). https://doi.org/10.1365/s40702-018-00468-4
38. Z. Zheng, S. Xie, H.N. Dai, X. Chen, H. Wang, Blockchain challenges and opportunities: A survey. Int. J. Web Grid Serv. **14**, 352–375 (2018). https://doi.org/10.1504/IJWGS.2018.095647
39. Y. Chen, Y. Wang, S. Nevo, et al., IT capability and organizational performance: The roles of business process agility and environmental factors. Eur. J. Inform. Syst. **23**, 326–342 (2014). https://link.springer.com/article/10.1057/ejis.2013.4. Accessed 23 May 2021
40. A. Arora, S. Bansal, C. Kandpal, et al., Measuring social media influencer index-insights from facebook, Twitter and Instagram. J. Retail. Consum. Serv. **49**, 86–101 (2019). https://doi.org/10.1016/j.jretconser.2019.03.012
41. J. Brauer, B. Linnala Eriksson, Blockchain's influence on digital marketing: An exploratory study examining Blockchain in relation to big data and digital marketing, in *DIVA*, (2020), http://www.diva-portal.org/smash/record.jsf?pid=diva2%3A1446822. Accessed 15 Feb 2020
42. R. Felix, P.A. Rauschnabel, C. Hinsch, Elements of strategic social media marketing: A holistic framework. J. Bus. Res. **70**, 118–126 (2017). https://doi.org/10.1016/j.jbusres.2016.05.001
43. R. Kher, S. Terjesen, C. Liu, Blockchain, Bitcoin, and ICOs: A review and research agenda. Small Bus. Econ. **56**, 1699–1720 (2020). https://doi.org/10.1007/s11187-019-00286-y
44. J.A. Fehrer, S. Benoit, L. Aksoy, et al., Future scenarios of the collaborative economy. J. Serv. Manag. **29**, 859–882 (2018). https://doi.org/10.1108/josm-04-2018-0118

45. H. Min, Blockchain technology for enhancing supply chain resilience. Bus. Horiz. **62**, 35–45 (2019). https://doi.org/10.1016/j.bushor.2018.08.012
46. M. Pilkington, Blockchain technology: Principles and applications, in *Research Handbook on Digital Transformations*, (Edward Elgar Publishing, Cheltenham, 2016), pp. 225–253. https://doi.org/10.4337/9781784717766.00019
47. T.T. Dinh, R. Liu, M. Zhang, et al., Untangling Blockchain: A data processing view of Blockchain systems. IEEE Trans. Knowl. Data Eng. **30**, 1366–1385 (2018). https://doi.org/10.1109/tkde.2017.2781227
48. D.E. O'Leary, Open information enterprise transactions: Business intelligence and wash and spoof transactions in Blockchain and social commerce. Intell. Syst. Account. Finance Manag. **25**, 148–158 (2018). https://doi.org/10.1002/isaf.1438
49. T. Klein, H.P. Thu, T. Walther, Bitcoin is not the new gold: A comparison of volatility, correlation, and portfolio performance. SSRN Electron. J. (2018). https://doi.org/10.2139/ssrn.3146845
50. A.R. Mathew, Cyber security through Blockchain technology. Int. J. Eng. Adv. Technol. **9**, 3821–3824 (2019). https://doi.org/10.35940/ijeat.A9836.109119
51. J. Astill, R.A. Dara, M. Campbell, et al., Transparency in food supply chains: A review of enabling technology solutions. Trends Food Sci. Technol. **91**, 240–247 (2019). https://doi.org/10.1016/j.tifs.2019.07.024
52. P.C. Verhoef, T.H.A. Bijmolt, Marketing perspectives on digital business models: A framework and overview of the special issue. Int. J. Res. Mark. **36**, 341–349 (2019). https://doi.org/10.1016/j.ijresmar.2019.08.001
53. S. Saberi, M. Kouhizadeh, J. Sarkis, L. Shen, Blockchain technology and its relationships to sustainable supply chain management. Int. J. Prod. Res. **57**, 2117–2135 (2019). https://doi.org/10.1080/00207543.2018.1533261
54. J.L. Zhao, S. Fan, J. Yan, Overview of business innovations and research opportunities in Blockchain and introduction to the special issue. Financ. Innov. **2**(28), 1–7 (2016). https://doi.org/10.1186/s40854-016-0049-2
55. PricewaterhouseCoopers. Blockchain is here. What's your next move? in *PwC*, (2018), https://www.pwc.com/gx/en/industries/technology/Blockchain/Blockchain-in-business.html. Accessed 12 Jan 2021.

Blockchain Technology: Unlocking the Business Model Maze for Evolving Businesses and Start-Up

Richa Chauhan and Vidhi Kaul

1 Introduction

Emerging new technologies often enable transformations on social, economic and business platforms [1]. Globally, digitalisation of economy is in process since 1960s. Earlier, technologies based on digitalisation were utilised to enhance procedural aspects of business via conveying the same output in a swift and secured manner. However, Blockchain technology offers an exchange of value which is completely distinct from previous technologies. As per 2018 technology trends, blockchain occupied a place among the top five such specific technologies, which are capable of revolutionising the business world [2].

International Data Corporation (IDC) anticipated expenditure on Blockchain to reach new heights of approximately 12 billion USD by 2022. Fundamentally blockchains' potential to revolutionised financial service industry has been exposed more; however, the influence of blockchain technology surpasses the financial sector and encircles any business practice that acts via mediator and draws economic gains as an intermediary in the value chain [3]. Hence, it is forecasted that blockchain will question traditional business models and furnish new opportunities of value creation. Constructive guidance with respect to various existing blockchain solution and their probable impact on businesses and business models is quite minuscule.

Despite the attention that Bitcoin receives, it is far from the only variant of blockchain technology. While it is quite serene to locate mediums that prop up the prospects of blockchain as disruptive technology, for example mobile, Internet, social media, email, etc., it is profoundly strenuous to perceive the literature regarding the

R. Chauhan (✉)
Banasthali Vidyapith, Niwai, Rajasthan, India

V. Kaul
Independent Researcher, Rajasthan, India

S. M. Idrees, M. Nowostawski (eds.), *Transformations Through Blockchain Technology*, https://doi.org/10.1007/978-3-030-93344-9_13

range of blockchain based technology and their capabilities to add value to businesses and business models [4].

1.1 Business Models

Lately, Business Model is the most talked about concept of management studies. The first mention of the concept can be traced back to 1957 and often confused with the concept of business strategy, however not until the end of the twentieth century, the business model earned the interest of researchers [5]. Although in the recent times there is a significant development in the theory of business models, still researchers are unable to conceptualise the definition of BM that could capture its true essence and which is globally acceptable [6].

As a specific definition of BM yet to be found, it has been recognised as the 'story that explains how an enterprise works' or also as system for executing business. As per Osterwalder, theoretically a business model is an instrument which constitutes four major sectors of a business namely, Product: What is the business's product and what is their value postulation for the market? Consumer interface: Who are their consumers, how do they provide value to them, and how do they establish and sustain relationships with them? The infrastructure management: What infrastructure is required to provide optimal customer value? The financial aspects: What are the costs associated with the business, and how will the business, price its products or services to achieve a profit? [7].

Various constituents of business models and their elucidation are expressed with help of Fig. 1.

Elements	Constituents of Business Model	Elucidation
Product	Value Postulation	The value postulation is a transparent method of explaining what pains, concerns and needs, the enterprise resolve for its consumers.
Consumer Interface	Target audience	Describes the segment of consumers company wants to deliver value to
	Channel of Distribution	Describes the various means to get in touch with the consumers
	Consumer Association	Explains the kind of association the company wishes to establishes with various consumer groups
Infrastructure management	Value layout	Explains the configuration of resources and activities.
	Core competancy	Elaborates the competencies required to deliver the business model
	Partner Tieups	Presents the matrix of cooperative tieups with other companies to efficiently propose and monetise value
Financial aspects	Cost structure	Sumarises the monetory consequences of the method deployed in business model. The cost of delivering value.
	Revenue model	The way earning is made through variety of revenue flows.

Fig. 1 Constituents of business model

These four elements are further segregated among nine sub-segments, namely customer relationships, value propositions, revenue streams, channels, customer segments, key resources, key activities, key partnerships, cost structure as shown in Fig. 2 which are completely interrelated and capable of putting influence on a business model [8].

Nevertheless, value generation is the focal point in the interpretations presented by various academicians. As the definitions of business model were evolving, gradually the distinction between business strategy and business models also become more and more comprehensible. Where, business model explains the logic of how a company generates value, and business strategy's main objective is to enhance competitiveness by being distinct. Business model focuses on the process of consumers' estimation and value premise [6]. Where, framing of business strategy is a top management approach defining and explain modus operandi in competition; operational levels bring about business model explaining how to execute the business strategy [9].

To earn commercial success and substance in the market, the company needs to have a proper blend of both business strategy and business model. Logical and simple business model is the excellent one [10]. It should take into consideration human resources and physical resources as well.

Development of business model often faces three major hurdles. Firstly, there are compatibility issues between business model and real market situations; secondly, most of the time business model tend to constitute only a fragment of business strategy; lastly the main aim of business model is value hypothesising rather than capturing the true value [11]. New ventures show more vulnerability towards these hurdles. An appt business model could facilitate a start-up to align of its resources according to consumers' needs and wants.

Besides the concepts of value creation, business process and strategic approach, now a days BMs delineated a concrete and strong connect with technology [12]. Particularly, the era of early 2000 shows the advent of designing many BM supportive tools [13]. Business Model Canvas (BMC) is among the industry's best known and commonly used tools for modelling, designing and analysing business models.

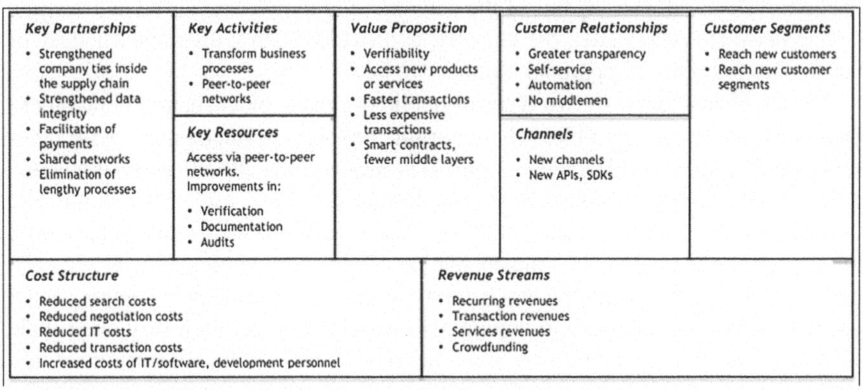

Fig. 2 Sub-segments of business model canvas

In BMC, a quick description of the business model is provided in the form of short, informative texts, often in the form of 'post-it' notes. Using the BMC is a highly agile way of discovering challenges and opportunities, making it possible to render and change the business model as the company grows. The design and purchase organisations need to understand fundamental relationship between BMs and IT [14]. Technological advancements and innovations alone cannot secure and assure the any firms. In order to gain competitive advantage, BMs have to arbitrate between economic value creation, ever growing technologies and firm's performance [15]. It is necessary to apprehend the value created by an innovation and assure its success commercially [16]. The analysis of relationship between technological advancements and BMs becomes more relevant, when introduction of technological innovations and advancements are industry specific. Inertly going by longstanding belief system and not incorporating technological innovations and advancements in BMs of companies have proven to bring consequences that are fatal for businesses [17].

1.2 Blockchain Technology

The genesis of blockchain begins when the white paper was written by Satoshi Nakamoto in 2008. Nakamoto inaugurated bitcoin as a peer-to-peer kind of electronic cash that permits payments to be sent amongst parties and do not need any centralised financial institution as intermediaries. Nakamoto also conceived the ledger while implementing bitcoin, which he named a 'chain of blocks' [18]. This sequence of blocks assists the novel kind of electronic cash and subsequently came to be known as blockchain. Initially blockchain implementations were intended to deliver an alternative trust mechanism between transacting parties based on cryptography [19]. Further, blockchain technology facilitates a collective bookkeeping system (ledger), which permits participants to reach an agreement on the acceptance of a transaction via a mathematical function (hash function). Each transaction is recorded in a separate 'block'. In a chronological order, these blocks are added to the computers of all participants of the network after they are reviewed and verified by the network. Afterwards, the network is provided with a distributed ledger of verified transactions of a particular unit.

Mainly blockchain is of two types of public or private blockchain. Public blockchain also known as open blockchain permits anybody to interconnect with other party while transacting. The identity of two parties is either entirely anonymous (i.e. parties who are entering into the transaction do not know each other before the transaction) or even pseudonymous [20]. A public blockchain implies very little privacy in transactions, which means that each and every participant can see each and every transaction. A public blockchain also needs a considerable computational capacity which is a requisite to maintain a distributed ledger on a very large scale [21]. In order to attain consensus in public blockchains, each and every node in a matrix requires to resolve a resource-intensive cryptographic, complex complexity known as proof of work to make sure that each and every node of the blockchain is properly in sync. Public blockchain comprises of Litecoin (a cryptocurrency designed to be faster and more efficient in comparison of Bitcoin), Bitcoin and Ethereum (which is processed unlike Litecoin and Bitcoin) and is fundamentally utilised for smart contracts.

Closed or private blockchains permit only those groups of individuals or individuals which are pre-validated to access the ledger for the purpose of entering and viewing data. In closed blockchain, identities of users are known to each other prior to entering into transaction. Consortium or federated prototype is a variant in which the closed blockchain work under the guidance of a group. In this variant of closed blockchain, permission regarding uses of blockchain network by new entrants varies, only already existing members can choose the future entrants. There is a regulatory authority or a consortium which gives users licenses to new members and make decisions regarding their participation.

Private blockchain tends to provide greater privacy to the transaction, involving critical and sensitive data (e.g. the transfer of financial or medical data). The authority to read the private blockchain might be allowed for some or completely no permission to even members. Closed blockchains are capable of cutting down costs, enjoy feature of significant transactional throughput and easy to scale up. In addition, it also comprises of merits like added reliability and security and greater trust, because only those members are allowed to commence new node in the blockchain which are pre-verified [22]. For examples Linux-based Hyperledger, which provide the collaborative and collective evolution of blockchains and tools in supply chain, IoT, manufacturing technology, finance, banking and R3.

A distributed ledger technology company R3 guides a consortium of over 200 plus firms and engaged in development of various applications for commerce and finance [20].

Though the differences between open and closed blockchains are many, both of them also have some common benefits to offer. Both the types of blockchains offer decentralised peer-to-peer networks, within which every member sustains a replication of a shared transaction of append-only ledger with digitally signatures; both the blockchains keep the duplicates which are properly synchronised through a protocol called consensus; private and public blockchains facilitate certain level of guaranteed stability of the ledger, when even some network members are engaged in malicious or faulty activities [22].

Today, blockchain has been put forward as a solution for a wide range of transactions, from transferring funds across currencies (remittances, micro payments) to real-time payments between two parties (swift settlement and without the requirement of a bank account), and digital assets (records of ownership stored in digitalised manner). The influence of blockchain technology may go far beyond than just bringing simple modifications to processes and introduction of some new services and products. A number of market researchers and academicians believe that the repercussions of blockchain technology might be far reaching and could affect the entire setup of BMs [23]. Blockchain's influence on BMs proves to be exemplary comprehensive potential of IT. Blockchain's distributed ledger technology (DLT), has sparked great attraction amongst Information Systems (IS) practitioners and communities in context of cryptographic features and trust, procedures and their suggestions, and various matters in relation to virtual currencies as well [17].

2 Influence of Blockchain on Business Models

Traditional incumbents are threatened directly by blockchain technologies, which offer the possibility of creating entirely new businesses. Conventional business models that assume an organisation acts as an intermediary between two transacting parties should reflect on whether and how such technologies put influence on their efforts of value propositions, how they operate in the market and face competition. Several pilot projects are undergoing in number of industries like, application of blockchain to track goods within industrial logistics, smart contracts application to ensure swift, secure and reasonable transactions regarding property and evaluate blockchain's efficiency to allow consumers to send funds overseas without the involvement of any unnecessary time-consuming activities or immoderate exchange fees. Today business organisations need to consider and evaluate the influence of booming blockchain applications on their business models. This section chapter outlines blockchain technologies influence on eight components of business model canvas.

2.1 Customer Segments

According to Osterwalder and Pigenur, a customer segment refers to various classes of consumers or organisations which an enterprise looks out to serve. Blockchain technology permits organisations to concentrate on existing market segments. An experimental pilot program backed by ChromaWay will allow people to purchase and sell real estates in Sweden using blockchain technology. Blockchain is ideally suited to serve diversified, niche and mass consumer markets, same as traditional enterprises could. Although there is one unique aspect of blockchain technology that makes it possible for businesses to reach a previously inaccessible market, at the same time creating new customer section [24].

2.2 Value Proposition

A value proposition embraces all those activities of the firms which are designed to create a value for its consumers [25]. Theodore Levitt famously claimed: 'People don't want to buy a quarter-inch drill, they want a quarter-inch hole'. Specifically, consumers do not buy products; they purchase a solution. Customers will derive more value by emphasising the significance of a task and by being satisfied with the alternatives currently available, and their price [26]. Through blockchain technology, consumers can gain access to products and services previously unavailable or only accessible after spending a significant amount of time and money. Furthermore, blockchain-based transactions are commonly faster and less expensive than traditional business transactions.

2.3 Channels

Channels are the processes through which a business communicates with and attain its consumer segments in order to carry out its value proposition. Channels may comprise the company's website, sales force, or stores of the company's wholesalers or partners. Application of blockchain-based options provide ease to business execution. By eliminating stipulated time and personnel needed for a validity check or transaction, this can be accomplished. Within the company, new types of channels can be introduced as a result of utilising common codes [27].

2.4 Consumer Relationship

An organisation's customer relationship building block explains the variety of relationships it has with specific segments of its customers. A company might build these relationships for a variety of reasons, such as to gaining and keep consumers, or boosting of sales. A number of categories of relationships exist, such as dedicated and personal assistance, automated and self-services, creating communities, etc.

2.5 Revenue Streams

By 2023, blockchain projects are expected to generate $10.6 billion in revenue, primarily from sales of software and services [28]. Companies which are administering services of professional level via blockchain-related technology generate revenue from agreement level services with enterprises or fees for providing software-as-a-service (SaaS) contracts. However, maximum earning of revenue from blockchain technology is obtained from crypto crowdfunding, via utilisation of initial coin offerings (ICOs). In contrast to private equity firms, venture capital and banks that are well known for providing equity and capital debt, ICO's raise depends on the utilisation of cryptocurrencies and trading based on blockchain to raise funds. An ICO offers tokens instead of shares as a reward for early investors. An aftermarket for these tokens is available, and all transactions are verifiable on a blockchain platform.

2.6 Key Resources and Activities

Key resources are those resources or assets which are critical to fabricate a business model. These resources are the one which define a value proposition, markets outreach, nurture relationships with customers, and generate revenue. Resources can be financial, human, intellectual or physical. An organisation's key activities include all those undertakings which are essential to deliver value, which means all those processes which result in the metamorphosing of resources and assets into value. Companies need to review the key resources that comprise their business model in order to incorporate blockchain technologies. There are several facets of blockchain technology that put their impact on activities and resources. Firstly, there is a possibility of increasing resource flexibility, enabling firms to resettle from conventional ownership structures and acquire resources only when they are needed. Specifically, relevant opportunity is the implementation of public blockchain technologies, where anyone can enter into transaction and negotiate with other member on a peer-to-peer network. Sometimes, a firm may not need to invest in building IT infrastructure or maintenance since public blockchains provide these resources and processes. Further, both the private and public/federated blockchain applications allow businesses to automate processes which were previously handled manually, for example audit reporting, documentation and verification, allowing human resources to concentrate on other value-added tasks. Secondly, blockchain technologies can influence organisation's resources and activities, when users assist in providing various key resources and processes and employ blockchain technologies to facilitate resource exchanges, for example while transacting in real estate segment via smart contracts, resources such as human capital (e.g. experience, knowledge, skills) and physical capital like assets are provided by the parties involve in transactions, while peer-to-peer exchange of these resources are aided by blockchain technologies [25].

2.7 Key Partnerships

A business model is built around a network of partners and suppliers. Partnerships can take the form of joint ventures, strategic alliances, or purchase and supplier relationships to secure authentic supply. A potential application of blockchain could reduce the role of conventional intermediaries (currency exchanges, banks, notaries, etc.) or modify financial institutions (credit card companies, etc.). Blockchain can ease up the joining of new members, like IT companies which create application programming interface (API) and software development kits (SDK) and sustain the transaction algorithms. In addition, blockchain allows businesses to form peer-to-peer partnerships, thus supporting and expanding supply chains.

2.8 *Cost Structure*

The cost structure explains each and every cost which is associated with operating a business model. By eliminating the costs of intermediaries and negotiation costs, blockchain implementations can reduce transaction costs. According to Gregario, blockchain is expected to enable cost savings of $15–20 billion in the financial services industry by 2022 [29]. The savings were achieved via the reduction of cost of information technology infrastructure and the abolishing non-value appending manually performed processes. A blockchain implementation for managing financial transfers could reduce the authorisation holds currently used in banking and credit card processing. Funds can take several days to get clearance from holds of authorisation body. With public blockchain protocols, these holds can be reduced to a matter of minutes. A private blockchain reduces these holdings to milliseconds [30]. With blockchain technology, data aggregation, amendment, and sharing are simplified, and regulatory reporting and audit materials are produced with less manual steps. Consequently, employees can focus on tasks that contribute to greater revenue generation, while consumers save time and money.

3 Segments Facilitating Blockchain Development and Implication

In this chapter, an attempt has been made to discuss eight specific segments where blockchain use was found, namely crypto currency services, asset management services, crowd funding service, digital identity services, other financial services, contract management services, blockchain development services, traceability services healthcare, supply chain, energy, banking, which are extensively described below.

3.1 *Crypto Currency Services*

The fundamental technology that supports cryptocurrencies is blockchain. Cryptocurrency is an alternative form of monetary setup which is virtual and beyond any monetary authority regulated by the government such us Litecoin, Ethereum, Bitcoin, etc. The cryptocurrency talked about a novel economy which is not definite and therefore beyond the influence of any legal and political construct and geographic areas as well. Cryptocurrency is a decentralised ledger, permitting peer to peer transfer of funds without requiring bank as third party.

3.1.1 Implication and Start-Ups

Blockchain technology enables generation of cryptocurrency which is authentic and legally tradable in nature. The first and foremost acknowledged application of blockchain was with Bitcoin cryptocurrency. As the time passes by, many more cryptocurrencies have come up besides Bitcoin, namely Gridcoin, Ethereum, Ripple etc., by far till today Bitcoin enjoys the lion's share in the world. Cryptocurrencies can also be an incentive apparatus for suggesting proposals performed by cross-functional group. Smart contracts are performed to complete the whole process and groups that turn up with the finest proposals with pre-established offerings of digital coins are decently rewarded automatically [31].

3.2 Asset Management Services

The concept of asset management services controls the ownership of an asset or property (non-physical like company share and physical like house, cars, etc.) as a smart property via smart contracts using blockchain. Blockchain technology encodes property and convert it into smart property which is transactable through smart contracts and further thus enables trust-less trading and lending of property.

3.2.1 Application and Start-Ups

The applicability of block chain in the segment of asset management has been via supervising smart property through smart contracts, and SwanCoin is one such start-up which deals with smart property where 121 artworks of physical world, fabricated on 30 × 30 cm varnished plywoods are available for buying purpose, and monetary transfer happen through the Bitcoin blockchain [4]. Another start-up, namely Ownage, an Ethereum-based platform, collects, distributes and trades content of digital game. Smart property setup could also be used with several other blockchain application, for example digital identity which facilitates easy access of identity credentials to use it as in the case of a smartphone or to operate the doors of physical assets like homes and cars. Coloured coins are one of its kind implementations in smart property where some explicit bitcoins are tagged or coloured in correspondence to a certain asset.

3.3 *Crowd-Funding Services*

Platforms based on blockchains allow start-ups to raise funds through their websites. They can also create digital currencies and sell 'cryptographic' shares to early backers, and investors receive a token representing their share of the company they support [4].

3.3.1 Applications and Start-Ups

Another application is in crowd-funding services to raise funds for blockchain-based projects using blockchain technology. FunderGrowth is a start-up that offers investors the opportunity to invest in Blockchain start-ups using blockchain technology. The start-up Swarm is another great example of a cryptocurrency incubator that has given birth to various funded projects extending from the development of a decentralised cryptocurrency workplace to smart personal drone matrix [4].

3.4 *Digital Identity Services*

Permanence and security features of Blockchain make it a great technology to be applied for securing and verifying one's digital identity. Its benefits are proven by the fact that all the users of 33 cryptocurrencies have their personalised e-wallet, and their wallet addresses could be further utilised to verify the user if at all required.

3.4.1 Application and Start-Ups

One of the basic applications of blockchain in digitalised identity services has been the availability of identity verification online. It incorporates an identity verification of an individual to a website employing verification via blockchain. BitHandle and OneName are two famous start-ups facilitating such services. These start-ups facilitate decentralised and trustless services so that one's digitalised identity goes beyond the control of central institution, that's how these start-ups secure more authentication to websites than social media sites for the same. Another successful application of the concept is in insurance sector where any asset or property which is registered under blockchain concept of smart property can be used to verify and provide owner's digital authentication. Blockchain-based digital identity services are also used in verifying and checking other blockchain-based transactions like smarty property, cryptocurrency and smart contracts in order to substantiate the parties taking part in the transactions [4].

3.5 Other Financial Services

Associating cryptocurrencies with financial market and conventional banking system is considered to be the major area for blockchain applications. Blockchain empowers several financial services with speedy international fund transfer, that too with quick transfer at less transfer cost.

3.5.1 Applications and Start-Ups

The main application of blockchain in financial service segment is in facilitating quick and low-cost payment in cryptocurrency, covering the entire blockchain across several business enterprises. One of the major start-ups in this financial service segment is Ripple, which permits direct foreign exchange and fund transfer transaction amongst banks. Remitsy is another such start-up which provide funds transfer internationally. There are start-ups which are assisting payments in cryptocurrency to other conventional market and financial payments applications like GoCoin and Coinbase and PayPal in the case of Bitpay which is a processor of payment for.

Cryptocurrency trade is an additional application in cryptocurrency. Cryptocurrency trade obtains cryptocurrency from customers and then trade it with other parties as per current cost of the currency, as it happens in solutions, namely Coinbase and Bitpay. Start-ups by the name Kraken also provide banking assistance in lending services and saving services of digitalised currency for bitcoin. TeraExchange and BTCiam start-ups are assisting users in getting same kind of services.

Chain.com is another application of blockchain that assists NASDAQ in exercising private equity exchange, which is not only fast with efficient tracking system but also far more effective in terms of stock value creation compared with conventional ways of trading in stocks.

Bitshare, Blockstream and Medici are some start-ups which uses blockchain solutions in providing securities exchange services.

3.6 Contract Management Services

A smart contract is a process of entering into the agreements with people through the blockchain. In smart contract, an individual using the main blockchain decides whether or not a particular blockchain operation, such as payments, should be allowed. The smart contract permits transacting members to design self-regulating and self-evaluating contracts codes, exhibiting an autonomous status once they have been placed in the blockchain.

3.6.1 Applications and Start-Ups

The application of blockchain in contract management services is in the form of smart contracts, in which member party can come up with self-regulating and self-evaluating contract code, Ethereum being the leading start-up involved with smart contracts. It is a stateful user-created virtual machine based on a blockchain platform. For the creation and publication of distributed ledger, Ethereum is a foundational programming language and a complete infrastructural platform. Ethereum functions like amalgamated development platform for all blockchains and protocols as well. The blockchain-based Ethereum platform is already being utilised in a wide array of early applications via smart contracts such as trading and settlement of financial derivatives, keyless access, governance, crowdfunding, autonomous banks [32].

3.7 Blockchain Development Services

Many blockchain development service projects are initiated which are not only facilitating development of blockchain protocols for their own application and processes but provide help in creating the same for others.

3.7.1 Applications and Start-Ups

This service provides opportunity to projects that are involved in the development of new protocol to come up with such blockchain based technology which is widely acceptable and popular. Start-ups namely Ethereum, Ripple, etc. are the ones which are involved in the development of protocol for their own blockchain and corresponding application, whereas start-ups namely, NXT, open transactions, etc. create cover blockchain platforms for bitcoin. Start-ups for example Stellar and chain and Blockchain.info propose development of application programming interface protocol for them.

3.8 Traceability Services

Blockchain facilitates authenticity of entity and anti-counterfeiting via fabricating immutable ledger and sharable consensus base and enables tracing of the origin and the modification an entity has undergone by generating a formal registry which in turn help in the tracking and identification of entity ownership.

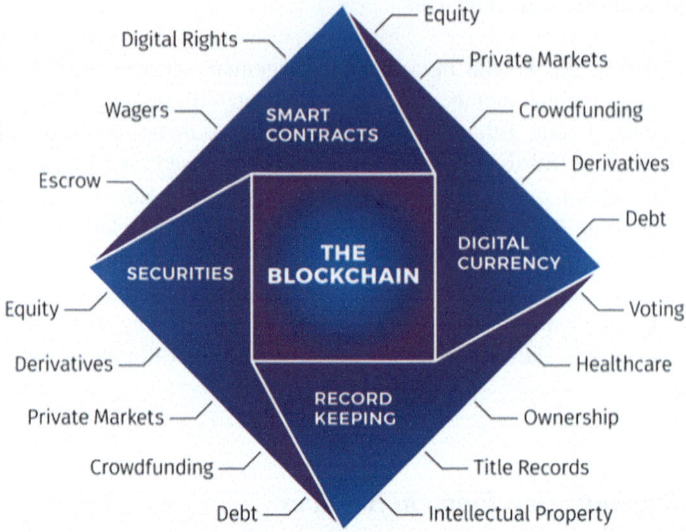

Fig. 3 Implementation of blockchain in business

3.8.1 Applications and Start-Ups

Among the many applications for this technology, the most prominent one is in the supply chain management process. Blockchain-based start-up Everledger develops a permanent ledger for the certification of diamonds and records the history of transacted diamond.

Similarly, BlockVerify is an anti-counterfeit startiup that is using blockchain technology for its anti-counterfeit solutions in luxury products, diamonds, pharmaceutical and the electronic sector. In addition to traceability services, blockchain can be used as proof-of-existence. Blockchain can be envisioned to be such digital platform where all legal documents, notaries, private securities, health records, etc., can be placed [33]. Several other implications of blockchain technology are shown in Fig. 3.

4 Discussion

Blockchain adoption has not yet reached a critical mass. Most blockchain projects have not progressed beyond the pilot stage. A recent Gartner study shows that only 1% of responding CEOs are using blockchain, and only 8% are planning short-term and pilot projects [34]. Deloitte was quite hopeful. As per Deloitte report, although 74% of respondents said their companies are dubious about the merits of blockchain technology, but not less than 34% of respondents admits that their company has already started deployment of blockchain based technology, Deloitte also observed that there is a presence of many stumbling blocks which hinder the mainstream acceptance of blockchain technology:

- The blockchain's operating system is considered slow. In spite of its capability to provide remarkable gains in comparison to delayed clearance by banks and credit card companies, still on public distributed ledger system performing operations based on consensus is a time taking exercise. In order to keep data confidential, additional obfuscation and encryption are required. This puts negative impact on the creation of consumer value, as business organisations and consumers both expect almost instant service.
- Blockchain applications require specialised developers and complex integrations: the cost of developing blockchain applications to customer specifications is high.
- Managers have a quite discouraging viewpoint regarding data security feature of blockchain platforms, as there are events of contravention of data on cryptocurrency platforms, and corporates require data security in a wide range of blockchain-supported platforms.
- There is no standardisation of blockchain construct. GitHub in 2018 listed over 6500 blockchain based projects, covering a variety of protocols, consensus protocols, privacy protocols, and coding languages.
- Due to the lack of standardisation in blockchain construct, it is difficult to establish business relations between firms by using them, since this requires integrating a variety of constructs and protocols.
- Finally, a critical mass of users is required in order for blockchain technology to be widely adopted. The blockchain is being used to address the needs by projects like Everest's humanitarian initiatives, which was performed on large platform. Utilisation of the blockchain for disenfranchised like in Everest's humanitarian initiatives could help accelerate the use of the technology throughout the world [35].

Nevertheless, contemporary evolution in the blockchain constructs, regulatory framework and alliances amongst organisations resulted in the removal of many of such obstacles. Figure 4 represents the progressive picture of blockchain technology adding value to businesses in coming future.

- Blockchain-based technologies namely Hyperledger, R3, Ripple and Stellar are using new mechanisms of achieving consensus which in turn facilitates gaining higher throughput, easy implementations and better performance, and also allowing processing time to be reduced from minutes to milliseconds. Blockchain consensus refers to the mutual acknowledgment of the validity of recorded transactions.
- The process of standardisation continues. The number of blockchain consortiums initiating projects is now over 60. In these consortiums, hundreds of companies and government agencies are interested in exploring the possibilities of blockchain applications. Blockchain matrix developers create applications and develop technologies, as well as develop use cases and standards. Others provide

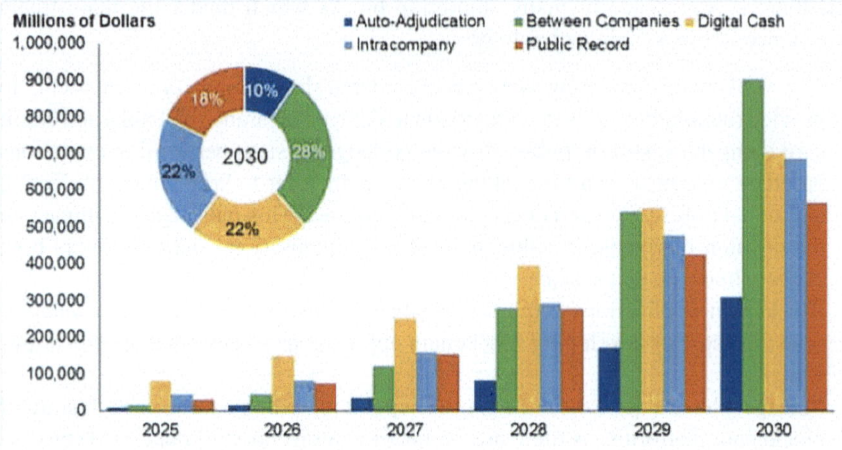

Fig. 4 Forecasted business value add of blockchain

their members with educational opportunities or research. Deloitte explained that the increasing number of users shall increase the value of a network. The Hyperledger Foundation with more than 250 organisations and Enterprise Ethereum Alliance with 600 plus organisations are amongst this industry's consortiums. Collaboration between companies outside the consortia is growing as well.

- There has been a decline in both the complexity and the cost of blockchain implementations. Microsoft, IBM, and Amazon provide cloud-based implementations of blockchain technology and templates that are less expensive than specialised development. Utilising these templates will enable organisations to minimise costs of these initiatives and shorten establishment times. The regulatory environment of many countries is improving. A number of states in the USA have passed legislation facilitating the adoption of blockchain technology for certain medical applications [36].

5 Conclusion

As industries capitalise on digitalisation, the fourth industrial revolution (Industry 4.0) is revolutionising how business is conducted across business value chains [37]. Industrial sector is in the process of becoming increasingly 'smart' with the use of data exchange intensives, predictive analytics Internet of Things (IoT) technologies, etc. [38]. Optimisation and automation of business processes have several merits like reduction in overhead costs, enhancement of productivity, production at high pace, and as a matter-of-fact major reduction of errors [39]. The majority of industry experts consider this as a positive development, as recent PwC and BCG reports predict that Industry 4.0 will deliver 15–20% efficiency boosts and generate 20% of revenue growth in the next coming 5 years. There are substantial opportunities to innovate business models using digital technologies in a business to business (B2B) setup, and simultaneously generating new value and revenue opportunities [40]. Companies which are able to enhance their efficiency of digital modification make better use of big data analytics will excel more than their peers, on the parameters of operating productivity and revenue growth [41]. Many organisations in order to gain perceived opportunities and benefits of blockchain technology are experimenting with innovational business models based on the same technology [42]. The recent developments in blockchain technology are attributed to the growth of collaborations and the formation of consortia. In order to demonstrate proofs of concepts, organisations are launching pilot projects to evaluate the blockchain movement. Moreover, blockchain tokens are being issued and sold by entrepreneurs and at the same time reconstructing entrepreneurship and bringing innovations in ways of raising fund, community building, investments and open sourcing [43]. In order to carry out blockchain projects, companies must decide which blockchain model they want: public or private? Both the types of blockchains enjoy different market preposition. While private blockchains are advantageous in terms of time and cost savings, public blockchains may disrupt the financial setup of an industry through disintermediation, for example a case of cryptocurrencies and Bitcoin, or by offering and creating completely new business models [44]. Identification, selection, and execution of customised blockchain innovations that are beneficial to a company's operations is a key challenge as discussed above for many companies. Furthermore, the need to analyse, customise, sell or buy intangible products or services is another challenge. This is apparent from business models via blockchain supportive advanced service, where the offer consists not of a product or a service but of an assurance to provide a specific outcome to consumers [45]. In order to be competitive and provide long-term value to consumers, blockchain technology and innovations in business model are navigating the need for continuous improvement.

References

1. Idrees, S. M., Agarwal, P., & Alam, M. A. (Eds.). (2021). Blockchain for Healthcare Systems: Challenges, Privacy, and Securing of Data. CRC Press.
2. J. Kietzmann, From hype to reality: Blockchain grows up. Bus. Horiz. **62**(3), XXX (2019)
3. A. Hughes, A. Park, C. Archer-Brown, J. Kietzmann, Beyond bitcoin: What blockchain and distributed ledger technologies mean for firms. Bus. Horiz. **62**(3), XXX (2019)
4. M. Swan, *Blockchain: Blueprint for a New Economy*, 1st edn. (O'Reilly Media, Sebastopol, 2015). https://doi.org/10.1017/CBO9781107415324.004
5. B.W. Wirtz, A. Pistoia, S. Ullrich, V. Göttel, Business models: Origin, development and future research perspectives. Long Range Plan. **49**(1), 36–54 (2016)
6. C. Zott, R. Amit, L. Massa, The business model: Recent developments and future research. J. Manag. **37**, 1019 (2011)

7. A. Osterwalder, Y. Pigneur, C.L. Tucci, Clarifying business models: Origins, present, and future of the concept, in *Communications of the Association for Information Systems*, vol. 16 (2005)
8. W. Mougayar, V. Buterin, *The Business Blockchain: Promise, Practice, and Application of the Next Internet Technology* (Wiley, New York, 2016)
9. A. Onetti, A. Zucchella, M.V. Jones, P.P. McDougall-Covin, Internationalization, innovation and entrepreneurship: Business models for new technology-based firms. J. Manag. Gov. **16**(3), 337–368 (2012)
10. M. Morris, M. Schindehutte, J. Allen, The entrepreneur's business model: Toward a unified perspective. J. Bus. Res. **58**(6), 726–735 (2005)
11. S.M. Shafer, H.J. Smith, J.C. Linder, The power of business models. Bus. Horiz. **48**(3), 199–207 (2005)
12. D. Veit, E. Clemons, A. Benlian, P. Buxmann, T. Hess, D. Kundisch, J. Leimeister, P. Loos, M. Spann, Business models. Bus. Inf. Syst. Eng. **6**(1), 45–53 (2014)
13. M.M. Al-Debei, D. Avison, Developing a unified framework of the business model concept. Eur. J. Inf. Syst. **19**(3), 359–376 (2010)
14. M.W. Johnson, C.M. Christensen, H. Kagermann, Reinventing your business model. Harv. Bus. Rev. **86**(10), 57–68 (2008)
15. C. Baden-Fuller, S. Haefliger, Business models and technological innovation. Long Range Plan. **46**(6), 419–426 (2013)
16. H. Chesbrough, R.S. Rosenbloom, The role of the business model in capturing value from innovation: Evidence from Xerox Corporation's technology spin-off companies. Ind. Corp. Chang. **11**(3), 529–555 (2002)
17. R. Beck, J. Stenum Czespluch, N. Lollike, S. Malone, Blockchain—The gateway to trust-free cryptographic transactions, in *ECIS 2016 Proceedings* (2016)
18. S. Nakamoto, Bitcoin: A peer-to-peer electronic cash system (2008), https://bitcoin.org/bitcoin.pdf
19. Rather, I. H., & Idrees, S. M. (2021). Blockchain Technology and Its Applications in the Healthcare Sector. In Blockchain for Healthcare Systems (pp. 17-25). CRC Press.
20. W. Vaughn, Open vs closed blockchains: Let's end this madness (2015), https://medium.com/@WayneVaughan/open-vs-closed-blockchains-let-s-end-thismadness-8313e4095ead
21. P. Jayachandran, *The Difference Between Public and Private Blockchain* (IBM, Somers, 2017)., https://www.ibm.com/blogs/blockchain/2017/05/the-difference-betweenpublic-and-private-blockchain/
22. J. Coburn, Public vs. private blockchains: Understanding the differences. Blocks Decoded (2018), https://blocksdecoded.com/public-private-blockchains/
23. D. Tapscott, A. Tapscott, *Blockchain Revolution: How the Technology Behind Bitcoin Is Changing Money, Business and the World* (Penguin, New York, 2016)
24. G. Larios-Hernandez, Blockchain entrepreneurship opportunity in the practices of the unbanked. Bus. Horiz. **60**(6), 865–874 (2017)
25. A. Osterwalder, Y. Pigneur, *Business Model Generation: A Handbook for Visionaries, Game Changers, and Challengers* (Wiley, Hoboken, 2013), p. P38
26. M.W. Johnson, C.M. Christensen, H. Kagermann, Reinventing your business model. Harv. Bus. Rev. **86**(12), 50–59 (2008)
27. M. Montecchi, K.A. Plangger, M. Etter, It's real, trust me! Establishing supply chain provenance using blockchain. Bus. Horiz. **62**(3), XXX (2019)
28. L. Mearian, Q&A: Walmart's Frank Yiannas on the use of blockchain for food safety. Computerworld (2018), https://www.computerworld.com/article/3309656/emerging-technology/qa-walmarts-frank-yiannas-on-theuse-of-blockchain-for-food-safety.html
29. M.D. Gregorio, Blockchain: A new tool to cut costs. PricewaterhouseCoopers (2017), https://www.pwc.com/m1/en/media-centre/articles/blockchain-newtool-to-cut-costs.html
30. M. Vukolic, *Behind the Architecture of Hyperledger Fabric* (IBM, Somers, 2018)., https://www.ibm.com/blogs/research/2018/02/architecture-hyperledgerfabric/

31. M. Pilkington, Blockchain technology: Principles and applications, in *Research Handbooks on Digital Transformations*, (University of Burgundy/Edward Elgar Publishing Ltd., Burgundy/Cheltenham, 2016), pp. 225–253. https://doi.org/10.4337/9781784717766
32. E. Kane, Is blockchain a general purpose technology? SSRN Electron. J. **68**, 1–27 (2017). https://doi.org/10.2139/ssrn.293258
33. M. Crosby, V.S. Nachiappan, V. Kalyanaraman, Blockchain technology beyond bitcoin. Blockchain Technologies (2015), pp. 1–27, http://www.blockchaintechnologies.com/blockchain-definition
34. Gartner, Blockchain status 2018: Market adoption reality (2018), https://www.gartner.com/doc/3869693/blockchain-status–market-adoption
35. Deloitte, Blockchain technology use cases in organizations worldwide as of April 2018*. Statista (2018), https://www.statista.com/statistics/878732/worldwideuse-cases-blockchain-technology
36. D. Schatsky, A. Arora, A. Dongre, Blockchain and the five vectors of progress. Deloitte (2018), https://www2.deloitte.com/insights/us/en/focus/signals-for-strategists/value-of-blockchain-applicationsinteroperability.html
37. M.E. Porter, J.E. Heppelmann, How smart, connected products are transforming companies. Harv. Bus. Rev. **93**, 96–114 (2015)
38. S. Lenka, V. Parida, J. Wincent, Digitalization capabilities as enablers of value co-creation in servitizing firms. Psychol. Mark. **34**, 92–100 (2017)
39. T. Grubic, I. Jennions, Remote monitoring technology and servitised strategies—Factors characterising the organisational application. Int. J. Prod. Res. **56**, 2133–2149 (2018)
40. Gardener Glossary. Digitalization (2018), https://www.gartner.com/it-glossary/digitalization/. Accessed 14 Nov 2018
41. IBM, *Innovative Analytics (GBE03664-USEN-02)* (IBM, Somers, 2015)
42. T. Baines, A. Ziaee Bigdeli, O.F. Bustinza, V.G. Shi, J. Baldwin, K. Ridgway, Servitization: Revisiting the state-of-the-art and research priorities. Int. J. Oper. Prod. Manag. **37**, 256–278 (2017)
43. Y. Chen, Blockchain tokens and the potential democratization of entrepreneurship and innovation. Bus. Horiz. **61**, 567–575 (2018). https://doi.org/10.1016/j.bushor.2018.03.006, https://www.sciencedirect.com/science/article/pii/S0007681318300037
44. D. Tamayo, IBM blockchain explained (2017), https://www.slideshare.net/DiegoDiaz49/1-ibmblockchain-explained
45. A. Ziaee Bigdeli, T. Baines, O.F. Bustinza, V. Guang Shi, Organisational change towards servitization: A theoretical framework. Compet. Rev. Int. Bus. J. **27**, 12–39 (2017)

Transformation of Logistics Value Chain for Enhancing Cross-Border Trade Using Blockchain and IoT

Gulpreet Kaur Chadha, Aakarsh Shrivastava, Deepti Mehrotra, and Renuka Nagpal

1 Introduction

S. Nakamoto in 2008 [1] brought Blockchain into the picture through his bitcoin white paper. The development of blockchain has become rapid because of its applicability that is not only limited to the Bitcoin network but also in various other domains due to the features like being highly secure and robust. It has already received the attention of the big companies, and they have started implementing it in financial services and supply chain management. This technology can further prove to be a great help in the logistics value chain. Logistics value chains from one country to another generally perform a very complex process in which they are bound to perform all the manual processes mandated by regulatory authorities like custom departments, airport authorities, etc. These manual processes require a lot of documentation which makes it much tougher for both the exporter and the importer as well. It is also tough to keep track of the status of shipment, and the process can be delayed as many stakeholders are involved that can cause friction in global trade. Companies and organizations exploit user's data for optimization of their business processes which makes data an asset in our economy [2]. Hence, it is important to keep the data safe in this entire process of the logistics value chain. Blockchain can be used to solve this problem as firstly the data required by every regulatory authority can be stored in the block so that no time is wasted in manual documentation and information of exporter, or the importer gets shared only with the relevant

G. K. Chadha · A. Shrivastava · D. Mehrotra (✉)
Department of IT, Amity School of Engineering and Technology, Amity University
Uttar Pradesh, Noida, India

R. Nagpal
Department of CSE, Amity School of Engineering and Technology, Amity University
Uttar Pradesh, Noida, India
e-mail: rnagpal1@amity.edu

S. M. Idrees, M. Nowostawski (eds.), *Transformations Through Blockchain Technology*, https://doi.org/10.1007/978-3-030-93344-9_14

stakeholder in the entire process. This access control can be created using permissioned blockchain. For keeping a track of the shipment, IoT can be used further. So, implementation of both the technologies together might help to develop a more robust and transparent logistics value chain. The aim is to develop an Ethereum-based smart contract system to implement a supply chain system for delivery of a package.

There are many problems faced by the importer and exporter while dealing with international trading. Different countries have different norms, and these norms provide various restrictions. The various areas that lead to problems are- mode of payment, transportation and geography, legal norms, language barriers, finding the right importer and adapting the company for international trade [3]. The most common problem is delay in the shipment process is due to unforeseen reasons, like vehicular breakdown, bad weather, shipping delay, etc. Loss of cargo could also occur during mishaps like losing the packages in sea or during an accident. There could be a typing error or mistake in the paperwork. This could lead to confusion at later stages of the process. Government policies keep on changing due to political or economic reasons. This has a direct impact on international trade. Transport and logistics services have been facilitating international trade for a long time now and have a vital role in the growth and development of rural as well as urban economy. The efficiency and quality of logistics and its services matters for international trade. On the other hand, weak and faulty infrastructure of logistics and operational processes could act as a big hindrance towards global trade and its integration. In contrast, a new and improved trade logistics when combined with a liberalized economic environment can increase the quantity of trade and economies of scale and scope in distribution and production activities [4]. Figure 1 indicates the various industries where the proposed system could be adopted. If working on a larger scale, the system can be used to implement international courier services and e-commerce and trade. However, on a smaller scale, it can also be applied to food delivery and movers and packers.

In this chapter, a smart contract for cross-border trade is proposed using the blockchain technology. Extensive literature review of blockchain technology is done in Sect. 2. Overview of smart contract and blockchain technology used in this study is given in Sect. 3. Solution is proposed in Sect. 4 followed by their implementation in consequent Sect. 5. This chapter is concluded in Sect. 6 followed by references used in the chapter.

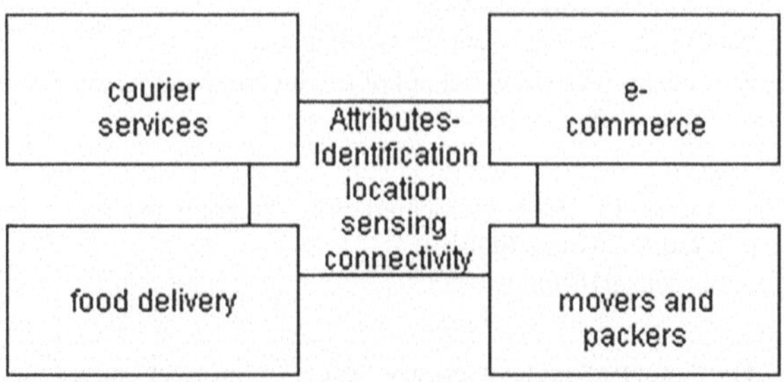

Fig. 1 Target areas for the proposed system

2 Literature Review

Brought together to the world, by one or more individuals under the pseudonym Satoshi Nakamoto [1], bitcoin, the cryptocurrency and its underlying technology, blockchain (BCT) successfully created a huge impact in the electronic payment systems by using peer to peer approach [2, 5]. On a large scale, the blockchain and its underlying technology provides the suitable environment which makes it possible to implement secure and direct transactions between participants without any third party or intermediary ('internet of value') [4]. George Foroglou et al. [6] researched the scope and usage of the blockchain technology and got into depth of concepts like currency, contract, voting, intellectual property rights, smart property and finance. The main reason why blockchain is attracting so much attention is due to the attributes of security, data integrity and anonymity without any intermediate or third party that controls the organization and creates interesting research areas, from the perspective of technical limitations and challenges [7]. Being a shared ledger, it provides the apt infrastructure for secure and direct exchange of values between participants [4] which are confirmed with a series of literature reviews [3, 8–12]. Apart from the above applications, blockchain has also been known to revolutionize the Internet of Things.

Various research works have been put forward as an attempt to understand blockchain technology, especially in ways that blockchain can be used with supply chain management and logistics. Various studies also highlight how data security can be enhanced by optimizing blockchain and how the integration of blockchain can be done with the IoT. Nir Kshetri [13] pointed out how blockchain can have an impact on the supply chain management and its objectives like cost, quality, speed, risk reduction, sustainability, speed and flexibility with the incorporation of IoT. Idrees S. M. et al. [14–17] have emphasized upon the secure nature of blockchain technology and also suggested that blockchain technology can provide secure solutions in supply chain management. K. Francisco et al. [18] reported that blockchain had the potential to bring transparency in supply chain management to a new level, but adoption of the same has been limited because of our understanding. Zyskind Guy et al. [19] proposed a method where blockchain could work as an access control manager without requiring any third party for the same. Idrees S. M. et al. [15] pointed out in a study that blockchain can help to implement privacy, security and trust in the twenty-first century. Kshetri Nir (2017) [20] in his study pointed out the potential of blockchain-based identity and access management systems to address various key challenges that are associated with security in the case of IoT. Haoyan Wu et al. [21] proposed a framework consisting of a set of private distributed ledgers and a single blockchain public ledger where every private ledger allowed the private sharing of custody events amongst the trading partners in the given shipment. AlTawy Riham et al. (2017) [22] proposed a model for a physical delivery system called Lelantos which promises to offer customer anonymity, unbiased transactions and unlinkability between merchant and customer. The system was based on 'onion routing', a term which was introduced to establish anonymous delivery of messages. Lelantos claimed to be dependent on the principles of decentralization and pseudonymity of the blockchain and the consensus mechanisms that are distributed. This was done to ensure that there are fair and irrefutable transactions between distrustful contractual parties. This idea was a good attempt to execute an anonymous delivery system. How blockchain is best represented by state machine replication is proposed by Haseeb et al. [23]. It focused on the various applications of blockchain and threw light on how Hyperledger fabric is a standout amongst numerous ventures that are used as techniques to implement and run the blockchain technology. A theoretical design and model is put forward by Janssen M. et al. [24] that could be used to adopt the main idea behind blockchain technology and understand its relations with technical, market and institutional factors. The model that was proposed outlined the possible outcomes and results that could be possible and that the whole process of changing is focal, as it can shape how the blockchain and its applications work. So, the model put forward can hence be adopted by various organizations as a reference point and can even be used to expand, refine and evaluate the research into the complete idea of blockchain. It can be used by various establishments to recognize the various domains where blockchain can be used. It categorizes the requirements and importance of understanding institutional and organizational aspects that can outline the way blockchain applications can run and how it can bring about an evolution in the market scenario.

Chang Y. et al. [25] proposed a research work where the author studied the challenges of global supply chain and trade operations, along with capabilities and potential of the use of blockchain. The inference blockchain can have on the government agencies and custom agencies. The corresponding challenges that enable the wide scale deployment of blockchain are also crucial points of discussion. The key capabilities of blockchain were listed as adoptability, smart contracts, verifiability, automation, immutability, transparency and disintermediation. However, use of blockchain in various domains faced issues like complexity, scalability, and the huge size of the network as depicted in Table 1.

Tijan et al. [26] researched how the unique technology blockchain can be incorporated in distinguished systems to perform the basic idea of decentralization and storing data with utmost security. The various options for using the blockchain technology for supply chain management and logistics have also been put forward. The advantages of using blockchain in finance has gained huge attention as of today's scenario. Blockchain technology has been explored a lot lately, especially in the financial sector, but it has also been observed that many challenges prevail in the path of adopting blockchain in logistics. Order delay, damage to goods, errors and multiple data entry are few challenges that could be reduced by incorporating blockchain into our already existing technologies. The research work presented an elaborate learning about the current popular trends of the blockchain and how it can be used with logistics and supply chain management. The IBM blockchain uses cloud and its underlying technology and thus aims to deliver end-to-end services, as demanded by the client, who should be able to activate, develop, operate, govern and secure their business networks. Blockchain, as a technology, contributes to increased sustainability, reduces and eliminates frauds and errors, improves

Table 1 Applications of blockchain

S. no	Domain/area	Name of the organization/ platform/ application	Feature/proposed feature	Problem faced
1	Entertainment/ media	Kodak	Project to track intellectual property rights and images clicked by photographers	Complexity
		UjoMusic	Application to create a record for tracking royalties given to musicians	Data
2	Transport/ tourism	Arcade City	Application to track moving ride sharing in taxi services	Network size
		La'Zooz	Online platform to track empty seats for the passengers in need of a lift	Data
		Webjet	Travel portal to monitor empty rooms of hotels and to tackle networks of middlemen	Scalability
3	Real estate	Ubiquity	Company creating an application to track complex legal systems for real estate transfer. This start-up is creating a blockchain driven system	Lack of awareness in the sector
4	Financial	Bank Hapoalim	Collaboration between Microsoft and Israel to manage bank guarantees	–
		Maersk	A project to keep a record of marine insurance	–
5	Charity	BitGive	Online platform to make the charity process more transparent in nature	Lack of awareness in the sector
6	Administration	Various countries	For creating a data Centre for keeping public and confidential records	Scalability
7	Manufacturing	Reliance Jio	Trying to improve the supply chain of logistics by making their own cryptocurrency JioCoin	Complexity
		Hijro	Framework for prototyping and proof of concept	Complexity
		Transactive Grid	A community project to locally produce and distribute energy	Scalability
8	Healthcare	Gem	To create an efficient response towards a disaster due to disease outbreak	Scalability
		MedRec	A project by MIT for sharing and safety of medical records	Scalability

inventory management, minimizes courier costs and reduces delays caused by paperwork, wasting and identifying issues faster. Smart contracts may be implemented over a blockchain based system in cases where the specific conditions are being met, and a third party is not needed. Smart contracts and blockchains received an increasing amount of attention in recent years, especially in academic circles [27]. With the fast-paced development of the blockchain technology, smart contracts can be implemented as computer programs that can run on nodes of the blockchain and could be issued among untrusted and anonymous parties, without involving third parties. The initial successful blockchain-based smart contract was the Bitcoin script, which was based on simple predefined commands, and not a not-Turing-complete language. Some simple examples of smart contracts are pay-to-public-key-hash (P2PKH) and pay-to-script-hash (P2SH), and both of these were defined with bitcoin script [28]. There are also a few platforms that enable us to implement complex contractual functionalities and flexibilities, for example Ethereum [29] adopts a Turing-complete language for smart contracts. Newer blockchain platforms like Neo and Hyperledger fabric allow smart contracts to be written in various high-level languages. Ethereum can easily be understood to be a blockchain-based ecosystem that provides a complete in-built environment to code, run and deploy smart contracts. Writing and implementing smart contracts before using Ethereum was very challenging. It required using unconventional programming paradigms, due to inherent characteristics of blockchain. Also, the bugs and errors of the contracts may lead to consequences. Therefore, it is profitable to have a solid platform and code pattern to simplify implementation of smart contracts [30].

3 Overview of Blockchain and Smart Contracts

The combination of blockchain with smart contracts can prove to be of high importance to many technical collaborations. Smart contracts can be implemented on blockchain-based systems even when all specified conditions are not being met and third parties are not required. Smart contracts and blockchain, both have gained a lot of popularity in the recent past [31]. Catering to the fast-growing popularity of the cryptocurrency and its underlying BTC, Ethereum and Hyperledger platforms have started supporting smart contracts. Basically, smart contracts can be thought of as computer protocols that are intended to digitally facilitate, verify and enforce the performance or negotiation of a particular contract. These smart contracts are known to have a broad range of applications, like in financial services, prediction markets, the Internet of things (IoT), etc. However, there are still many challenges related to security issues and privacy disclosures that need to be further studied in-depth in future [32].

3.1 Blockchain Architecture

Chained Architecture Blockchain is known to offer a distributed ledger that exhibits tamperproof transactions in a decentralized network [33]. It can be rightly called a database in the decentralized network to validate and store all transactions in a consensus that is agreed upon by all nodes in the network, without any central authority. The complete and valid transactions are added to the distributed ledger with a timestamp and other details [34]. Hence, the blockchain technology can exchange tangible and intangible data among the participants of the public ledger. Each stakeholder owns a copy of the public transaction ledger of the blockchain, and this helps avoid system failure and loss of data [25]. During changes, like addition of a block or updating of information, all the copies of the ledger are updated in real time, and the records are registered [2]. These changes thus create a chain. Figure 2 shows the chained architecture of the blockchain network.

Every block in the chain has information about a particular transaction on the whole network. This block is linked to another block through a hash pointer. Every block header comprises a Merkle root through which other blocks originate. Each block also stores the hash pointer of the next block it is linking to [35].

Permissionless and Permissioned Blockchain In the current industry, many projects are being implemented on public and private blockchains. Blockchain is an evolving technology that is in its nascent stage. There are a lot of confusions in the market with reference to the state-of-art-technology blockchain. One of these is regarding one of the biggest divisions in blockchain technology—Permissionless and Permissioned Blockchain. A permissioned blockchain needs to have approval in advance of beginning the functioning regarding who can use and access the blockchain. On the other hand, a permissionless blockchain lets anyone participate in the system. The two systems are similar in some respects and differ as well on a few points (Table 2). The usage of both these is quite different with respect to the targeted application areas. For instance, cryptocurrencies will not prefer to use a permissioned blockchain. Table 2 shows the difference between the two types of blockchains [36].

Hyper-Ledger Fabric This type of fabric is a distributed operating system for permissioned blockchains, which can implement distributed applications that are written in general purpose programming languages. A fabric network is known to support many blockchains that are connected to the same ordering service. Every sub-blockchain is called a channel and has different peers as its members. These

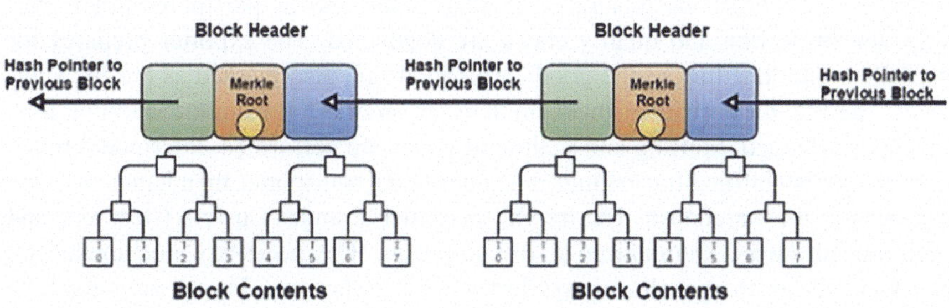

Fig. 2 Representation of chaining in a blockchain

Table 2 Difference between permissioned and permissionless blockchain

Permissioned blockchain	Permissionless blockchain
It provides transparency and anonymity, being a distributed ledger	There is a user-incentivizing token that can increase or decrease in value, based on relevance
It provides a varying degree of decentralization	These are bound to be transparent
Governance is decided by the members of the business network	These are decentralized as there is no central entity that shuts down the network, changes the protocol or edits the ledger

channels can be used to partition the blockchain network; however, consensus across these channels is not coordinated.

4 Proposed Solution

Excluding the part of the process where negotiation is taking place, we will apply blockchain in the rest of the process. Use of Ethereum-based blockchain is proposed. The reason for using blockchain is that the blockchain will increase transparency in between all the stakeholders involved in the entire process and will make it a quicker process by reducing the manual documentation process that has to be done at ports. For implementing blockchain, smart contracts are used. The process of a logistic value chain majorly involves two entities that are an importer and an exporter. An importer identifies the need of the product and floats a query in the international market. Then the exporter and importer interact to discuss the terms and conditions of each party. Importers shortlist quotes and get in touch with the most suitable exporter. Importer and exporter negotiate, and the contract is fixed. Payment, insurance, delivery, transportation and timeline are finalized. The exporter prepares and sends Performa Invoice to the importer. The PI is a preliminary cost of the sale that is sent to buyers before the shipment or delivery of goods. The importer prepares and sends an export order and purchase order to the exporter. The importer and exporter arrange for finance and bank loans. The exporter arranges for manufacturers of required goods. Insurance is also established as per the contract. Pre-shipment inspection and quality check are conducted. The exporter prepares for packaging and labelling of the goods. The goods are also marked as per the shipment. The Freight Forwarder picks up delivery order (DO) from the shipping line. Pickup is arranged. Stuffing and sealing of goods are performed. Intermodal transportation is established for loading. The intermodal transporter then hands over the shipment to the warehouse. The freight forwarder arranges Custom Clearance and performs physical examination of the shipment. The necessary documents are checked and the charges due are paid to suitable authorities. The workers load the goods to the scheduled goods carriers for further transportation. The exporter mails the bill of loading and other documents to the consignee or importer. Once the

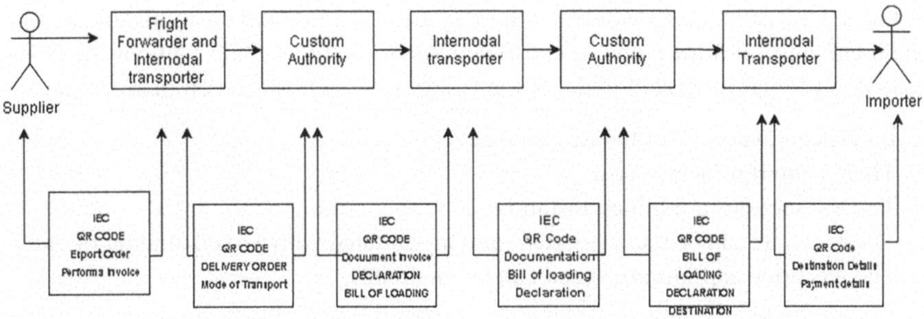

Fig. 3 Flowchart of the proposed system

goods have been received, the customs agent presents the documents for import clearance. The transportation of goods to the destination is also arranged. The custody of shipment is handed over to the intermodal transporter. The intermodal transporter transports the goods to the local warehouse or the assigned place. The importer arranges for vacating the container in which the goods arrived. Once the importer receives the goods as expected, the smart contract initiates the fund transfer to the exporter. The importer then may apply for government incentives. The detailed architecture is given in Fig. 3. It highlights the flow of events. Along with using blockchain for implementing the proposed system, authors also propose using a GPS tracker to track the location of the product that must be transported. The delivery parcel that must be transported across an international border relates to a GPS tracker, and the live location of the parcel is tracked at regular intervals of time. This location is stored in a block of the blockchain that is being developed. This tracked location helps cross-check the location and know where the parcel is at every point of time.

5 Implementation

Ethereum aims to dig away bitcoin's structure, in this case, with the goal that professionals may implement applications or programs to have advancements and set new standards in the technological aspects, etc. The main aim of Ethereum's Turing finished language was to allow the engineers to build more projects that use blockchain and its underlying technologies. The underlying design of the Ethereum's blockchain is the same as the one of bitcoin, where it shows a mutual record of the whole exchange history. Every hub on the system stores a duplicate of the transaction history. The structure of the Ethereum blockchain is fundamentally the same as bitcoin's, in that it is a mutual record of the whole exchange history. The Ethereum hubs store the latest condition of each savvy contract and do not withstand the entirety of other exchanges. Ethereum is also known to utilize accounts. Like finance reserves, ether tokens are present in the wallet and can be ported to other records.

Ganache CLI is the up-to-date version of Test RPC: a fast and customizable blockchain emulator. It allows one to call functions of the blockchain without the overhead of an actual Ethereum node. Some features of the Ganache cli platform are:

- It gives an overview of the test-chain events.
- There is no transaction cost.
- The transactions are mined instantly.
- The accounts can be recycled, reset and instantiated with a fixed amount of ether.
- The gas price and mining speed can be modified.
- A convenient GUI can give you an overview of test chain events.

While implementing the proposed system, the following categories are assigned:

Port Authority The set order function will require the manager to insert the basic details of the product like product ID and weight of the product. He will also be able to enter the price of the product and product description for further description. The set order flow function (Fig. 4) will help him to design the flow of delivery and define it. He will be able to set the flow and identify the stakeholders and allot them a certain address.

Port Authority Function There will be two port authorities in every case importer port authority and exporter port authority (Fig. 5), and both will be able to use the port authority function.

The struct Place will help authorities to store the information where this package is supposed to be sent. It will also be scanning the package using Scan structure that will be created while the final packing is done. Here the sender will be able to enter final package details by create package function storing information about package owner and other details. They can also get the same info by getpackage by ID function, getScanIDs, etc.

Customs Customs will be also at both the ends that are custom authorities at importer end and custom authorities at exporter end. So, the main work to do is that there is a set limit for every product by weight that can be delivered by a specific person or can be imported by specific person (Fig. 6).

Here there are the two functions, the checkweight function will check the weight of the product as per weight on the port by customs and customs can approve or

Fig. 4 Set flow function of port authority

```
function setflowoforder( uint orderId,
    address _manufacturer,
    address _exLandtrasport,
    address _excustoms,
    address _exportAuthority,
    address _shipping,
    address _importAuthority,
    address _imcustoms,
    address _imLandtransport,
    address _distrubuter){
```

```
contract PortAuthority {
    struct Place {
        bool exists;
        string country;
        string location;
    }

    struct Package {
        address sentBy;
        string destination;
    }

    struct Scan {
        address placeOwner;
        uint packageId;
        uint timestamp;  // block number
        string country;
        string location;
    }
```

Fig. 5 Port authority function

Fig. 6 Customs authority function

```
contract Customs is main {

    event passed(uint orderId);
    event unpassed(uint orderId);
    event underweight(uint orderId);
    event overweight(uint orderId);
```

Fig. 7 Transportation contract

```
contract Transportation {
    enum PackageStatus {
        inTransit,
        received,
        cancelled,
        arrived,
        booked
    }
```

reject this order by using the approve in category function. This contract can be used at both the ends by the customs officer.

Transportation: There will be one importer land transporter, one importer land transporter and one shipment process. The transportation contract (Fig. 7) will be called four times, and it has three stakeholders so it will be as follows:

- Exporter–Exporter Land Transporter–Exporter Customs
- Importer Customs–Importer Land transportation–Importer
- Exporter Port Authorities–Shipping–Importer Port Authorities
- Exporter–Delivery Company (if any)–Importer

The Sender will be able to call addPackage function where he will enter the details of the package and the address to where it must be transported. The transporter will be able to use the pickPackage function, and the receiver will be able to call two functions, one will be the receivedPackage function and the other will be to cancelOrder function. The transporter will be able to call the arrivedPackage function as soon as he reaches the port.

As a result, we were able to get transactions when we tested the transactions on the Ethereum test network using Blockchain. The user was able to call the functions mentioned above using the Remix IDE. The manager or the exporter was able to use a setOrderFlow function and setOrder function. In set order, he was able to allot address to the respective stakeholders. The stakeholders included are:

- Exporter (manufacturer)
- Exporter land transporter
- Exporter custom authority
- Exporter port authority
- Shipping
- Importer (manufacturer)
- Importer land transporter
- Importer custom authority
- Importer port authority
- Distributer (importer)

The manufacturer (manager) can also call setOrder to define what has to be delivered and how much. The custom was to call the weightPackage and approvePackage. Customs of both the ends can call the same feature. The port authorities were able to use various features like createPackage, scanPackage, setPlace. Both the importer and exporter port authority can call this function. There will be one importer land transporter, one exporter land transporter and one shipment process. The transportation contract will be called four times, and it has three stakeholders so it will be as follows:

- Exporter–Exporter Land Transporter–Exporter Customs
- Importer Customs–Importer Land transportation–Importer
- Exporter Port Authorities–Shipping–Importer Port Authorities
- Exporter–Delivery Company (if any)–Importer

The Sender will be able to call addPackage function where he will enter the details of the package and the address to where it must be transported. The transporter will be able to use the pickPackage function, and the receiver will be able to call two functions one will be to insert the function; one will be the receivedPackage function and the other will be to cancelOrder function. The transporter will be able to call the arrivedPackage function as soon as he reaches the port.

In this particular contract, the functions available for the sender is an addPackage. The transporter can use the pickPackage and arrivedPackage functions. The receiver can use two functions: cancelOrder and receivedPackage.

Fig. 8 Ethereum chain

The backend of this blockchain system uses truffle as a framework. The folder structure of this framework will be as follows:

1. Build
2. Contracts
3. Migrations
4. Tests
5. Compile.js
6. Deploy.js
7. Package-lock.json
8. Package.json
9. Truffle-congif.js

As soon as any contract is deployed, or any function is called, the number of transactions will increase on the respective account and can be verified in Ganache GUI. The amount deducted and left for a particular account can be monitored in the same way.

The final transaction can also be checked for each account like above and simultaneously a chain of transactions will be called, which can be called as Blockchain or Ethereum Chain (Fig. 8).

5.1 Invalid Transactions

Figure 9 shows four transactions. The first transaction is successful, as indicated by the green tick and no error message. The second transaction depicts a failed transaction. This happens when the contract is breached and when someone tries to tamper with the blockchain. This is an important reason for blockchains popularity and robustness of the entire system.

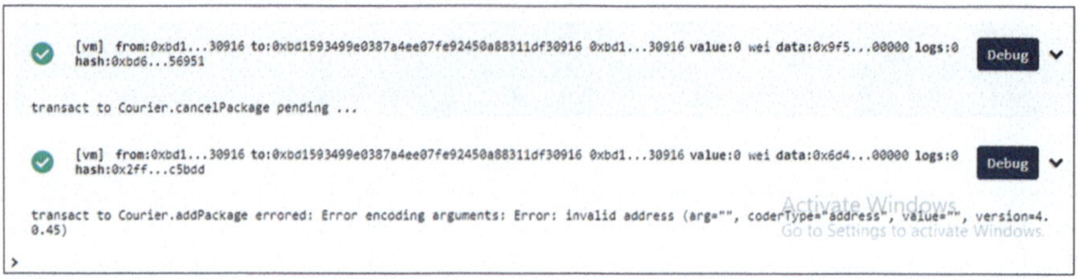

Fig. 9 Examples of invalid transactions

Table 3 Comparison between Ethereum and Blockchain-based systems

Parameters	Ethereum based	Blockchain based
Cost	Cost is directly dependent on the developer and smart contract	Cost is high anyway considering the price of a bitcoin
Speed of process	Faster	Slower
Speed of computation	Slower	Faster
Rigidity	More rigid	Less rigid
Security	Highly secure	Comparatively less secure

5.1.1 System Based on Blockchain or Ethereum

Blockchain and Ethereum are somewhat similar but however use different algorithms and different functionalities for function (Table 3). A comparative has been drawn if this value chain logistics system would have been on blockchain and not on Ethereum. Factors on which comparison is being drawn are as follows:

- *Cost*: As far as the cost is considered, the blockchain architecture would be more when only the cryptocurrency is involved which is bitcoin as bitcoin has more value than ethers. So, let us say if some amount of cryptocurrency has to be given to a miner for a particular transaction, then the blockchain system would cost more. But there is also a different angle to this. In Ethereum, the concept of smart contract might make it more expensive than the blockchain system. As every time the contract is called, some amount of cryptocurrency will be expended, and there will be multiple smart contracts and hence more ethers would be spent which might increase its cost than the normal blockchain system. Also, every mathematical computation, memory costs some amount of money. Like addition, subtraction or using memory would also cost ether. Hence it also depends upon the developer how much of the cost is spent because it will depend directly on the code structure and quality of the code.
- *Speed of Process*: The process can be faster in the case of Ethereum as no verification has to be done in the case the process can be made in such a manner that no verification is required manually hence if the smart contract is hit sequentially, it will be much faster and reliable.

- *Speed of Computation*: Faster a transaction, faster is the computation. Blockchain has more community than that of Ethereum that is it has more no of miners. In addition to that, Ethereum focuses on serializing the record and hence focuses more on security hence is comparatively slow than normal blockchain.
- *Rigidity*: When it comes to the rigidity the Ethereum system is more rigid that means that a particular contest will only be called only when a particular contract is called. Hence in Ethereum, it is not just about the log creation but also following the sequence of events. Hence in Ethereum the process cannot be manipulated.
- *Security*: As Ethereum focuses on serializing, the record according to nodes of the system makes it more secure. Secondly the rigidity of the entire process, as no one can manipulate the process, hence the Ethereum is more secure.

6 Conclusion

In this work, it has been identified how we can incorporate blockchain-based architecture to implement a supply chain management system. This promises a solution for the problems that were faced by exporters and importers in the traditional export–import process. Innovations are being carried out across the globe to enhance the system by automating it through its integration with the field of Information Technology. IoT and Blockchain based mechanism that will be able to track the live location of the delivery parcel at regular intervals of time. This would enable the user to verify the position of the parcel. Further this system will provide the relevant data to the relevant stakeholder which will ensure data security. Blockchain can be used in the proposed system in two ways: as a permissioned blockchain or as two parallel blockchains (one public and the other private). A permissioned blockchain can be implemented by using a Hyperledger-based Sawtooth framework. Hyperledger fabric is basically an inbuilt framework that enables the user to implement a distributed ledger and exhibits a modular and versatile design with increased scalability and preserving the privacy of the data at the same time. The model as two aspect, the core side and the application side. The core part handles the basic set of rules and functionality, and the application part accommodates user-specific changes. Thus, by using a Hyperledger framework, parallel transactions can be made in both public and private aspects. Another way to implement the system is by using an entangled blockchain that also lets us perform the functions of both public and private blockchain. The last option to implement the same idea is by using a public Ethereum blockchain in parallel with a private Ethereum blockchain. Ethereum's personal network comprises a totally private blockchain that is isolated from the main Ethereum network. This network was created by various organizations to reduce the permissions given to a blockchain. The nodes of this network would not be connected to the main network but will only be restricted to a private blockchain. It can thus act as a distributed database, containing private data, having private access rights and conducting free transactions and buying virtual ethers. An important future development in this paper would be to integrate the existing system with IoT-based devices, like GPS tracker that would enable us to track the live location of the package that is out for delivery. The location tracked in the real-time system would be stored in blocks, thus making them completely secure from hackers, etc.

References

1. S. Nakamoto, Bitcoin: A peer-to-peer electronic cash system. Decentral. Bus. Rev., 21260 (2008)
2. K.R. Lakhani, M. Iansiti, The truth about blockchain. Harv. Bus. Rev. **95**(1), 119–127 (2017)
3. C. Catalini, How blockchain applications will move beyond finance. Harv. Bus. Rev. **2** (2017)
4. D. Tapscott, A. Tapscott, How blockchain will change organizations. MIT Sloan Manag. Rev. **58**(2), 10 (2017)
5. C. Mulligan, J.Z. Scott, S. Warren, J.P. Rangaswami, Blockchain beyond the hype: A practical framework for business leaders, in *White Paper of the World Economic Forum*, (2018)
6. G. Foroglou, A.L. Tsilidou, Further applications of the blockchain, in *12th Student Conference on Managerial Science and Technology*, (2015), pp. 1–8
7. J. Yli-Huumo, D. Ko, S. Choi, S. Park, K. Smolander, Where is current research on blockchain technology?—A systematic review. PLoS One **11**(10), e0163477 (2016)
8. M. Swan, *Blockchain: Blueprint for a New Economy* (O'Reilly Media, Newton, MA, 2015)
9. Y. Yuan, F. Wang, Development status and prospect of blockchain technology. J. Autom. **42**(4), 481–494 (2016)
10. Y. Guo, C. Liang, Blockchain application and outlook in the banking industry. Finan. Innov. **2**(1), 1–12 (2016)
11. G. Ishmaev, Blockchain technology as an institution of property. Metaphilosophy **48**(5), 666–686 (2017)
12. A. Tapscott, D. Tapscott, How blockchain is changing finance. Harv. Bus. Rev. **1**(9), 2–5 (2017)
13. N. Kshetri, Blockchain's roles in meeting key supply chain management objectives. Int. J. Inf. Manag. **39**, 80–89 (2018)
14. S.M. Idrees, M. Nowostawski, R. Jameel, A.K. Mourya, Security aspects of Blockchain technology intended for industrial applications. Electronics **10**(8), 951 (2021)
15. S.M. Idrees, I. Aijaz, R. Jameel, M. Nowostawski, Exploring the Blockchain technology: Issues, applications and research potential. Int. J. Online Biomed. Eng. **17**(7) (2021)
16. S.M. Idrees, M. Nowostawski, R. Jameel, Blockchain-based digital contact tracing apps for COVID-19 pandemic management: Issues, challenges, solutions, and future directions. JMIR Med. Inform. **9**(2), e25245 (2021)
17. S.M. Idrees, M. Nowostawski, R. Jameel, A.K. Mourya, Privacy-preserving infrastructure for health information systems, in *Data Protection and Privacy in Healthcare*, (CRC, Boca Raton, FL, 2021), pp. 109–129
18. K. Francisco, D. Swanson, The supply chain has no clothes: Technology adoption of blockchain for supply chain transparency. Logistics **2**(1), 2 (2018)
19. G. Zyskind, O. Nathan, Decentralizing privacy: Using blockchain to protect personal data, in *2015 IEEE Security and Privacy Workshops*, (IEEE, Piscataway, NJ, 2015), pp. 180–184
20. N. Kshetri, Can blockchain strengthen the internet of things? IT Profess. **19**(4), 68–72 (2017)
21. H. Wu, Z. Li, B. King, Z. Ben Miled, J. Wassick, J. Tazelaar, A distributed ledger for supply chain physical distribution visibility. Information **8**(4), 137 (2017)
22. R. AlTawy, M. ElSheikh, A.M. Youssef, G. Gong, Lelantos: A blockchain-based anonymous physical delivery system, in *2017 15th Annual Conference on Privacy, Security and Trust (PST)*, (IEEE, Piscataway, NJ, 2017), pp. 15–1509

23. M. Haseeb, K.R. Sekhar, Y.V. Spandana, M. Syam, A fabric architecture towards block chain application using hyper ledger. Int. J. Eng. Technol. **7**(2.20), 83–85 (2018)
24. M. Janssen, V. Weerakkody, E. Ismagilova, U. Sivarajah, Z. Irani, A framework for analysing blockchain technology adoption: Integrating institutional, market and technical factors. Int. J. Inf. Manag. **50**, 302–309 (2020)
25. Y. Chang, E. Iakovou, W. Shi, Blockchain in global supply chains and cross border trade: A critical synthesis of the state-of-the-art, challenges and opportunities. Int. J. Prod. Res. **58**(7), 2082–2099 (2020)
26. E. Tijan, S. Aksentijević, K. Ivanić, M. Jardas, Blockchain technology implementation in logistics. Sustainability **11**(4), 1185 (2019)
27. M. Heutger, M. Kueckelhaus, *Blockchain in Logistics* (Bonn, DHL, 2018)
28. A.M. Antonopoulos, *Mastering Bitcoin: Unlocking Digital Cryptocurrencies* (O'Reilly Media, Newton, MA, 2014)
29. G. Wood, Ethereum: A secure decentralised generalised transaction ledger, in *Ethereum Project Yellow Paper*, vol. 151, (2014), pp. 1–32
30. M. Wohrer, U. Zdun, Smart contracts: Security Patterns in the Ethereum Ecosystem and Solidity, in *2018 International Workshop on Blockchain Oriented Software Engineering (IWBOSE)*, (IEEE, Piscataway, NJ, 2018), pp. 2–8
31. M. Alharby, A. Van Moorsel, Blockchain-based smart contracts: A systematic mapping study. In *arXiv preprint arXiv:1710.06372* (2017)
32. S. Wang, Y. Yuan, X. Wang, J. Li, R. Qin, F.Y. Wang, An overview of smart contract: Architecture, applications, and future trends, in *2018 IEEE Intelligent Vehicles Symposium (IV)*, (IEEE, Piscataway, NJ, 2018), pp. 108–113
33. J. Zhang, C. Thomas, P. FragaLamas, T.M. Fernández-Caramés, Deploying blockchain technology in the supply chain, in *Computer Security Threats*, (IntechOpen, London, 2019)
34. C. Holotescu, Understanding blockchain technology and how to get involved, in *The 14th International Scientific Conference eLearning and software for Education, Bucharest, 19–20 April*, (2018)
35. G.K. Chadha, A. Singh, Bitcoin block-chain mining, in *2019 9th International Conference on Cloud Computing, Data Science & Engineering (Confluence)*, (IEEE, Piscataway, NJ, 2019), pp. 152–157
36. T.K. Sharma, *Permissioned and Permissionless Blockchains: A Comprehensive Guide* (Blockchain Certifications, 2020)

Rural Logistics Transformation Through Blockchain

Ritu Maheshwari

1 Introduction

Rural Logistics refers to the procurement, transportation, and storage of materials throughout the manufacturing and selling process. Different types of logistics activities include harvesting or purchasing raw materials; transporting those materials via truck, train, ship, plane, or a combination thereof; storing the materials at a warehouse; processing the materials into a saleable product; and transporting the finished product to the point of sale. The supply chain is tied to the complex processes of the creation and distribution of goods. Depending on the product, the supply chain can include many phases, multiple geographic locations, several accounts and payments, several individuals, entities, and means of transport. Therefore, procurement of supplies can be extended over several months. Because of the complexity and the lack of transparency of traditional supply chains, it is of great interest for the stakeholders involved in the logistics process to introduce and develop blockchain technology to enhance the logistics processes in the supply chain, making them more sustainable. Blockchain technology is most often mentioned and used in cryptocurrencies, but the extent of possible applications is significantly larger. Blockchain is a distributed book (ledger) with many potential applications. It can be used for any data exchange, whether it is contracting, tracking of shipments, and financial exchanges (payments) [14]. Each action is captured in the block, and the data is distributed over many nodes (computers), making the system transparent. Every block connects to the one before and after, which makes the system safer. Blockchain can increase the efficiency and transparency of the supply chain and

R. Maheshwari (✉)
Atal Bihari Vajpayee Institute of Good Governance and Policy Analysis (A Registered Body of Government of M.P. Under Public Service Management Department), Bhopal, Madhya Pradesh, India
e-mail: ritu.maheshwari14@mp.gov.in

S. M. Idrees, M. Nowostawski (eds.), *Transformations Through Blockchain Technology*, https://doi.org/10.1007/978-3-030-93344-9_15

positively affect all logistic processes, from storage to delivery and payment. In addition to increased transparency and security achieved through blockchain, it is possible to speed up the physical flow of goods. Tracking goods through blockchain can improve the decision-making process with the end result being a more satisfying service for the end-user. Blockchain technology possesses the potential for the creation of new logistics services, as well as new business models. As a relatively new technology, blockchain is designed to achieve decentralization, real-time peer-to-peer operation, anonymity, transparency, irreversibility, and integrity in a widely applicable manner. The verification of every transaction requires the acknowledgment of every node in the network, which will take substantially more time than the centralized system. It should be emphasized that no available literature sources have been found that comprehensively explains the exclusive application of blockchain technology principles in logistics [2, 11].

2 Related Literature Review

The concept of rural logistics encompasses transport, distribution, storage, material handling, and the packaging of goods in rural areas, as well as the flow of information and funds in support of rural production and consumption. Rural logistics encompasses more than just the outbound flow of agricultural products from rural areas. It includes the movement of agricultural inputs and consumer products into rural areas, as well as the movement of light industrial goods produced in villages. Rural logistics also carries a regional connotation, as it focuses on logistics activities in administrative regions at or below the county level, among them county-level cities, autonomous county-level administrative units, townships, and villages. Rural logistics and supply chains are affected by a number of complex and interrelated factors and issues that include: land tenure, farm size, market structure (including the market power of various actors in the chain), information flows, the availability and cost of finance and banking facilities, available logistics infrastructure and services, government policies (legislative, regulatory and fiscal environment), and levels of public and private participation. Many of these factors lie outside of the scope of issues normally dealt with by transport ministries. They do, however, illustrate the need for a multi-sectored approach to rural logistics and supply chains. A supply chain consists of all parties involved, directly or indirectly, in fulfilling a customer request. The supply chain includes not only the manufacturer and suppliers, but also transporters, warehouses, retailers, and even customers themselves. Within each organization, such as a manufacturer, the supply chain includes all functions involved in receiving and filling a customer request. These functions include, but are not limited to, new product development, marketing, operations, distribution, finance, and customer service [6, 19].

Tribal areas were characterized by a lifestyle distinct from agrarian communities, and their agriculture practices are distinct. Today, the tribal majority areas, which overlap with the country's major forest areas, are also areas with the highest concentrations of poverty. Agriculture is predominantly rainfed and mono-cropped. Horticulture is marginally developed in the tribal areas with the present area under fruits, vegetables, and spices accounting for only about 2.5% of the cultivated area. Women participate in all agricultural operations excepting ploughing and sowing of rice seed, contributing between 70% and 80% of the total labor. Despite favorable resource conditions, tribal regions perform poorly in terms of infrastructure, returns from agriculture and almost all human development indicators.

- *Policy Intervention*

- Usually, the market outcomes are determined by the free forces of demand and supply of a product. However, some products may require government intervention to either facilitate the market for the good or to restrict the market for the good. In such cases, the market outcomes differ from the equilibrium outcome like price ceiling, price floor, taxes, and subsidy.
- *Program Intervention*
- To implement new processes, change is required to the way that organizations work together and how individuals operate. Making those changes is hard and many projects fail. The bi-model business model is here to stay, but it is a challenge; carrying out changes while the supply chain continues to operate is like "doing heart surgery while running a marathon". In the supply chain, change involves multiple departments and organizations. Sales, manufacturing, logistics, purchasing, and finance teams within an organization all play a part. Externally, suppliers, customers, service providers, and IT providers also have a role. There is a tendency to optimize in silos because this avoids political issues.
- *Management Intervention:* It is required for improved reporting, improved data use, improved supply chain infrastructure, management practices, and outcomes.

Inventory management is one of the most important business processes during the operation of a manufacturing or production company as it relates to purchases, sales, and logistic activities. It is concerned with the control of stocks throughout the whole supply chain. Inventory control sits at the data level where the day-to-day business is organized. Activities here are data-driven and are primarily concerned with short-term planning and recording of events. Inventory control is concerned with maintaining the correct level of stock and recording its movement. Inventories are essential for keeping the production wheels moving, keep the market going and the distribution system intact. They serve as lubrication and spring for the production and distribution systems of organizations. Management is very critical about any shortage of inventory items required for production. Any increase in the redundancy of machinery or operations due to shortages of inventory may lead to production loss and its associated costs. In a manufacturing system, even for a very complicated and flexible process of different products, the physical arrangement and capabilities of machines usually determine how production control should operate. The machines are not specified only for one task, and different tools need to be set up whenever the work changes. Managing supply chain costs is the most important aspect of an organization; to achieve this, an organization has to employ a qualified professional who understands inventory management techniques [12].

Warehousing is one of the main spheres of logistics. The very broad meaning of it is storage of finished goods or materials (raw and components) for manufacturing, agricultural, or commercial purposes. In fact, warehousing contains numerous functions, like acceptance of products (loading, unloading), inspection, and "proper storage." It is the whole system (warehouse management system) that includes warehouse infrastructure, tracking systems, and communication "between product stations." One of the most sustainable trends in storage solutions is the "Just in Time technique." It means product delivery directly from supplier to producer without warehousing. But this system has quite limited application as the distances between intermediaries are growing with the globalization process of the world economy. Modern logistics cannot survive without warehousing services, but various sustainable modifications of warehousing infrastructure can be introduced [13] (Fig. 1).

3 Challenges and Issues

3.1 Customer Service

Logistics customer service is a part of a firm's overall customer service offering, customer service elements that are specific to logistics operations including fulfilment, speed, quality, and cost. The term fulfilment process has been described as the entire process of filling the customer's order. The process includes the receipt of the order, managing the payment, picking and packing the goods, shipping the package, delivering the package, providing customer service for the end-user, and handling the possible return of the goods.

Fig. 1 Warehousing in Madhya Pradesh

3.2 *Transportation Cost Control*

Transportation cost determines how much a firm will charge for the goods and services that it makes available in the supply chain. Pricing affects the behavior of the buyer of the good or service, thus affecting supply chain performance. For example, if a transportation company varies its charges based on the lead time provided by the customers, customers who value efficiency will likely order early and customers who value responsiveness will be willing to wait and order just before they need a product transported. Differential pricing provides responsiveness to customers that value it and low cost to customers that do not value responsiveness as much. Any change in pricing impacts revenues directly but could also affect costs based on the impact of this change on the other drivers. Our definitions of these drivers attempt to delineate logistics and supply chain management. Supply chain management includes the use of logistical and cross-functional drivers to increase the supply chain surplus. Cross-functional drivers have become increasingly important in raising the supply chain surplus in recent years [4, 23].

3.3 *Planning and Risk Management*

A supply chain risk management plan is a strategy to speed response to as many circumstances as can be predetermined to minimize disruptions to the supply chain if they were to occur. Supply chain risks can come from anywhere, as we have learned. Some of the most common are natural disasters and weather-related risks, perhaps from global warming. COVID-19 is still a threat to supply chains while some risks happen more frequently than others, even those that appear less emergent can be equally destructive. These may include changing market conditions, the competition gaining market share or operating at lower costs or evolving customer tastes. Once again, supply chain management can either help or hurt a business.

3.4 *Lack of Awareness*

Lack of visibility into your supply chain and in rural logistics can cause companies to lose track of their supply network. If you do not have the technology that permits you to forecast demand, you could be left with not enough or too much inventory at different points in your supply chain.

3.5 Lack of Education

It is a fact that education and access to better technology are two major factors contributing to fighting poverty and lifting communities out of a place of constant strife. If vast amounts of people around the globe simply do not have the knowledge needed to help themselves increase production and make a climb up the economic ladder, then the responsibility to help them lies with the people who are already above them.

3.6 Supplier/Partner Relationships

A supplier relationship is important because a long-term relationship between your organization and its suppliers allows for the free flow of feedback and ideas. Over time, this will create a more streamlined, effective supply chain that will have a positive impact on costs and customer service.

3.7 Government and Environmental Regulations

The impact of the environmental regulations on green supply chain management. Specifically, what firms must address to comply with regulations and how firms deal with requests or requirements from their customers, as well as the impositions they place on their suppliers.

3.8 Slow Digital Transformation

Digital transformation in logistics has allowed small companies as well to make a global reach. Over the few years ahead, this race will be mostly to work upon customer's experience in logistics and to make way to win over other companies. It is helping with logistics control, reducing operation cost, improving efficiency, with it, it is allowing the company to satisfy customer's same-day delivery needs too. New delivery capabilities are also evolving, allowing logistics to enjoy and harness technology in more efficient and effective ways [8].

3.9 Lack of Actionable Data on Insight

Big supply chain analytic uses data and quantitative methods to improve decision-making for all activities across the supply chain. In particular, it does two new things. First, it expands the data set for analysis beyond the traditional internal data held on Enterprise Resource Planning (ERP) and supply chain management (SCM) systems. Second, it applies powerful statistical methods to both new and existing data sources. This creates new insights that help improve supply chain decision-making, all the way from the improvement of front-line operations, to strategic choices, such as the selection of the right supply chain operating models.

3.10 The Limited Granularity of Data

The single most important design issue facing the data warehouse developer is determining the proper level of granularity of the data that will reside in the data warehouse. Granularity is also important to the warehouse architect because it affects all the environments that depend on the warehouse for data. Granularity affects how efficiently data can be shipped to the different environments and determines the types of analysis that can be done. The trade-off in choosing the right levels of granularity centers around managing the volume of data and storing data at too high a level of granularity, to the point that detailed data is so voluminous that it is unusable. In addition, if there is to be a truly large amount of data, consideration must be given to putting the inactive portion of the data into overflow storage.

3.11 Product Cost

Supply chain costs are defined as costs that constitute a considerable percentage of the total sales price of a product or service. Manufacturers usually define supply chain costs using the total cost of ownership. The total cost of ownership is defined as the combination of the purchase or acquisition price of a good or service. To this, they add the additional costs incurred before or after the product or service delivery. Applying the total cost of ownership analysis to the supply chain implies identifying all direct, indirect, and other associated costs.

3.12 Low-Frequency Visit of APMC (Agricultural Produce Market Committee)

The frequency of farmers visits the APMC market is 1–2 times a month only. The visit to the market by the farmer is low as most of the produce is sold to village aggregators and it is uneconomical to transport a small lot of produce to the market which is at a distance. Most of the farmers sell their produce to traders outside of RMC yards. The next large category is comprised of farmers who sell their produce to village-level aggregators. Only a few farmers sell their produce to traders/wholesalers sitting inside the RMC yards.

3.13 Non-Participation of the Buyer at RMC Markets (Ready Mixed Concrete)

The markets under RMC are devoid of any transaction as there are no market channels in a function. Like sellers, the traders also face uncertainty in trading in RMC markets as there is a risk of inconsistent supply of low marketable lots and poor quality of the produce. Buyers have no motivation to trade in the market as farmer producers do not visit these markets to sell the produce. However, buyers are willing to engage in markets provided sellers have sufficient produce to trade. The three major reasons that emerged from the farmer survey behind not utilizing the RMC platform are the location of RMC, the absence of traders at RMC, and transportation problems. The other reasons which were cited by the farmers were small surplus with farmers and uncertainty about getting a deal and price.

4 Why Blockchain in Rural Logistics

Blockchain is a universally acclaimed innovation based on a distributed ledger technology, which originated from the efforts of anonymous developers to create a secure digital currency. Digital currencies that are based on a blockchain are defined as cryptocurrencies since they rely on cryptographic mathematical tools. Big companies such as IBM and Microsoft have started to consider offering blockchain services. Supply chain managers can use blockchain for transparent and controlled transactions. Blockchain results in faster transparent settlements as ledger is updated automatically. It has pre-approved transaction fees for procurement, digital payments and contracts, provenance and traceability, logistics, etc. Blockchain can store relevant data from all the partners involved in supply chain with a 360-degree view of the total volume of purchases, regardless of who managed the purchase activity. The ledger can also be programmed to trigger transactions automatically. For Governance, public permissioned blockchain and private permissionless blockchain is preferable [15, 16].

5 Research Methodology

5.1 *The Objective of the Study* "Digital Interventions in Rural Logistics and Its Solutions Through Blockchain Covers"

1. Adequate management of **rural logistics** in districts of MP.
2. **Initiatives and challenges** in rural logistics.
3. **Digital Interventions** in rural logistics.
4. **Its Solutions through emerging technology—BLOCKCHAIN.**

5.2 *Scope of the Study*

It is a live use case and a technology-based project that focuses on building up Blockchain-enabled smart rural logistics in MP which is traditionally having poor and disintegrated management system for Traffic and Vehicle Systems/Transportation/Material/Goods/Marketing/Finance/Crisis and Insurance leading to poor delivery of low-quality rural logistics of product lagging behind in time/cost/space/quality/technology domain and failing in better, smart and environment friendly rural logistics performance.

Blockchain-enabled smart rural logistics will provide Mobility as a Service (MaaS) feature in rural logistics for real-time traffic and vehicle management and smart transaction/payment systems for cost transparency between stakeholders, confidential information system, immutability and transparency of records, traceability, auditability and verifiability of integrated smart rural logistics systems with proper smart contracts to reduce risks of frauds [5].

Traditional Rural Logistics suffer from many drawbacks, lacking areas, and disadvantages in its processes. Rural logistics management is required to carry out its functions through emerging technology Blockchain in the following areas of rural logistics that lacks in the management of the following in an integrated manner [3, 22]:

- Traffic, Vehicle, and Transportation.
- Material and Market.

This research paper will provide infrastructure requisites/strategy and framework for building up emerging technology-based, Blockchain-enabled Smart Rural Logistics in MP with proper management of the aforementioned sub-domains of Rural Logistics using an integrated mobility system.

The survey has been fielded in seven (07) Districts of Madhya Pradesh, viz. Bhopal, Hoshangabad, Sagar, Sehore, Dewas, Khandwa, and Ujjain. The semi-structured questionnaire survey had been distributed to Sr. Government officials and senior technical Executives at Jila Panchayat, Janpad Panchayat, Mandi Board, Collectorate, Public Service Management, Digital Center, Common Service Center, and Stakeholders like Warehouses, Transporters, Traders, Suppliers, Farmers, Shopkeepers, and Manufacturers. The respondent includes Sr. Government officials and senior technical Executives of Government Departments. The data has been tabulated, analyzed, and presented.

Government must focus on Blockchain policy and its technological implementation about rural logistics in Madhya Pradesh to help reduce time, cost, and risk in the areas of regulatory compliance, contract management, identity management, and citizen services. Strategy and framework have been designed for stakeholders in the area of rural logistics. The decentralization process requires the involvement of different actors and those actors can play a very important role in providing the necessary data intended to gather from this study. Therefore, the target populations for the study are the respondents based on their participation and experience in the decentralization process. This target population is a sample representative of the whole of Madhya Pradesh and stakeholders involved in the current decentralization process has been drawn from the following departments/concerned respondents:

- Jila Panchayat.
- Janpad Panchayat.
- Mandi Board.
- Collectorate.
- Public Service Management.
- Digital Center, Common Service Center.
- Stakeholders like Warehouses, Transporters, Traders, Suppliers, Farmers, Shopkeepers, and Manufacturers.

5.3 *Research Design*

It is a mixed-method-based research methodology that follows an inductive approach and includes descriptive and non-experimental case study-based research design. This research design is a cross-sectional study used to observe the current IT ecosystem of the government departments at a given time. It adheres to both pragmatism and realism philosophy.

5.4 *Sampling Method and Techniques for Data Collection*

Probability sampling method has been used that utilizes a combination of two sampling techniques, viz. non-uniform systematic cluster sampling and stratified random sampling. The population of Madhya Pradesh has been targeted. Seven clusters (Districts) have been chosen systematically based on the Rural Logistics activities pattern. Districts are Bhopal, Sehore, Sagar, Hoshangabad, Khandwa, Ujjain, and Dewas. Stratas has been designed as per the need of the Rural Logistics. Sixty samples have been gathered for its data analysis and predictions. Sampling frame includes Jila Panchayat, Janpad Panchayat, Mandi Board, Collectorate, Public Service Management, Digital Center, Common Service Center, and Stakeholders like Warehouses, Transporters, Traders, Suppliers, Farmers, Shopkeepers, and Manufacturers.

5.5 Data Collection Techniques

Qualitative and quantitative methods of data collection have been followed based face to face interactions, in-depth interviews, focused group discussions, and close-ended questionnaire survey forms as tools for primary data collection. The permission letters had been obtained from the government departments to facilitate data collection from its structures and partners/stakeholders. Secondary data has been collected through websites, international research publications, national and international journals, books, online pdfs, etc.

- *Quantitative Close-Ended Questionnaire Survey*: Four distinct sections:
  - Baseline: Questions involving queries of age, gender, and education level achieved.
  - Disruptive Technology: General questions regarding respondents' knowledge and awareness of Blockchain technology innovations and any concerns they may have concerning them.
  - Blockchain General: Introductory questions on Blockchain to establish if there is an awareness of Blockchain among respondents.
  - Blockchain Specific: Using specific examples of how Blockchain can be used.

5.6 Data Analysis Techniques

- *Pivot Charts using Excel* for Graphical Analysis of Data.

6 Data Findings and Outcomes

This Primary Research Data has been collected from the questionnaire survey. Out of the survey, the total valid questionnaires were kept as benchmarked outcomes. Total Expected Standard Outcomes depict the results out of the questionnaire without using blockchain technology. Percentage of the total expected standard outcomes is the expected results from the districts' stakeholders in terms of rural logistics without using any blockchain technology. Percentage of the received valid outcomes depicts the actual results from the stakeholder's district wise those have shown the outcomes in rural logistics without Blockchain. Difference in the current outcome and the Blockchain enabled outcome is the scope for this blockchain technology to be implemented in rural logistics. Finally ranking of the districts, depicts the performance based on the digital interventions in rural logistics and its scope for improvement through blockchain technology. Table 1 and graphs show our unique and proposed work for required improvement in rural logistics using Blockchain.

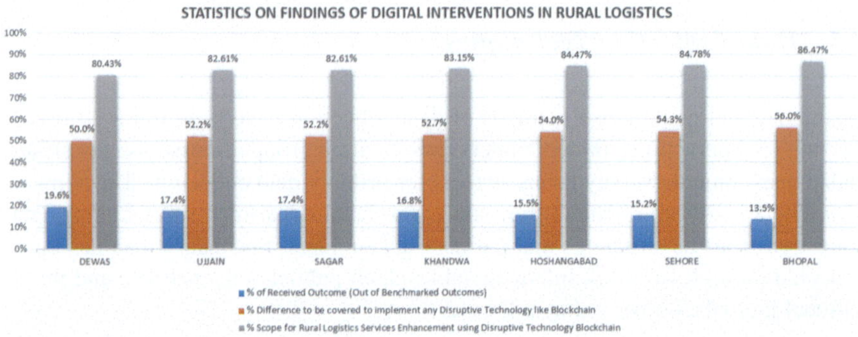

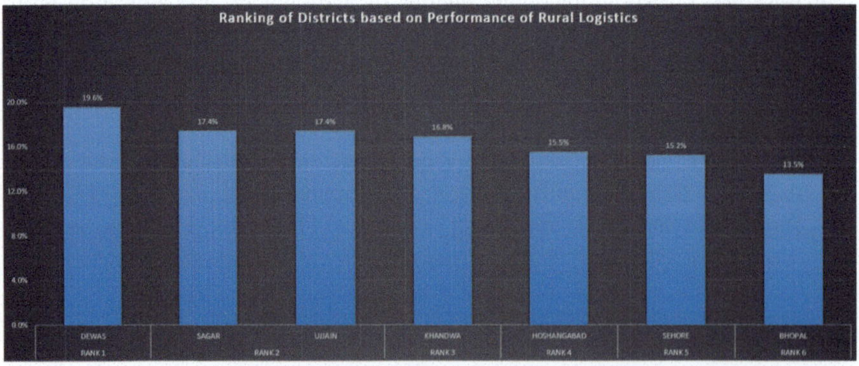

Rural logistics aims to have ICT-based transit management and traffic management system which shall add value to citizens, city authorities, and society in general by bringing down travel time, reducing travel-related energy consumption's, increasing comfort and safety of travel, establishing efficient and effective management procedures, and working toward environmental sustainability. Benefits include maximization of the traffic flow by reducing the average queue lengths and average wait times using dynamic traffic flow data read from the adaptive traffic control system.

A robust intelligent traffic management system infrastructure can be developed that would be reliable and efficient leading to effective traffic management in the state to improve journey time reliability, to increase operational efficiency, to improve customer services, and to create a blockchain-enabled platform to share traffic information [7, 17].

Table 1 Statistics on findings of digital interventions in rural logistics

No.	District	Total questions	Benchmarked outcomes (using any disruptive technology)	Total expected standard outcomes (without using any disruptive technology)	% of Total expected standard outcomes (without using any disruptive technology)	Received valid outcomes	% of Received outcome (out of benchmarked outcomes)	Difference to be covered to implement any disruptive technology like blockchain	Not received outcomes (out of benchmarked outcomes)	% Scope for rural logistics services enhancement using disruptive technology blockchain	Ranking of districts based on performance of rural logistics
6	DEWAS	50	46	32	69.6%	9	19.6%	50.0%	37	80.43%	RANK 1
7	UJJAIN	50	46	32	69.6%	8	17.4%	52.2%	38	82.61%	RANK 2
5	SAGAR	100	92	64	69.6%	16	17.4%	52.2%	76	82.61%	RANK 2
4	KHANDWA	200	184	128	69.6%	31	16.8%	52.7%	153	83.15%	RANK 3
2	HOSHANGABAD	350	322	224	69.6%	50	15.5%	54.0%	272	84.47%	RANK 4
3	SEHORE	300	276	192	69.6%	42	15.2%	54.3%	234	84.78%	RANK 5
1	BHOPAL	450	414	288	69.6%	56	13.5%	56.0%	358	86.47%	RANK 6

Fig. 2 Rural logistics road infrastructure

Installation of traffic signals, surveillance camera, and integration of the same with police control room and command control center is required with the purpose to reduce congestion and friction in traffic.

- *Rural logistics road infrastructure must include*:

Traffic control system, surveillance camera, automatic number plate recognition system, red light violation detection system, variable message sign boards, public address system, emergency call box system, traffic junctions, entry-exit points, location for speed detection, etc. (Fig. 2).

The solution must include hardware installation provisioning, repair, and replacement service of terminals. Vehicle tracking system must involve attaching a GPS device to a vehicle where the device gathers information about the location of a vehicle and records the position of the vehicle at regular intervals, sending all this information to a centralized location. A person at the center location can monitor the information to know the exact location of a vehicle at a given time, and the route followed by the vehicle can be known [1, 20].

This system would apply the information, communication, and control technologies built on a blockchain platform to improve the operation of logistics networks. It will save time and lives, will improve the quality of logistics and environment, and will also improve the productivity of logistics activities. Integrated system required [21]:

- Low road network speeds.
- Higher travel times.
- The increased cost of travel.
- Increased energy consumption.

- Increased pollution (both noise and air).
- The decline in human productivity.
- Decrease in the overall competitiveness of the city.
- The decline in overall quality of life.

Integrated Vehicle Tracking System will help in [2]:

- Collection of data on traffic conditions in real-time from field equipment (about the transportation network highways, state and local roads, traffic signal system).
- Analysis of data and control of traffic management systems (traffic signals, variable signs used for dynamic re-routing, etc.)
- Monitoring of information on traffic conditions and incidents and emergencies by operators.
- Timely interventions to rectify traffic incidents and/or system failures.
- Real-time dissemination of accurate information to travelers and other stakeholders (like media, emergency response teams, etc.) Table 2 below show the stakeholder's current state comparision with the future state.

Table 2 Stakeholders' Current State vs. Future State

Stakeholders	Current state	Future state
Manufacturer	Has limited visibility and control into the system date and processes	Complete provenance trail of every asset
	No insight into lost/spilled goods	Visibility into process flow and claim data via POS integrations
	Dependency on the paper trail	Losses are reported in real-time via Tol sensors
		Real-time shipment acknowledgments
Warehouse	No visibility of incoming shipments	Complete provenance trail of every asset
	Real-time stock and sales data are not available	Visibility into process flow and claim data via POS integration
	Slow and isolated processes	Losses are reported in real-time via lot sensors
		Real-time shipment acknowledgments
Retailer	Fertilizer quality is not guaranteed	Fertilizer quality can be traced back to the manufacturing source and B2 certificate
	No visibility of incoming shipments	Tol devices can help identify pilferage sources
	Fertilizer losses along the way	
Government Agencies	Auditing inventory and sales data is complex	Holistic participant data view for each participant
	Isolated process structures and inconsistent soiled data	Consensus and immutability ensure data is valid and can be trusted
		Minimizes the need for accounting, auditing

7 Suggestions

7.1 Online Platform to Organize Logistics

With increasing digitization, platform-based business models will connect new players, wash away inefficient old ones, and harness the cloud. Multi-party coordination of this asset-intensive industry adds to the overall complexity. A blockchain-based platform for managing global shipments involving multiple stakeholders. Events across the shipping life cycle—credit checks, contract signing, arrival at the port, and payment—can be recorded publicly. On trade lens, event data and document information can be written on the blockchain, which creates a single source of truth that all can see. [9] Contracts can also execute automatically on the blockchain. When a specific event is recorded—say a delivery at a port—the corresponding contracts encoded in the blockchain are automatically activated, removing human errors, delays, and lost documentation [18].

7.2 Rural Logistics Collaboration Index

"Knowledge is power," as they say, and in rural logistics collaboration, sharing of information leads to enhanced knowledge across the chain that allows you to achieve [10]:

- Lower inventory levels and higher inventory turns.
- Lower transportation and warehousing costs.
- Lower out-of-stock levels.
- Shorter lead times.
- Improved customer service metrics.
- Visibility into customer demand and supplier performance.
- Earlier and quicker decision-making.

Transparency and collaboration can be difficult to execute, but well worth the effort when you consider the potential reduction in risk and costs, and improvement in customer satisfaction and loyalty. Our system has the flexibility to allow everyone, at every location across the globe, to understand what is happening when and where, in real-time, via an easy-to-navigate dashboard. The platform is able to manage massive amounts of current inventory specifics and has the power to analyze past activities, anticipate future needs through "what if" modeling, understand, and plan replenishment cycles, manage accounting functions, track productivity and collaborate throughout the entire supply management process. The system also has the transparency and interoperability to integrate completely with customers' systems, even if it is a patchwork of legacy systems.

8 Recommendations

- Development of Multi-Modal logistics parks in Madhya Pradesh.
- Development of Integrated Mobile App mLogistics.
- Amendment of e-Governance model for information sharing.
- Need of logistics synchronization.
- Requirement for enhancement of incentive alignment.
- Eradication of rural logistics interventions in tribal areas.
- Government regulation.
- Environment safety is a necessity.
- Proper management of logistics cost.
- Need of redesigning of logistics network structure.
- Need of enforcement of flexibility principle, simplification principle, and gravity principle in material handling.
- Need of sufficient warehouses.
- Need to develop a Blockchain-enabled policy framework for transportation.
- Need of a robust packaging system.
- Need of enhancement of measured services like availability and service reliability.
- Need of robust rural infrastructure.
- Eradication of lack of awareness and education on rural logistics management system.
- Requirement of sufficient actionable data.
- Need to improve the low frequency visit of APMC (Agricultural Produce Market Committee).
- Need to eradicate Non-participation of buyer at RMC markets (Ready Mixed Concrete).
- Need of Mandi Board operations to be taken up on Blockchain Platform: Mandi Board.
- Need of integration of Agricultural Marketing Information System Network (Agmarknet) with Mandi Board.
- Need of integration of database of Krishinet, Panchlekha, e-Gram Suvidha, Parakh, and Web-based Rural Soft monitoring System onto single Blockchain Platform.
- Need of emerging technologies like Blockchain, Internet of Things, Artificial Intelligence, Drone Delivery, Driverless Vehicles.
- Need of reverse rural logistics.
- Need of rural logistics COLLABORATION INDEX.
- Need of state level blockchain enabled rural logistics policy making.

9 Conclusion

- Dewas has scored the highest outcome out of **19.6%** out of 69.6% minimum valid outcome without using technology.
- Ujjain and Sagar both have scored an equal **17.4%** out of 69.6% minimum valid outcome without using technology.
- Khandwa has scored **16.8%** out of 69.6% minimum valid outcome without using technology.
- Hoshangabad has scored **15.5%** minimum valid outcome without using technology.
- Sehore has scored **15.2%** minimum valid outcome without using technology.
- Bhopal has scored the lowest outcome of **13.5%** minimum valid outcome without using technology.

District Dewas has 80.4% scope for improvement using blockchain technology, Ujjain and Sagar has 82.6% scope for improvement using blockchain technology, Khandwa has 83.1% scope for improvement using blockchain technology, Hoshangabad has 84.4% scope for improvement using blockchain technology, Sehore has 84.7% scope for improvement using blockchain technology, and finally Bhopal has 86.4% scope for improvement using blockchain technology. Blockchain technology offers an innovative platform for a new decentralized and transparent transaction mechanism in industry and business. Features of this technology increase confidence through transparency within any transaction of data, goods, and financial resources. Blockchain technology can easily provide secure business operations in logistics. In the logistics sector, blockchain technology could dramatically reduce time delays, added costs, and human errors. Finally, by using blockchain technology, the challenges encountered by the logistics sector can be minimized or even eliminated, sustainability can be greatly increased, based on previous research, and appropriate conclusions are derived about possible obstacles and advantages in blockchain technology implementation. This technology can facilitate logistics tasks: it can be used to track purchase orders, order changes and freight documents, and it can help in information sharing about manufacturing process and delivery. Blockchain technology has huge potential for development and application in the logistics sector and supply chain, presenting challenges for further research.

Some of the major benefits of adopting an integrated Blockchain platform are as follows:

- The citizen benefits because there is transparency, efficiency and integrity in his dealings with the government; furthermore, there is easy information access.
- The government benefits because it reduces redundancy and duplication. The processes of data collection, analysis, and audit are made much easier. Decision-making gets expedited.
- The business community benefits because e-Governance can become a catalyst and a channel for e-Business, furthermore a web-based government will enable tax-paying online, reduce corruption and bending of laws.

We in Madhya Pradesh envision the commitments made to the citizens and work for nurturing our dreams to come true using blockchain technology.

References

1. L. Zhu et al., Big data analytics in intelligent transportation systems: A survey. IEEE Trans. Intell. Transp. Syst. **20**(1), 383–398 (2015)
2. K. Witkowski, Internet of things, big data, industry 4.0—Innovative solutions in logistics and supply chains management. Proc. Eng. **182**, 763–769 (2017)
3. Marsh & McLennan Companies, *Drones—A View Into the Future for the Logistics Sector?* (Marsh & McLennan Companies, New York, 2015)
4. M. Makridis et al., *Connected and Automated Vehicles on a Freeway Scenario. Effect on Traffic Congestion and Network Capacity* (Transport Research Arena, Vienna, 2018)
5. R. Baker et al., *Disruptive Technologies and Transportation* (Texas A&M Transportation Institute, PRC 15–45 F, Dallas, TX, 2016)
6. P.M. Bosch et al., Cost-based analysis of autonomous mobility services, in *Working Paper 1225*, (Institute for Transport Planning and Systems, Zurich, 2017)
7. A. Das, P. Dash, B. Mishra, *An Innovation Model for Smart Traffic Management System Using Internet of Things (IoT)* (Springer International, Berlin, 2018), pp. 364–365
8. Deloitte, *Transport in the Digital Age, Disruptive Trends for Smart Mobility* (Deloitte, London, 2015)
9. P. Kumar, M. Anas, An ABC-analysis for the multiple-products inventory management. Int. J. Res. Eng. Adv. Technol. **1**(5), 1–6 (2013)
10. Council of Logistics Management, Definition of logistics, from http://www.cscmp.org/, in *Tilanus, B. (1997). Information Systems in Logistics and Transportation*, (Elsevier Science, Amsterdam, 1991), 350 p
11. Idrees, S. M., Aijaz, I., Agarwal, P., & Jameel, R. (2021). Blockchain-Based Smart and Secure Healthcare System. In *Transforming Cybersecurity Solutions using Blockchain* (pp. 165-183). Springer, Singapore.
12. O. Ozer, *Inventory sManagement: Information, Coordination and Rationality* (Stanford University, Stanford, CA, 2014)
13. Government of the People's Republic of India, Ministry of Transport; Ministry of Agriculture; State Post Bureau; and India Co-op, *Several Opinions on Promoting the Healthy Development of Rural Logistics Cooperatively to Better Serve Agricultural Modernization* (Beijing, 2014)
14. Idrees, S. M., Aijaz, I., Jameel, R., & Nowostawski, M. (2021). Exploring the Blockchain Technology: Issues, Applications and Research Potential. International Journal of Online & Biomedical Engineering, 17(7).
15. I. Weber, X. Xu, R. Riveret, G. Governatori, A. Ponomarev, J. Mendling, Untrusted business process monitoring and execution using blockchain, in *International Conference on Business Process Management*, (2016)
16. G. Wood, Ethereum: A secure decentralized generalized transaction ledger—Homestead draft, in *Technical Report*, (2016)
17. G. Zhao, S. Liu, C. Lopez, H. Lu, S. Elgueta, H. Chen, B.M. Boshkoska, Blockchain technology in Agri-food value chain management: A synthesis of applications, challenges and future research directions. Comput. Ind. **109**, 83–99 (2019)
18. D. Tse, B. Zhang, Y. Yang, C. Cheng, H. Mu, Blockchain application in food supply information security, in *International Conference on Industrial Engineering and Engineering Management (IEEM)*, (IEEE, Piscataway, NJ, 2017), pp. 1357–1361
19. Idrees, S. M., Agarwal, P., & Alam, M. A. (Eds.). (2021). Blockchain for Healthcare Systems: Challenges, Privacy, and Securing of Data. CRC Press.

20. P.K. Sharma, S.Y. Moon, J.H. Park, Block-VN: A distributed blockchain-based vehicular network architecture in smart city. J. Inf. Process. Syst. **13**, 184–195 (2017)
21. M. Friedlmaier, A. Tumasjan, I.M. Welpe, Disrupting industries with blockchain: The industry, venture capital funding, and regional distribution of blockchain ventures, in *Proceedings of the 51st Hawaii International Conference on System Sciences, Waikoloa Village, HI, USA, 3–6 January 2018*, (2018)
22. N. Hackius, M. Petersen, Blockchain in logistics and supply chain: Trick or treat? in *Proceedings of the Hamburg International Conference of Logistics (HICL), Hamburg, Germany, 12–13 October 20*, (2020)
23. C. Carter, L. Koh, *Blockchain Disruption in Transport: Are you Decentralized Yet?* (Catapult Transport Systems, Milton Keynes, 2018)

Index

S. M. Idrees, M. Nowostawski (eds.), *Transformations Through Blockchain Technology*, https://doi.org/10.1007/978-3-030-93344-9

The manufacturer's authorised representative in the EU is Springer Nature Customer Service Centre GmbH, Europaplatz 3, 69115 Heidelberg, Germany. If you have any concerns regarding our products, please contact ProductSafety@springernature.com

Printed and bound by CPI Group (UK) Ltd, Croydon, CR0 4YY
30/09/2026
02225496-0013